LONDON MATHEMATICAL SOCIETY LECTURE NOTE SERIES

Managing Editor: Professor Endre Süli, Mathematical Institute, University of Oxford, Woodstock Road, Oxford OX2 6GG, United Kingdom

The titles below are available from booksellers, or from Cambridge University Press at www.cambridge.org/mathematics

361 Words: Notes on verbal width in groups, D. SEGAL
362 Differential tensor algebras and their module categories, R. BAUTISTA, L. SALMERÓN & R. ZUAZUA
363 Foundations of computational mathematics, Hong Kong 2008, F. CUCKER, A. PINKUS & M.J. TODD (eds)
364 Partial differential equations and fluid mechanics, J.C. ROBINSON & J.L. RODRIGO (eds)
365 Surveys in combinatorics 2009, S. HUCZYNSKA, J.D. MITCHELL & C.M. RONEY-DOUGAL (eds)
366 Highly oscillatory problems, B. ENGQUIST, A. FOKAS, E. HAIRER & A. ISERLES (eds)
367 Random matrices: High dimensional phenomena, G. BLOWER
368 Geometry of Riemann surfaces, F.P. GARDINER, G. GONZÁLEZ-DIEZ & C. KOUROUNIOTIS (eds)
369 Epidemics and rumours in complex networks, M. DRAIEF & L. MASSOULIÉ
370 Theory of p-adic distributions, S. ALBEVERIO, A.YU. KHRENNIKOV & V.M. SHELKOVICH
371 Conformal fractals, F. PRZYTYCKI & M. URBAŃSKI
372 Moonshine: The first quarter century and beyond, J. LEPOWSKY, J. MCKAY & M.P. TUITE (eds)
373 Smoothness, regularity and complete intersection, J. MAJADAS & A. G. RODICIO
374 Geometric analysis of hyperbolic differential equations: An introduction, S. ALINHAC
375 Triangulated categories, T. HOLM, P. JØRGENSEN & R. ROUQUIER (eds)
376 Permutation patterns, S. LINTON, N. RUŠKUC & V. VATTER (eds)
377 An introduction to Galois cohomology and its applications, G. BERHUY
378 Probability and mathematical genetics, N. H. BINGHAM & C. M. GOLDIE (eds)
379 Finite and algorithmic model theory, J. ESPARZA, C. MICHAUX & C. STEINHORN (eds)
380 Real and complex singularities, M. MANOEL, M.C. ROMERO FUSTER & C.T.C WALL (eds)
381 Symmetries and integrability of difference equations, D. LEVI, P. OLVER, Z. THOMOVA & P. WINTERNITZ (eds)
382 Forcing with random variables and proof complexity, J. KRAJÍČEK
383 Motivic integration and its interactions with model theory and non-Archimedean geometry I, R. CLUCKERS, J. NICAISE & J. SEBAG (eds)
384 Motivic integration and its interactions with model theory and non-Archimedean geometry II, R. CLUCKERS, J. NICAISE & J. SEBAG (eds)
385 Entropy of hidden Markov processes and connections to dynamical systems, B. MARCUS, K. PETERSEN & T. WEISSMAN (eds)
386 Independence-friendly logic, A.L. MANN, G. SANDU & M. SEVENSTER
387 Groups St Andrews 2009 in Bath I, C.M. CAMPBELL *et al* (eds)
388 Groups St Andrews 2009 in Bath II, C.M. CAMPBELL *et al* (eds)
389 Random fields on the sphere, D. MARINUCCI & G. PECCATI
390 Localization in periodic potentials, D.E. PELINOVSKY
391 Fusion systems in algebra and topology, M. ASCHBACHER, R. KESSAR & B. OLIVER
392 Surveys in combinatorics 2011, R. CHAPMAN (ed)
393 Non-abelian fundamental groups and Iwasawa theory, J. COATES *et al* (eds)
394 Variational problems in differential geometry, R. BIELAWSKI, K. HOUSTON & M. SPEIGHT (eds)
395 How groups grow, A. MANN
396 Arithmetic differential operators over the p-adic integers, C.C. RALPH & S.R. SIMANCA
397 Hyperbolic geometry and applications in quantum chaos and cosmology, J. BOLTE & F. STEINER (eds)
398 Mathematical models in contact mechanics, M. SOFONEA & A. MATEI
399 Circuit double cover of graphs, C.-Q. ZHANG
400 Dense sphere packings: a blueprint for formal proofs, T. HALES
401 A double Hall algebra approach to affine quantum Schur–Weyl theory, B. DENG, J. DU & Q. FU
402 Mathematical aspects of fluid mechanics, J.C. ROBINSON, J.L. RODRIGO & W. SADOWSKI (eds)
403 Foundations of computational mathematics, Budapest 2011, F. CUCKER, T. KRICK, A. PINKUS & A. SZANTO (eds)
404 Operator methods for boundary value problems, S. HASSI, H.S.V. DE SNOO & F.H. SZAFRANIEC (eds)
405 Torsors, étale homotopy and applications to rational points, A.N. SKOROBOGATOV (ed)
406 Appalachian set theory, J. CUMMINGS & E. SCHIMMERLING (eds)
407 The maximal subgroups of the low-dimensional finite classical groups, J.N. BRAY, D.F. HOLT & C.M. RONEY-DOUGAL
408 Complexity science: the Warwick master's course, R. BALL, V. KOLOKOLTSOV & R.S. MACKAY (eds)
409 Surveys in combinatorics 2013, S.R. BLACKBURN, S. GERKE & M. WILDON (eds)
410 Representation theory and harmonic analysis of wreath products of finite groups, T. CECCHERINI-SILBERSTEIN, F. SCARABOTTI & F. TOLLI
411 Moduli spaces, L. BRAMBILA-PAZ, O. GARCÍA-PRADA, P. NEWSTEAD & R.P. THOMAS (eds)
412 Automorphisms and equivalence relations in topological dynamics, D.B. ELLIS & R. ELLIS

413 Optimal transportation, Y. OLLIVIER, H. PAJOT & C. VILLANI (eds)
414 Automorphic forms and Galois representations I, F. DIAMOND, P.L. KASSAEI & M. KIM (eds)
415 Automorphic forms and Galois representations II, F. DIAMOND, P.L. KASSAEI & M. KIM (eds)
416 Reversibility in dynamics and group theory, A.G. O'FARRELL & I. SHORT
417 Recent advances in algebraic geometry, C.D. HACON, M. MUSTAŢĂ & M. POPA (eds)
418 The Bloch–Kato conjecture for the Riemann zeta function, J. COATES, A. RAGHURAM, A. SAIKIA & R. SUJATHA (eds)
419 The Cauchy problem for non-Lipschitz semi-linear parabolic partial differential equations, J.C. MEYER & D.J. NEEDHAM
420 Arithmetic and geometry, L. DIEULEFAIT *et al* (eds)
421 O-minimality and Diophantine geometry, G.O. JONES & A.J. WILKIE (eds)
422 Groups St Andrews 2013, C.M. CAMPBELL *et al* (eds)
423 Inequalities for graph eigenvalues, Z. STANIĆ
424 Surveys in combinatorics 2015, A. CZUMAJ *et al* (eds)
425 Geometry, topology and dynamics in negative curvature, C.S. ARAVINDA, F.T. FARRELL & J.-F. LAFONT (eds)
426 Lectures on the theory of water waves, T. BRIDGES, M. GROVES & D. NICHOLLS (eds)
427 Recent advances in Hodge theory, M. KERR & G. PEARLSTEIN (eds)
428 Geometry in a Fréchet context, C.T.J. DODSON, G. GALANIS & E. VASSILIOU
429 Sheaves and functions modulo p, L. TAELMAN
430 Recent progress in the theory of the Euler and Navier–Stokes equations, J.C. ROBINSON, J.L. RODRIGO, W. SADOWSKI & A. VIDAL-LÓPEZ (eds)
431 Harmonic and subharmonic function theory on the real hyperbolic ball, M. STOLL
432 Topics in graph automorphisms and reconstruction (2nd Edition), J. LAURI & R. SCAPELLATO
433 Regular and irregular holonomic D-modules, M. KASHIWARA & P. SCHAPIRA
434 Analytic semigroups and semilinear initial boundary value problems (2nd Edition), K. TAIRA
435 Graded rings and graded Grothendieck groups, R. HAZRAT
436 Groups, graphs and random walks, T. CECCHERINI-SILBERSTEIN, M. SALVATORI & E. SAVA-HUSS (eds)
437 Dynamics and analytic number theory, D. BADZIAHIN, A. GORODNIK & N. PEYERIMHOFF (eds)
438 Random walks and heat kernels on graphs, M.T. BARLOW
439 Evolution equations, K. AMMARI & S. GERBI (eds)
440 Surveys in combinatorics 2017, A. CLAESSON *et al* (eds)
441 Polynomials and the mod 2 Steenrod algebra I, G. WALKER & R.M.W. WOOD
442 Polynomials and the mod 2 Steenrod algebra II, G. WALKER & R.M.W. WOOD
443 Asymptotic analysis in general relativity, T. DAUDÉ D. HÄFNER & J.-P. NICOLAS (eds)
444 Geometric and cohomological group theory, P.H. KROPHOLLER, I.J. LEARY, C. MARTÍNEZ-PÉREZ & B.E.A. NUCINKIS (eds)
445 Introduction to hidden semi-Markov models, J. VAN DER HOEK & R.J. ELLIOTT
446 Advances in two-dimensional homotopy and combinatorial group theory, W. METZLER & S. ROSEBROCK (eds)
447 New directions in locally compact groups, P.-E. CAPRACE & N. MONOD (eds)
448 Synthetic differential topology, M.C. BUNGE, F. GAGO & A.M. SAN LUIS
449 Permutation groups and cartesian decompositions, C.E. PRAEGER & C. SCHNEIDER
450 Partial differential equations arising from physics and geometry, M. BEN AYED *et al* (eds)
451 Topological methods in group theory, N. BROADDUS, M. DAVIS, J.-F. LAFONT & I. ORTIZ (eds)
452 Partial differential equations in fluid mechanics, C.L. FEFFERMAN, J.C. ROBINSON & J.L. RODRIGO (eds)
453 Stochastic stability of differential equations in abstract spaces, K. LIU
454 Beyond hyperbolicity, M. HAGEN, R. WEBB & H. WILTON (eds)
455 Groups St Andrews 2017 in Birmingham, C.M. CAMPBELL *et al* (eds)
456 Surveys in combinatorics 2019, A. LO, R. MYCROFT, G. PERARNAU & A. TREGLOWN (eds)
457 Shimura varieties, T. HAINES & M. HARRIS (eds)
458 Integrable systems and algebraic geometry I, R. DONAGI & T. SHASKA (eds)
459 Integrable systems and algebraic geometry II, R. DONAGI & T. SHASKA (eds)
460 Wigner-type theorems for Hilbert Grassmannians, M. PANKOV
461 Analysis and geometry on graphs and manifolds, M. KELLER, D. LENZ & R.K. WOJCIECHOWSKI
462 Zeta and L-functions of varieties and motives, B. KAHN
463 Differential geometry in the large, O. DEARRICOTT *et al* (eds)
464 Lectures on orthogonal polynomials and special functions, H.S. COHL & M.E.H. ISMAIL (eds)
465 Constrained Willmore surfaces, Á.C. QUINTINO
466 Invariance of modules under automorphisms of their envelopes and covers, A.K. SRIVASTAVA, A. TUGANBAEV & P.A. GUIL ASENSIO
467 The genesis of the Langlands program, J. MUELLER & F. SHAHIDI
468 (Co)end calculus, F. LOREGIAN
469 Computational cryptography, J.W. BOS & M. STAM (eds)
470 Surveys in combinatorics 2021, K.K. DABROWSKI *et al* (eds)

London Mathematical Society Lecture Note Series: 468

(Co)end Calculus

FOSCO LOREGIAN
Tallinn University of Technology

CAMBRIDGE
UNIVERSITY PRESS

University Printing House, Cambridge CB2 8BS, United Kingdom

One Liberty Plaza, 20th Floor, New York, NY 10006, USA

477 Williamstown Road, Port Melbourne, VIC 3207, Australia

314–321, 3rd Floor, Plot 3, Splendor Forum, Jasola District Centre, New Delhi – 110025, India

79 Anson Road, #06–04/06, Singapore 079906

Cambridge University Press is part of the University of Cambridge.

It furthers the University's mission by disseminating knowledge in the pursuit of education, learning, and research at the highest international levels of excellence.

www.cambridge.org
Information on this title: www.cambridge.org/9781108746120
DOI: 10.1017/9781108778657

First published 2021

A catalogue record for this publication is available from the British Library.

ISBN 978-1-108-74612-0 Paperback

A 'Εκάτερα
ambrosia pura
vestale di fuoco.

Φος

Contents

Preface

> No podemos esperarnos que ningún aspecto de la realidad cambie si seguimos usando los medios [...] de un lenguaje que lleva el peso de toda la negatividad del pasado. El lenguaje está ahí, pero tenemos que limpiarlo, revisarlo y, sobre todo, debemos desconfiar de él.
>
> J. Cortázar

What is (co)end 'calculus'. Coend calculus determines the behaviour of suitable universal objects associated to functors of two variables $T : \mathcal{C}^{\text{op}} \times \mathcal{C} \to \mathcal{D}$.

The intuition behind the process of attaching a special invariant to such a functor T can be motivated in many ways.

It is well-known that a measurable scalar function $f : X \to \mathbb{R}$ from a measurable space (X, Ω) can be integrated 'against' a measure μ defined on Ω to yield a real number

$$\int_X f(x) d\mu$$

(for example, when X is a smooth space, the measure can be legitimately thought to depend 'contravariantly' on x, as $d\mu$ is a volume form living in the top-degree exterior algebra of X). In a similar fashion, the evaluation map $V^\vee \otimes V \to k$ for a vector space V is a pairing $\langle \zeta, v \rangle = \zeta(v)$ between a vector v and a co-vector $\zeta : V \to k$, which becomes the sum $\sum_i \zeta_i v_i$ once a basis for V, and its dual basis, is chosen and the vector v has coordinates $(v_1, \dots, v_d)$, whereas ζ has coordinates $(\zeta_1, \dots, \zeta_d)$.

At the cost of pushing this analogy further than permitted, a functor $T : \mathcal{C}^{\text{op}} \times \mathcal{C} \to \mathcal{D}$ can be thought of as a generalised form of evaluation of an object of $\mathcal{C}$ against another; the 'quantity' $T(C, C')$ can then be 'integrated' to yield two distinct objects having dual universal properties:

C1. a *coend*, resulting from the symmetrisation along the diagonal of T, i.e. by modding out the coproduct $\coprod_{C \in \mathcal{C}} T(C, C)$ by the equivalence relation generated by the arrow functions $T(\_, C') : \mathcal{C}^{\mathrm{op}}(X, Y) \to \mathcal{D}(T(X, C'), T(Y, C'))$ and $T(C, \_) : \mathcal{C}(X, Y) \to \mathcal{D}(T(C, X), T(C, Y))$;

C2. an *end*, i.e. an object $\int_C T(C, C)$ arising as an 'object of invariants' or 'fixed points' for the same action of T on arrows; by dualisation, if a coend is a quotient of $\coprod_{C \in \mathcal{C}} T(C, C)$, an end is a subobject of the product $\prod_{C \in \mathcal{C}} T(C, C)$.

This also suggests a fruitful analogy with modules over a ring: if a functor $T : \mathcal{C}^{\mathrm{op}} \times \mathcal{C} \to \mathrm{Set}$ is a 'bimodule', which lets $\mathcal{C}$ act once on the left and once on the right on the sets $T(C, C')$, then the end $\int_C T(C, C)$ is the subspace of invariants for the action of $\mathcal{C}$, whereas the coend $\int^C T(C, C)$ is the space of orbits (or 'coinvariants') of said action.

In fact, a rather common way to employ coends is the following: consider a functor $F : \mathcal{C} \to \mathrm{Set}$ (a 'left module') and a functor $G : \mathcal{C}^{\mathrm{op}} \to \mathrm{Set}$ (a 'right module'), and tensor them together into a functor $(C, C') \mapsto GC \times FC'$; the symmetrisation of $F \times G$ yields a *functor tensor product* of F, G as the set

$$F \boxtimes G := \int^C FC \times GC.$$

Note that in this light, the analogy is meaningful: if $\mathcal{C}$ is a single-object category (so a monoid or a group G), such a pair of modules constitutes a pair (X, Y) of a left and a right G-set, and their functor tensor product can be characterised as the product $X \times_G Y$ obtained as the quotient of $X \times G$ for the equivalence relation $(g.x, y) \sim (x, g.y)$, so that $X \times_G Y$ is the universal G-*bilinear* product of sets, in that the 'scalar' $g \in G$ can pass left-to-right from $(g.x, y)$ to $(x, g.y)$ in the quotient. Of course, the terminology works better when X, Y are vector spaces carrying a *linear* representation of G.

Theorems involving ends and coends $\int^C T(C, C)$ and $\int_C T(C, C)$ can now be proved by means of the universal properties that define them; it is easily seen that given $T : \mathcal{C}^{\mathrm{op}} \times \mathcal{C} \to \mathcal{D}$ there exists a category $\bar{\mathcal{C}}$ and a functor $\bar{T} : \bar{\mathcal{C}} \to \mathcal{D}$ such that $\int^C T(C, C) \cong \mathrm{colim}_{\bar{\mathcal{C}}} \bar{T}$ and $\int_C T(C, C) \cong \lim_{\bar{\mathcal{C}}} \bar{T}$. In the example above, the tensor product of a left G-module and a right G-module (here 'module' means 'k-vector space') can be characterised as the coequaliser

$$\bigoplus_{g\in G} X \otimes_k Y \overset{\alpha}{\underset{\beta}{\rightrightarrows}} X \otimes_k Y \longrightarrow X \otimes_G Y \tag{0.1}$$

where $\alpha(g,(x,y)) = (g.x, y)$ and $\beta(g,(x,y)) = (x, g.y)$.

So, all (co)ends can be characterised as (co)limits; but they provide a richer set of computational rules than mere (co)limits. Often, establishing that an object has a certain universal property is a difficult task, because a direct argument tangles the reader into using elements. A general tenet of modern category theory is that cleaner, more conceptual arguments are preferred to element-wise proofs that are evil in spirit, if not in shape.

(Co)end calculus provides such a conceptualisation for many classical arguments of category theory: it is in fact possible to prove that two objects of a category, at least one of which is defined as a coend, are isomorphic by means of a chain of 'deduction rules'.

These rules are described in the first half of the book, but here we glimpse at what they look like.[1]

In order to make clear what this paragraph is about, let us consider the statement that *right adjoints preserve limits*; it is certainly possible to prove this by hand. Nevertheless, using little more than the Yoneda lemma it is possible to prove that if $R : \mathcal{D} \to \mathcal{C}$ is right adjoint to $L : \mathcal{C} \to \mathcal{D}$, there is a natural isomorphism of hom-sets $\mathcal{C}(C, R(\lim_{\mathcal{J}} D_J)) \cong \mathcal{C}(C, \lim_{\mathcal{J}} RD_J)$ for every object C, and every diagram $D : \mathcal{J} \to \mathcal{D}$ by arguing as follows:

$$\frac{\dfrac{\dfrac{\dfrac{\mathcal{C}(C, R(\lim_{\mathcal{J}} D_J))}{\mathcal{C}(LC, \lim_{\mathcal{J}} D_J)}}{\lim_{\mathcal{J}} \mathcal{C}(LC, D_J)}}{\lim_{\mathcal{J}} \mathcal{C}(C, RD_J)}}{\mathcal{C}(C, \lim_{\mathcal{J}} RD_J)}$$

[1] The somewhat far-fetched conjecture that permeates all the book is that coend calculus provides a higher dimensional version of a deductive system, suited for category theory (see [CW01] for some preliminary steps in this direction), having deduction rules similar to those of Gentzen's sequent calculus. We will never attempt to turn this enticing conjecture into a theorem, or even to make a precise claim; the interested reader is thus warned that their curiosity will not get satisfaction – not in the present book, at least. We record that the idea that coends categorify logical calculus comes from William Lawvere, and it was first proposed in [Law73].

where each step of this 'deduction' is motivated either by the fact that $L \dashv R$ are adjoint functors, or by the fact that all functors $\mathcal{C}(X, \_)$ preserve limits. Once this is proved, the Yoneda lemma (see A.5.3 for the statement) entails that there is an isomorphism $R(\lim D_J) \cong \lim RD_J$.

A similar argument is a standard way to prove that a certain object, defined (say) as the left adjoint to a certain functor, must admit an 'integral expansion' to which it is canonically isomorphic. For example, in the proof of what we call the *ninja Yoneda lemma* in 2.2.1, we carry out the following computation:

$$\frac{\frac{\frac{\frac{\mathrm{Set}\left(\int^{C\in\mathcal{C}} KC \times \mathcal{C}(X,C), Y\right)}{\int_{C\in\mathcal{C}} \mathrm{Set}(KC \times \mathcal{C}(X,C), Y)}}{\int_{C\in\mathcal{C}} \mathrm{Set}(\mathcal{C}(X,C), \mathrm{Set}(KC, Y))}}{[\mathcal{C}, \mathrm{Set}](\mathcal{C}(X, \_), \mathrm{Set}(K\_, Y))}}{\mathrm{Set}(KX, Y)}$$

where each step has to be interpreted as an application of a certain deduction rule that interchanges coends with ends, places them in and out of a hom functor, etc.

The reduction of proofs to a series of deduction steps embodies some sort of 'logical calculus', whose introduction rules resemble formulae such as

$$\mathrm{Cat}(\mathcal{C}, \mathcal{D})(F, G) \rightsquigarrow \int_C \mathcal{D}(FC, GC)$$

where the object $\mathrm{Cat}(\mathcal{C}, \mathcal{D})(F, G)$ is decomposed into an integral like $\int_C \mathcal{D}(FC, GC)$, and elimination rules look like

$$\int^C FC \times \mathcal{C}(C, X) \rightsquigarrow FX$$

where an integral is packaged into the object FX (of course, the symmetric nature of the canonical isomorphism relation makes all elimination rules reversible into introductions, and vice versa). Altogether, this allows us to derive the validity of a canonical isomorphism as a result of a chain of deductions, in a 'categorified' fashion.

The reader should not, of course, concentrate now on the meaning of these derivations at all; all notation will be duly introduced at the right time; when the statement of our Proposition 2.2.1 is introduced, the chain of deductions above will look almost tautological, and rightly so.

It is clear that, done in this way, category theory acquires an alluring algorithmic nature, and becomes (if not easy, at least) *easier to understand.*

Thus, the 'calculus' arising from these theorems encodes many, if not all, elementary constructions in category theory (we shall see that it subsumes the theory of (co)limits, it allows for a reformulation of the Yoneda lemma, it provides an explicit formula to compute pointwise Kan extensions, and it is a cornerstone of the 'calculus of bimodules', encoding the compositional nature of *profunctors*, the natural categorification of relational composition).

As it stands, (co)end calculus describes pieces of abstract and universal algebra [Cur12, GJ17], algebraic topology [Get09, May72, MSS02], representation theory [LV12], logic, computer science [Kme18], as well as pure category theory. The present book wishes to explore in detail a theory of (co)end calculus and its applications in detail.

So far, we have the motivations for the *topic* of this book. What about the motivation *for the book itself*? It shall be noted that (co)ends are not absent from the already existing literature on category theory: the topic is covered in [ML98], a statutory reading for every categorephile, and Mac Lane himself used coends to characterise a construction in algebraic topology as a 'tensor product' operation between functors in his [ML70]. Coends are mentioned (but not used as widely as they deserve) in Borceux's *Handbook* (in its first two tomes [Bor94a, Bor94b]). However, the topic lacks a treatment that is at the same time systematic, easy to read, and monographic.

As a result, (co)end calculus still lies just beyond the grasp of many people, and even of a few category theorists, because the literature that could teach its simple rules is a vast constellation of scattered papers, drawing from a large number of diverse disciplines.

This situation is all the more an issue because nowadays category theory has fruitfully contaminated with applied sciences. In the opinion of the author, it is of the utmost importance to provide his growing community with a single reference that accounts for the simplicity and unitary nature of category theory through (co)end calculus, thereby providing proof for its plethora of applications, and popularising this 'secret weapon' of category theorists, making it available to novices and non-mathematicians. The present endeavour is but a humble attempt to address this issue.

A brief history of (co)ends. Like many other pieces of mathematics, (co)end calculus was developed as a tool for homological algebra: the first definition of a universal object called '(co)end' was given in a paper studying the Ext functors, and the father of (co)end calculus is none other than Nobuo Yoneda. In [Yon60] he singled out most of the definitions we will introduce in the first five chapters of this book.

Having read Yoneda's original paper in order to write the present introduction, we find no better way than to quote the original text, untouched, just occasionally adding a few details here and there to frame Yoneda's words in a modern perspective, (but also in order to adapt them to our choice of notation).

There are multiple reasons for this strategy: [Yon60] is a mathematical gem, an enticing prelude of all the theory developed in the subsequent decades, and a perfect prelude to the story this book tries to tell. Even more so, in reporting Yoneda's words we believe we are also doing a service to the mathematical community, since the integral text of [Yon60] is somewhat difficult to find.

Our sincere hope is that this introduction, together with the whole book the reader is about to read, credits the visionary genius of Yoneda: category theory has few theorems, and one of them is a lemma. The *Yoneda lemma*, in its myriad incarnations, is certainly a cornerstone of structural thinking, way beyond category theory. If anything more was needed to revere Nobuo Yoneda, let this be (co)end calculus.

The paper [Yon60] starts introducing (co)ends in the following way:

> Let $\mathcal{C}$ be a category. By a *left $\mathcal{C}$-group* we mean a covariant functor M of $\mathcal{C}$ with values in the category Ab of abelian groups and homomorphisms. [...] Also by a $\mathcal{C}^*$-group (or a *right $\mathcal{C}$-group*) we mean a contravariant functor $K : \mathcal{C} \to \text{Ab}$. [...] Functors of several variables with values in Ab will accordingly be called $\mathcal{B}$-$\mathcal{C}$-groups, $\mathcal{B}^*$-$\mathcal{C}$-groups, etc.
>
> Let H be a $\mathcal{C}^*$-$\mathcal{C}$-group, and G an additive group. By a *balanced homomorphism* $\mu : G \Rightarrow H$ we mean a system of homomorphisms $\mu(C) : G \to H(C, C)$ defined for all objects $C \in \mathcal{C}$ such that for every

map $\gamma : C \to C'$ in $\mathcal{C}$ commutativity holds in the diagram

$$\begin{array}{ccc} G & \xrightarrow{\mu(C)} & H(C,C) \\ {\scriptstyle \mu(C')}\downarrow & & \downarrow{\scriptstyle H(C,\gamma)} \\ H(C',C') & \xrightarrow[H(\gamma,C')]{} & H(C,C'). \end{array}$$

Also by a balanced homomorphism $\lambda : H \Rightarrow G$ we mean a system of homomorphisms $\lambda(C) : H(C,C) \to G$ defined for all objects $C \in \mathcal{C}$ such that for every map $\gamma : C \to C'$ in $\mathcal{C}$ commutativity holds in the diagram

$$\begin{array}{ccc} H(C',C) & \xrightarrow{H(C',\gamma)} & H(C',C') \\ {\scriptstyle H(\gamma,C)}\downarrow & & \downarrow{\scriptstyle \lambda(C')} \\ H(C,C) & \xrightarrow[\lambda(C)]{} & G \end{array}$$

Of course, here a left/right '$\mathcal{C}$-group' is merely an Ab-enriched presheaf (covariant or contravariant) with domain $\mathcal{C}$. The above paragraphs define the fundamental notions we will use throughout the entire book: *wedges* and *cowedges*. These are exactly natural maps to/from a constant, which vary taking into account the fact that $H(C,C)$ depends both covariantly and contravariantly on C. There is a category of such (co)wedges, and a process dubbed *(co)integration* picks the initial and terminal objects of such categories:

> An additive group I together with a balanced $\theta : H \Rightarrow I$ is called *integration* of a $\mathcal{C}^*$-$\mathcal{C}$-group H if it is universal among balanced homomorphisms from H, i.e. if for any other balanced homomorphism $\lambda : H \Rightarrow G$ there is a unique morphism $\zeta : I \to G$ such that $\zeta \circ \theta(C) = \lambda(C)$ for every object $C \in \mathcal{C}$.
>
> [Integrations and cointegrations] are [...] given as follows: for a map $\gamma : C \to C'$ in $\mathcal{C}$ we put $H(\gamma) = H(C',C)$, $H(\gamma^*) = H(C,C')$, and define homomorphisms
>
> $$\partial_\gamma : H(\gamma) \to H(C,C) \oplus H(C',C')$$
> $$\delta_\gamma : H(C,C) \oplus H(C',C') \to H(\gamma^*)$$
>
> by
>
> $$\partial_\gamma(h') = h' \circ \gamma \oplus (-\gamma \circ h')$$
> $$\delta_\gamma(h \oplus h'') = \gamma \circ h - h'' \circ \gamma$$
>
> Denote by Σ_0 and Π^0 the direct sum $\sum_{C \in \mathcal{C}} H(C,C)$ and the direct product $\prod_{C \in \mathcal{C}} H(C,C)$ respectively. Also, denote by Σ_1 and Π^1 the

direct sum $\sum_{\gamma\in\hom(\mathcal{C})} H(\gamma)$ and $\prod_{\gamma\in\hom(\mathcal{C})} H(\gamma^*)$ respectively. Then $\partial_\gamma, \delta_\gamma$ are extended to homomorphisms

$$\partial : \Sigma_1 \to \Sigma_0 \qquad \delta_{CC'} : \Pi^0 \to \Pi^1.$$

Now [Yon60] proves that the 'integration' $\int_{\mathcal{C}} H$ and the 'cointegration' $\int_{\mathcal{C}}^* H$ of a $\mathcal{C}^*$-$\mathcal{C}$-group are given respectively by the cokernel of $\partial_{CC'}$, and by the kernel of $\delta_{CC'}$, for suitably defined maps $\partial_{CC'}$ and $\delta_{CC'}$.

To avoid confusion, we stress that in this definition Yoneda employed the *opposite* choice of terminology that we will introduce later on: an *integration* for H is a coend $\int^C H(C,C)$, but Yoneda denotes it $\int_C H(C,C)$; a *cointegration* is an end $\int_C H(C,C)$, but Yoneda denotes it $\int_C^* H(C,C)$. Always be careful if you consult [Yon60] or references from the same era.

The coend $\int^C H$ of a functor $H : \mathcal{C}^{\mathrm{op}} \times \mathcal{C} \to \mathcal{D}$ is a colimit (in the particular case of Ab-enriched functors, a cokernel) built out of H, and the end $\int_C H$ is a limit; precisely, the kernel of a certain group homomorphism. Once this terminology has been set up, given two functors $F, G : \mathcal{C} \to \mathcal{D}$, if we let H be the functor $(C, C') \mapsto \mathcal{D}(FC, GC')$, then a family of arrows $\alpha_C : FC \to GC$ forms the components of a natural transformation $\alpha : F \Rightarrow G$ if and only if the components $\alpha_C : FC \to GC$ lie in the kernel of a 'differential' $\bar{\delta} : \prod_{C\in\mathcal{C}} H(C,C) \to \prod_{\gamma\in\hom(\mathcal{C})} H(\gamma^*)$, obtained in the obvious way 'gluing' all the $\delta_{CC'}$ together.

> In dealing with functors of more variables, we shall often inscribe x (or y, z) to indicate the two entries to be considered in the (co)integration, namely we write
>
> $$\int_{X\in\mathcal{C}} H(\ldots, X, \ldots, X, \ldots). \tag{0.2}$$
>
> This is based on the following fact: let H, H' be $\mathcal{C}$-$\mathcal{C}^*$-groups, and let $\theta : H \to \int_{\mathcal{C}} H$, $\theta' : H' \to \int_{\mathcal{C}} H'$ be the integrations. Then a natural transformation $\eta : H \Rightarrow H'$ induces a unique homomorphism $\int_{\mathcal{C}} \eta : \int_{\mathcal{C}} H \to \int_{\mathcal{C}} H'$ such that $\left(\int_{\mathcal{C}} \eta\right) \circ \theta(C) = \theta'(C) \circ \eta(C,C)$. Thus if H is a $\mathcal{B}$-$\mathcal{C}^*$-$\mathcal{C}$-group, then $\int_{X\in\mathcal{C}}^{(*)} H(B, X, X)$ is a $\mathcal{B}$-group. On this account, for an $\mathcal{A}$-$\mathcal{B}^*$-$\mathcal{B}$-$\mathcal{C}$-$\mathcal{C}^*$-group H we have
>
> $$\int_{Y\in\mathcal{B}} \int_{X\in\mathcal{C}} H(A, Y, Y, X, X) = \int_{X\in\mathcal{C}} \int_{Y\in\mathcal{B}} H(A, Y, Y, X, X).$$

Here Yoneda introduces one of the pillars of coend calculus, the *Fubini rule*, i.e. the fact that the result of a (co)integration is the same regardless of the order of integration; this is ultimately just a consequence of the functoriality of the assignment sending a $\mathcal{C}$-$\mathcal{C}^*$-group H into its (co)integration. Of course, there is nothing special about the codomain of H being the category of abelian groups: any sufficiently (co)complete category $\mathcal{D}$ will do, as long as the (co)integrations involved exist.

In modern terms, the Fubini rule can be obtained as a consequence of a much deeper, and hopefully more enlightening, result: we prove it in our Theorem 1.3.1.

> Next for a $\mathcal{B}$-group M and a $\mathcal{C}$-group N, $M \otimes N : N(C) \otimes M(B)$ is a $\mathcal{B}$-$\mathcal{C}$-group, and $\underline{\hom}(M, N) = \hom(MB, NC)$ is a $\mathcal{B}^*$-$\mathcal{C}$-group. For an $\mathcal{A}$-group M and a $\mathcal{B}$-$\mathcal{C}$-$\mathcal{C}^*$-group H we have:
>
> $$\int_{X\in\mathcal{C}} MA \otimes H(B,X,X) = MA \otimes \int_{X\in\mathcal{C}} H(B,X,X)$$
> $$\int^*_{X\in\mathcal{C}} \hom(MA, H(B,X,X)) = \hom\left(MA, \int^*_{X\in\mathcal{C}} H(B,X,X)\right)$$
> $$\int^*_{X\in\mathcal{C}} \hom(H(B,X,X), MA) = \hom\left(\int^*_{X\in\mathcal{C}} H(B,X,X), MA\right)$$

As an immediate consequence of these statements we get another fundamental building-block of a coend 'calculus': given two functors $M : \mathcal{B} \to \mathrm{Ab}$ and $N : \mathcal{B}^{\mathrm{op}} \to \mathrm{Ab}$, they can be *tensored* by the integration $M \boxtimes N := \int_{B\in\mathcal{B}} N(B) \otimes_{\mathbb{Z}} M(B)$.

A rather interesting perspective on this construction is the following: the result remains true when the Ab-category $\mathcal{B}$ has a single object, so it is merely a ring B. In such a case, a functor $M : B \to \mathrm{Ab}$ is a left module, and a functor $N : B^{\mathrm{op}} \to \mathrm{Ab}$ is a right module. The integration (or in modern terms, the coend) $\int^B N \otimes_{\mathbb{Z}} M$ in this case is exactly the tensor product of B-modules: it has the universal property of the cokernel of the map

$$\bigoplus_{b\in B} M \otimes_{\mathbb{Z}} N \xrightarrow{\varrho} M \otimes_{\mathbb{Z}} N$$

defined by $\varrho(b, m, n) = b.m - n.b$.

As the reader might now suspect, few analogies are more fruitful than the one between modules over which a monoid object acts, and presheaves $\mathcal{C} \to \mathrm{Set}$.

Structure of the book. We shall now briefly review the structure of the book. In the first three chapters we outline the basic rules of (co)end calculus. After having defined (co)ends as universal objects and having proved that they can be characterised as (co)limits, we start denoting such objects as integrals $\int_C T(C,C)$ or $\int^C T(C,C)$. This notation is motivated by the fact that (co)ends 'behave like integrals' in that a *Fubini rule* of exchange holds, see Theorem 1.3.1.

Then in Chapter 2 we introduce the first rules of the calculus: the Yoneda lemma A.5.3 can be restated in terms of a certain coend computation, and pointwise Kan extensions can be computed by means of a (co)end.

After this, we study the particular case of left Kan extensions along the Yoneda embedding: in some sense, the theory of such extensions alone embodies category theory.

The subsequent chapters begin to introduce more modern topics described by means of (co)end calculus. The theories of *weighted (co)limits* (Chapter 4), of *profunctors* (Chapter 5) and *operads* (Chapter 6) are cornerstones of 'formal' approaches to category theory. Weighted (co)limits are the correct notion of (co)limit in an enriched or formal-categorical (see 2.4 and [Gra80]) setting; profunctors are a bicategory where one can re-enact all of category theory, and are deeply linked to categorical algebra and representation theory; operads, initially introduced as a technical means to solve an open problem in homotopy theory, now constitute the common ground where universal algebra and algebraic topology meet. The final point will be Theorem 6.4.7, where we draw a tight link between profunctors and operads.

Chapter 7 studies higher dimensional analogues of (co)ends; first, we study (co)ends in 2-categories; then we move up to infinity and study homotopy-coherent analogues of (co)ends in simplicial categories [Ber07], quasicategories [Lur09], model categories [Hov99] and derivators [Gro13].

Appendix A serves as a short introduction to category theory: it fixes the notation we employ in the previous chapters. A basic knowledge of elementary mathematics is a prerequisite, but we will introduce most categorical jargon from scratch.

Each chapter has a short introduction, in the form of a small abstract; this allows the interested reader to get a glimpse into the content and fundamental results of each chapter (often, one or two main theorems). We believe this format is easier to consult than a comprehensive survey of each chapter given all at once in the introduction, so we felt free to keep this introductory account of the content of the book pretty brief.

Several exercises follow each chapter of the book; there are questions of every level, sometimes easy, sometimes more difficult. Some of them make the reader rapidly acquainted with the computational approach to category theory offered by (co)end calculus; some others shed a new light on old notions. In approaching them, we advise you to avoid element-wise reasoning; instead, find either an abstract argument, or a 'deduction-style' one.

Some of the exercises are marked with a ◉◉ symbol (eyes wide with fear): this means they are more difficult and less well posed questions than the others. This might be deliberate (and thus part of the exercise is understanding what the question is) or not (and thus the question and its answer are not completely clear even to the author). In the second case, it is likely that a complete answer might result in new mathematics that the solvers are encouraged to develop.

Other kinds of 'small eyes' are present throughout the book: the paragraphs decorated with a ◔◔ contain material that can be skipped at first reading, or material that deepens a prior topic in a not-so-interesting detour; ◑◑ is used to signal key remarks and more generally important material that we ask the reader to digest properly and analyse in full detail.

Notation. Having to deal with many different sources for this exposition, to hope to maintain a coherent choice of notation throughout the whole book is wishful thinking; however, the author did his best to provide a coherent enough one, striving to make it at the same time expressive and simple.

In general, 1-dimensional category-like structures will be denoted as calligraphic letters $\mathcal{C}, \mathcal{D}, \dots$; objects of $\mathcal{C}$ are denoted $C, C', \dots \in \mathcal{C}$. In contrast, 2-categories are often denoted with a sans-serif case $\mathsf{Cat}, \mathsf{K}, \mathsf{A}, \dots$; an object of the 2-category of small categories is denoted $\mathcal{C} \in \mathsf{Cat}$, but an object of an abstract 2-category is denoted $A \in \mathsf{K}$.

Functors between categories are denoted as capital Latin letters such as F, G, H, K (although there can be small deviations from this rule); the category of functors $\mathcal{C} \to \mathcal{D}$ between two categories is almost always denoted as $\mathrm{Cat}(\mathcal{C}, \mathcal{D})$ (or less often $[\mathcal{C}, \mathcal{D}]$; this will be done especially when $[\mathcal{C}, \mathcal{D}]$ is regarded as the internal hom of the closed structure in Cat, or when it is necessary to save some space); the symbols $\_, =$ are used as placeholders for the 'generic argument' of a functor or bifunctor (they mark temporal precedence of saturation of a variable); morphisms in the category $\mathrm{Cat}(\mathcal{C}, \mathcal{D})$ (i.e. natural transformations between functors) are often written in lowercase Greek, or lowercase Latin alphabet, and collected in the set $\mathrm{Cat}(\mathcal{C}, \mathcal{D})(F, G)$.

The simplex category $\mathbf{\Delta}$ is the *topologist's delta* (as opposed to the *algebraist's delta* $\mathbf{\Delta}_+$ which has an additional initial object $[-1] := \varnothing$), having objects *non-empty* finite ordinals $[n] := \{0 < 1 \cdots < n\}$; we denote $\Delta[n]$ the representable presheaf on $[n] \in \mathbf{\Delta}$, i.e. the image of $[n]$ under the Yoneda embedding of $\mathbf{\Delta}$ in the category $\mathsf{sSet} = \widehat{\mathbf{\Delta}}$ of simplicial sets. More generally, we indicate the Yoneda embedding of a category $\mathcal{C}$ into its presheaf category with よ$_{\mathcal{C}}$, or simply よ, i.e. with the hiragana symbol for 'yo'.

Acknowledgments. Writing this book spanned the last five years of my life; it would probably take five more years to do it in the form the topic deserves. I did my best, and if I managed to be proud of the result, it is because I have been helped a lot.

The draft of this book accompanied me to all sorts of places: this has made the list of people who helped and guided me through it quite long. After all this time, it remains all the more true today that 'in some sense, I am not the only author of this note', as I used to write in the early stages of this project.

Several people participated and extensively commented on such drafts; each of them helped to make the present book a little better than it would have been if I had been alone.

I would like to thank S. Ariotta (for providing more help than I deserved, in the proof of the ∞-Fubini rule in 7.3), J. D. Christensen, I. Di Liberti, T. de Oliveira Santos (for an attentive proofreading when my eyes were bleeding), D. Fiorenza (a friend and an advisor I will never be able to refund for his constant support), A. Gagna, N. Gambino (for offering me the opportunity to discuss the content of this book in front of his students in Leeds; in just a few days I realised years of meditation were still insufficient to teach (co)end calculus), L. (for accompanying me there, and for being the best *mate* for a much longer trip), F. Genovese (for believing in (co)end calculus with all his heart, and for making dg-categories more understandable through it), A. Joyal (for letting me sketch the statement of 4.2.2 on the back of a napkin in a cafe in Paris), G. Marchetti, A. Mazel-Gee, B. Milewski, G. Mossa, D. Palombi (for sharing with me his truffles, his friendship, his monoidal categories, and his typographical kinks), D. Ravenel (for believing in this book's draft title, 'This is the (co)end, my only (co)friend' more than I will ever do), M. Román, G. Ronchi (for accepting my harsh ways while I was learning how to be a mentor, and because he Believes that mathematics is Beautiful), J. Rosický (for he trusted in me and gave me pearls of wisdom), E. Riehl, E. Rivas, T. Trimble, M. Vergura, C. Williams, K. Wright, N. Yanofsky, and countless other people who pointed out mistakes, faulty arguments, typos.

A separate thanks goes to S., P., G., C.: they opened their doors to me when I was frail and broken-hearted. Writing means, always, opening a wound, exposing your tendons to the blade of a sharp knife. During these years I have been protected by the warmth of your smile, of your houses, of your brilliance, of your friendship.

Various institutions supported me through the many years necessary to finish this book. I would like to thank the University of Western Ontario, Masaryk University, The Max Planck Institute for mathematics in Bonn for providing me time and money to keep writing, and the University of Coimbra, which added to the deal a *completely* distraction-free environment. Additional help came from Taltech, where I am working while this book approaches its final form, and from all Estonia: *aitäh!*

Ringrazio per ultimo mio padre, che non comprenderebbe questa dedica se non la scrivessi in italiano, e che non leggerà mai questo libro 'assorto com'è in altra e lontana scienza'. A sei anni mi fece sedere sulle sue ginocchia, e scrisse su un foglio il numero 1 seguito da nove zeri, per farmi capire quanto è grande un miliardo: si può dire tutto questo sia iniziato lì.

Grazie a tutti.

1

Dinaturality and (Co)ends

SUMMARY. Naturality of a family of morphisms $\alpha_C : FC \to GC$ defines the correct notion of map between functors F, G; yet it is not able to describe more subtle interactions that can occur between F and G, for example when both functors have a product category like $\mathcal{C}^{op} \times \mathcal{C}$ as domain. A transformation that takes into account the fact that F, G act on morphisms once covariantly and once contravariantly is called *dinatural*.

As ill-behaved as it may seem (in general, dinatural transformations cannot be composed), this notion leads to the definition of a *(co)wedge* and *(co)end* for a functor $T : \mathcal{C}^{op} \times \mathcal{C} \to \mathcal{D}$: a dinatural transformation having constant (co)domain, and a suitable universal property. This is in perfect analogy with the theory of (co)limits: universal natural transformations from/to a constant functor. Unlike colimits, however, (co)ends support a *calculus*, that is, a set of inference rules allowing mechanical proof of non-trivial statements as initial and terminal points of a chain of deductions.

The purpose of this chapter, and indeed of the entire book, is to familiarise its readers with the rules of calculus.

> Los idealistas arguyen que las salas hexagonales son una forma necesaria del espacio absoluto o, por lo menos, de nuestra intuición del espacio.
>
> ---
>
> J.L. Borges *La biblioteca de Babel*

1.1 Supernaturality

We choose to let the name 'supernaturality' describe the two sorts of generalisations of naturality for functors that we will investigate throughout the book: *di*naturality, in 1.1.1, and *extra*naturality, in 1.1.8.

1.1.1 Dinaturality

This first section starts with a simple example. We denote by Set the category of sets and functions, considered with its natural cartesian closed structure (see A.4.3.AD4). This means that we have a bijection of sets

$$\mathrm{Set}(A \times B, C) \cong \mathrm{Set}(A, C^B) \tag{1.1}$$

that is natural in all three arguments, where we let C^B denote the set of functions $f : B \to C$. The bijection above is defined by the maps

$$(f : A \times B \to C) \mapsto A \xrightarrow{\eta_{A,(B)}} (A \times B)^B \xrightarrow{f^B} C^B$$
$$(g : A \to C^B) \mapsto A \times B \xrightarrow{g \times B} C^B \times B \xrightarrow{\epsilon_{C,(B)}} C$$

by means of suitable *unit* and *counit* maps η and ϵ (see A.4.1) witnessing the adjunction. Let us concentrate on the counit map alone (a dual reasoning will yield similar conclusions for the unit): it is a natural transformations having components

$$\{\epsilon_{X,(B)} : X^B \times B \to X \mid X \in \mathrm{Set}\}. \tag{1.2}$$

This family of functions sends a pair $(f, b) \in X^B \times B$ to the element $fb \in X$, and thus deserves the name of *evaluation.*

For the purpose of our discussion, we shall consider this family of morphisms not only natural in X (as every counit morphism), but also mutely depending on the variable B in its codomain. This means that $X^B \times B$ is the image of the pair (B, B) under the functor $(U, V) \mapsto X^U \times V$, and X can be regarded similarly as the image of (B, B) under the constant functor in X. Both functors thus have 'type' $\mathrm{Set}^{\mathrm{op}} \times \mathrm{Set} \to \mathrm{Set}$.

The evaluation maps $\epsilon_{X,(B)}$ however do not vary naturally in the variable B; the most we can say is that for each function $f \in \mathrm{Set}(B, B')$ the following square is commutative:

$$\begin{array}{ccc} X^{B'} \times B & \xrightarrow{X^f \times B} & X^B \times B \\ {\scriptstyle X^{B'} \times f}\downarrow & & \downarrow{\scriptstyle \epsilon} \\ X^{B'} \times B' & \xrightarrow[\epsilon]{} & X. \end{array} \tag{1.3}$$

This relation does not resemble naturality so much, but it can be easily deduced from the requirement that the adjunction isomorphisms (1.1) are natural in the variable B. In fact, such naturality imposes the commutation of the square

$$\begin{array}{ccc} \mathrm{Set}(A, X^{B'}) & \longrightarrow & \mathrm{Set}(A \times B', X) \\ {\scriptstyle \mathrm{Set}(A,X^f)}\downarrow & & \downarrow{\scriptstyle \mathrm{Set}(A\times f,X)} \\ \mathrm{Set}(A, X^B) & \longrightarrow & \mathrm{Set}(A \times B, X) \end{array} \tag{1.4}$$

for an arrow $f : B \to B'$ (the horizontal maps are the adjunction isomorphisms $\_ \mapsto \epsilon \circ (\_ \times B)$), and this in turn entails that we have equations

$$\epsilon_{X,(B')} \circ (u \times B') \circ (A \times f) = \epsilon_{X,(B)} \circ (X^f \circ u) \times B \tag{1.5}$$

$$\epsilon_{X,(B')} \circ (X^{B'} \times f) \circ (u \times B) = \epsilon_{X,(B)} \circ (X^f \times B) \circ (u \times B) \tag{1.6}$$

for every $u : A \to X^{B'}$. But since this is an equality for every such u, the functions $\epsilon_{X,(B')} \circ (X^{B'} \times f)$ and $\epsilon_{X,(B)} \circ (X^f \times B)$ must also be equal.

So it would seem that there is no way to frame the diagram above in the usual context of naturality for a transformation of functors. Fortunately, a suitable generalisation of naturality (a 'supernaturality' condition) encoding the above commutativity is available to describe this and other similar phenomena.

As already said, the correspondence $(B, B') \mapsto C^B \times B'$ is a functor with domain $\mathrm{Set}^{\mathrm{op}} \times \mathrm{Set}$; it turns out that these functors, where the domain is a product of a category with its opposite, support a notion of *di*naturality besides classical naturality. This notion is more suited to capturing the phenomenon we just described: in fact, most of the transformations that are canonical, depending on two variables $(C, C') \in \mathcal{C}^{\mathrm{op}} \times \mathcal{C}$, but not natural, can be seen as dinatural.

Definition 1.1.1 (Dinatural transformation). Let $\mathcal{C}, \mathcal{D}$ be two categories. Given two functors $P, Q : \mathcal{C}^{\mathrm{op}} \times \mathcal{C} \to \mathcal{D}$, a *dinatural transformation* $\alpha : P \Longrightarrow Q$ consists of a family of arrows

$$\alpha_C : P(C, C) \to Q(C, C) \tag{1.7}$$

indexed by the objects of $\mathcal{C}$ and such that for any $f : C \to C'$ the following diagram commutes

$$\begin{array}{ccccc} P(C', C) & \xrightarrow{P(f,C)} & P(C, C) & \xrightarrow{\alpha_C} & Q(C, C) \\ {\scriptstyle P(C',f)}\downarrow & & & & \downarrow{\scriptstyle Q(C,f)} \\ P(C', C') & \xrightarrow[\alpha_{C'}]{} & Q(C', C') & \xrightarrow[Q(f,C')]{} & Q(C, C'). \end{array} \tag{1.8}$$

Remark 1.1.2. The notion of dinaturality takes into account the fact that a functor $P : \mathcal{C}^{\text{op}} \times \mathcal{C} \to \mathcal{D}$ maps at the same time two 'terms' of the same 'type' $\mathcal{C}$, once covariantly in the second component and once contravariantly in the first. On arrows $f : C \to C'$ the functor P acts as follows:

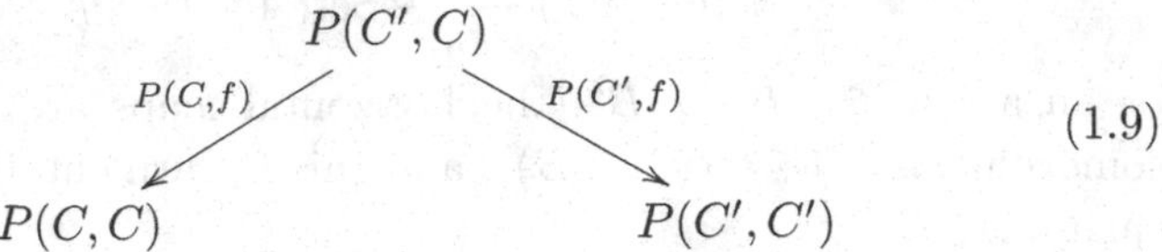

(1.9)

Given two such functors, say $P, Q : \mathcal{C}^{\text{op}} \times \mathcal{C} \to \mathcal{D}$, we can consider the two diagrams (1.9) and

$$
\begin{array}{ccc}
Q(C,C) & & Q(C',C') \\
& \searrow^{Q(C,f)} \quad {}^{Q(f,c)}\swarrow & \\
& Q(C,C') &
\end{array}
\tag{1.10}
$$

In the same way that a natural transformation $F \Rightarrow G$ can be seen as a family of maps that 'fill the gap' between $F(f)$ and $G(f)$ in a commutative square, a *dinatural* transformation between P and Q can be seen as a way to close the hexagonal diagram connecting the action on arrows of P to the action on arrows of Q:

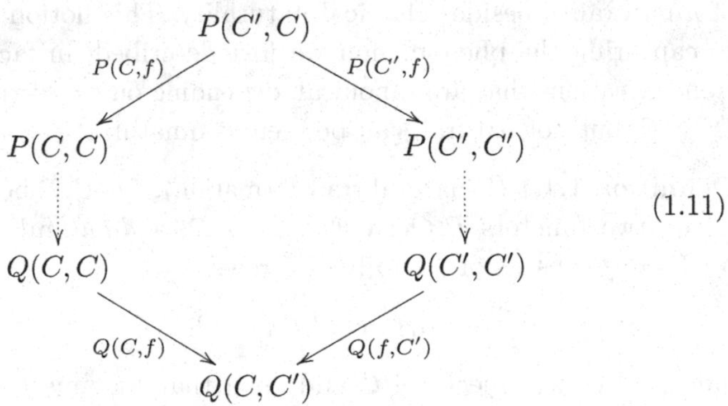

(1.11)

This is precisely the diagram drawn in (1.8).

Remark 1.1.3. If we let P_C be the functor $(U, V) \mapsto C^U \times V$, the counit components $\epsilon_{C(B)} : P_C(B, B) \to C$ of the cartesian closed adjunction form a dinatural transformation $\epsilon : P_C \Rightarrow \Delta_C$, where Δ_C is the constant functor at C.

Such dinatural transformations, having constant codomain, deserve a special name.

Definition 1.1.4 ((Co)wedge). Let $P : \mathcal{C}^{\text{op}} \times \mathcal{C} \to \mathcal{D}$ be a functor;

WC1. A *wedge* for P is a dinatural transformation $\Delta_D \Rightarrow P$ from the constant functor on the object $D \in \mathcal{D}$ (we often denote such a constant functor simply by the name of the constant, $D : \mathcal{C}^{\text{op}} \times \mathcal{C} \to \mathcal{D}$) defined by the rules $(C, C') \mapsto D$, $(f, f') \mapsto \text{id}_D$.

WC2. Dually, a *cowedge* for P as above is a dinatural transformation $P \Rightarrow \Delta_D$ having codomain the constant functor on the object $D \in \mathcal{D}$.

Remark 1.1.5. Wedges for a fixed functor P as above form the class of objects of a category $\text{Wd}(P)$, where a morphism of wedges is a morphism between their domains that makes an obvious triangle commute; given two wedges $\alpha : D \Rightarrow P$ and $\alpha' : D' \Rightarrow P$, a morphism of wedges consists of an arrow $u : D \to D'$ such that the triangle

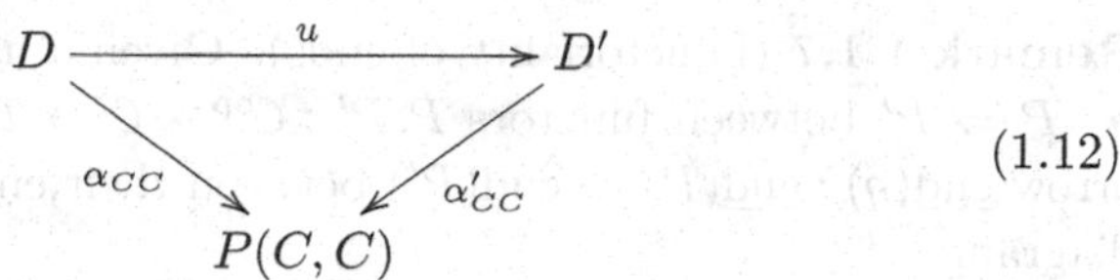

(1.12)

is commutative for every component α_{CC} and α'_{CC}. (Note the role of quantifiers: the same u makes (1.12) commute for *every* component of the wedges.)

Dually, there is a category $\text{Cwd}(P)$ of cowedges for P, where morphisms of cowedges are morphisms between codomains (of course there is a relation between the two categories: cowedges for P coincide with the opposite category of wedges for the opposite functor).

We now define the end of P as a terminal object in $\text{Wd}(P)$, and the coend as an initial object in $\text{Cwd}(P)$.

Definition 1.1.6 ((Co)end). Let $P : \mathcal{C}^{\text{op}} \times \mathcal{C} \to \mathcal{D}$ be a functor.

- The *end* of P consists of a terminal wedge $\omega : \underline{\text{end}}(P) \Rightarrow P$; the object $\underline{\text{end}}(P) \in \mathcal{D}$ itself is often called the *end* of the functor.
- Dually, the *coend* of P as above consists of an initial cowedge $\alpha : P \Rightarrow \underline{\text{coend}}(P)$; similarly, the object $\underline{\text{coend}}(P)$ itself is often called the coend of P.

Spelled out explicitly, the universality requirement means that for any other wedge $\beta : D \Rightarrow P$ the diagram

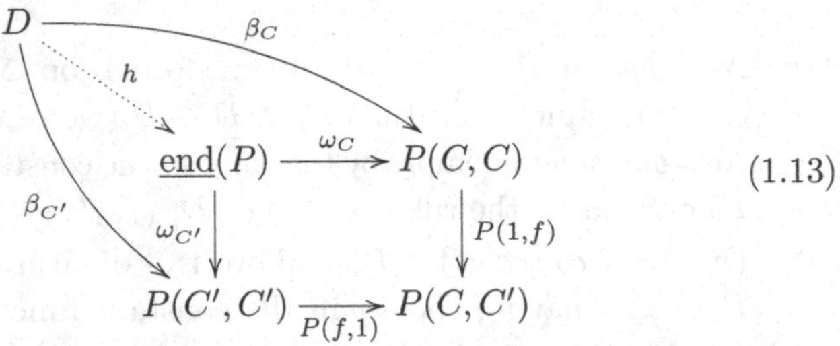

(1.13)

commutes for a unique arrow $h : D \to \underline{\mathrm{end}}(P)$, for every arrow $f : C \to C'$. Note again the role of quantifiers: the arrow h is the same for every component of the wedge. A dual diagram can be depicted for the coend of P.

Remark 1.1.7 (Functoriality of ends). Given a natural transformation $\eta : P \Rightarrow P'$ between functors $P, P' : \mathcal{C}^{\mathrm{op}} \times \mathcal{C} \to \mathcal{D}$ there is an induced arrow $\underline{\mathrm{end}}(\eta) : \underline{\mathrm{end}}(P) \to \underline{\mathrm{end}}(P')$ between their ends, as depicted in the diagram

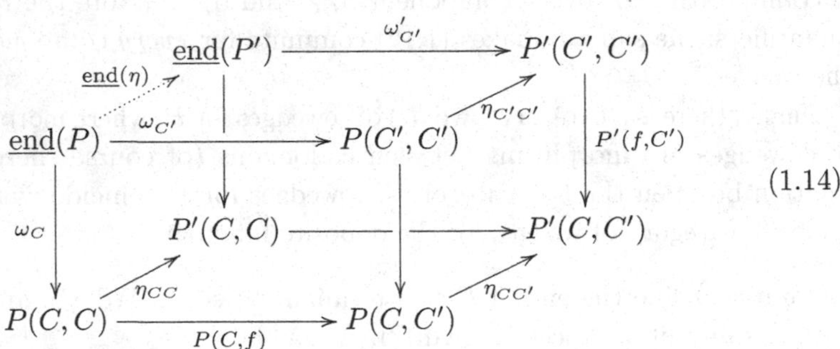

(1.14)

When all ends exist, sending a functor P into its end $\underline{\mathrm{end}}(P)$ is a (covariant) functor $\underline{\mathrm{end}} : \mathrm{Cat}(\mathcal{C}^{\mathrm{op}} \times \mathcal{C}, \mathcal{D}) \to \mathcal{D}$. $\underline{\mathrm{end}}$ preserves composition of morphisms, because the usual uniqueness argument using the universal property applies, and the arrow $\underline{\mathrm{end}}(\eta) \circ \underline{\mathrm{end}}(\eta')$ must coincide with $\underline{\mathrm{end}}(\eta \circ \eta')$ in a suitable pasting of cubes. Similarly, the unique arrow induced by $\mathrm{id}_P : P \Rightarrow P$ must be the identity of $\underline{\mathrm{end}}(P)$.

1.1.2 Extranaturality

A slightly less general, but better behaved[1] notion of supernaturality, which allows us again to define (co)wedges and thus (co)ends, is available: the notion is called *extra*naturality and it was introduced in [EK66b].

Definition 1.1.8 (Extranatural transformation). Let $\mathcal{A}, \mathcal{B}, \mathcal{C}, \mathcal{D}$ be categories, and P, Q be functors

$$P : \mathcal{A} \times \mathcal{B}^{\text{op}} \times \mathcal{B} \to \mathcal{D},$$
$$Q : \mathcal{A} \times \mathcal{C}^{\text{op}} \times \mathcal{C} \to \mathcal{D}.$$

An *extranatural transformation* $\alpha : P \Rightarrow Q$ consists of a collection of arrows

$$\alpha_{ABC} : P(A, B, B) \longrightarrow Q(A, C, C) \tag{1.15}$$

indexed by triples of objects in $\mathcal{A} \times \mathcal{B} \times \mathcal{C}$ such that the following hexagonal diagram commutes for every triple of arrows $f : A \to A'$, $g : B \to B'$, $h : C \to C'$, all chosen from the appropriate domains:

$$\begin{array}{ccccc} P(A, B', B) & \xrightarrow{P(f,B',g)} & P(A', B', B') & \xrightarrow{\alpha_{A'B'C}} & Q(A', C, C) \\ {\scriptstyle P(A,g,B)}\downarrow & & & & \downarrow{\scriptstyle Q(A',C,h)} \\ P(A, B, B) & \xrightarrow[\alpha_{ABC'}]{} & Q(A, C', C') & \xrightarrow[Q(f,h,C')]{} & Q(A', C, C'). \end{array} \tag{1.16}$$

Notice how this commutative hexagon can be equivalently described as the juxtaposition of three distinguished commutative squares, depicted in [EK66b]: the three can be obtained by letting f and h, f and g, or g and h respectively be identities in the former diagram, which thus collapses to

$$\begin{array}{ccc} P(A, B, B) & \xrightarrow{P(f,B,B)} & P(A', B, B) \\ {\scriptstyle \alpha_{ABC}}\downarrow & & \downarrow{\scriptstyle \alpha_{A'BC}} \\ Q(A, C, C) & \xrightarrow[Q(f,C,C)]{} & Q(A', C, C) \end{array} \qquad \begin{array}{ccc} P(A, B', B) & \xrightarrow{P(A,B',g)} & P(A, B', B') \\ {\scriptstyle P(A,g,B)}\downarrow & & \downarrow{\scriptstyle \alpha_{AB'C}} \\ P(A, B, B) & \xrightarrow[\alpha_{ABC}]{} & Q(A, C, C) \end{array}$$

[1] We say *better behaved* since extranaturality admits a graphical calculus translating commutativity requirements into the requirement that certain string diagrams can be deformed one into the other.

$$\begin{array}{ccc} P(A,B,B) & \xrightarrow{\alpha_{ABC}} & Q(A,C,C) \\ {\scriptstyle \alpha_{ABC'}}\downarrow & & \downarrow{\scriptstyle Q(A,C,h)} \\ Q(A,C',C') & \xrightarrow[Q(A,h,C')]{} & Q(A,C,C') \end{array} \tag{1.17}$$

Definition 1.1.9 (Mute functor)**.** Let $n \geq 1$ and $F : \mathcal{A}_1 \times \cdots \times \mathcal{A}_n \to \mathcal{B}$ be a functor; we say that F is *mute in its ith component* if F factors as

$$\begin{array}{ccc} \prod_i \mathcal{A}_i & \longrightarrow & \mathcal{B} \\ & \searrow & \nearrow_{\bar{F}} \\ & \prod_{k \neq i} \mathcal{A}_k & \end{array} \tag{1.18}$$

where π_i is the projection canceling out the ith factor of the product.

Remark 1.1.10. We can again define (co)wedges in this setting: if $\mathcal{B} = \mathcal{C}$ and in $P(A,B,B) \to Q(A,C,C)$ the functor P is the constant functor on $D \in \mathcal{D}$, and $Q(A,C,C) = \bar{Q}(C,C)$ is mute in A, then we get a wedge condition for $D \Rightarrow Q$; dually we obtain a cowedge condition for $P(B,B) \to Q(A,B,B) \equiv D'$ for all A, B, C.

It is worth mentioning that an extranatural transformation contains more information than a dinatural one, since in 1.1.8 we are given arrows

$$F(B,B) \xrightarrow{\alpha_{BB'}} G(B',B') \tag{1.19}$$

that are simultaneously a cowedge in B for each B', and a wedge in B' for all $B, B' \in \mathcal{B}$. We shall see in a while that extranaturality can be obtained as a special case of dinaturality.

Both dinatural and extranatural transformations give rise to the same notion of (co)end, defined as a universal (co)wedge for a bifunctor $F : \mathcal{C}^{\mathrm{op}} \times \mathcal{C} \to \mathcal{D}$. (More formally, the notion of dinatural (co)wedge is indistinguishable from the notion of extranatural (co)wedge, and thus the two give rise to the same notion of (co)end.)

We prefer extranaturality for a variety of reasons:

- it is less general (see Proposition 1.1.13), but it still makes (co)ends available;
- it gives rise to a fairly intuitive *graphical calculus* (see 1.1.11), and moreover it behaves better under composition (see Exercise 1.4);
- extranaturality is the correct notion in the enriched categorical setting (see 4.3.7 and the caveat right after).

Definition 1.1.11 (Graphical calculus for extranaturality). The graphical calculus for extranatural transformations depicts the components α_{ABC}, and arrows $f : A \to A'$, $g : B \to B'$, $h : C \to C'$, respectively as planar diagrams such as

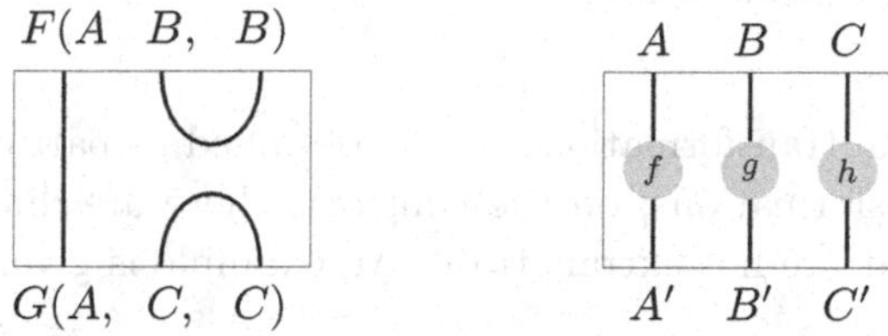

where wires are labelled by objects and must be thought of as oriented from top to bottom. The commutative squares of (1.17) become, in this representation, the following three string diagrams, whose equivalence is graphically obvious (the labels f, g, h are allowed to 'slide' along the wire they live on):

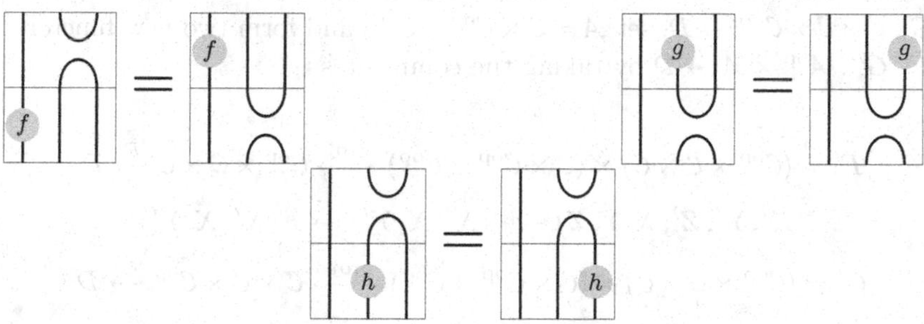

Remark 1.1.12. The notion of extranatural transformation can be specialised to encompass various other constructions. Simple naturality arises when F, G are both constant in their (co)wedge components, so the cap and cup in α_{ABC} vanish:

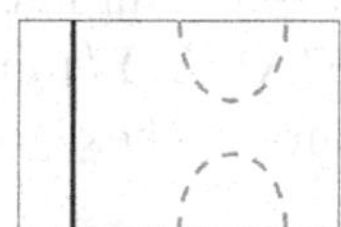

The wedge and cowedge conditions arise when either of F, G is constant, so that the straight line and one of the cup or the cap in α_{ABC} vanish:

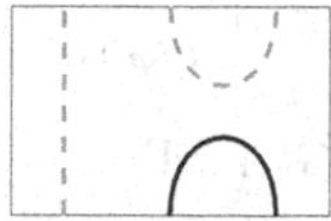

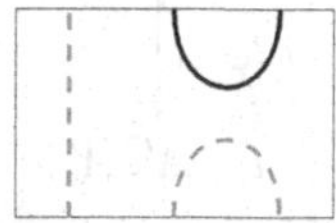

All the other mixed situations (a wedge-cowedge condition, naturality and a wedge, and others that do not have a specified name) admit a graphical representation of the same sort, and follow similar graphical rules of juxtaposition, when the boundaries of their associated cells agree in shape in the obvious sense.

All extranatural transformations can be obtained as particular cases of dinatural transformations; on the contrary, there are dinatural transformations that are not extranatural. An example is given in Exercise 1.5.

Proposition 1.1.13. ⚭ Extranatural transformations are particular kinds of dinatural transformations.

Proof (due to T. Trimble) Given functors $F : \mathcal{C}^{\mathrm{op}} \times \mathcal{C} \times \mathcal{C} \to \mathcal{D}$ and $G : \mathcal{C} \times \mathcal{C} \times \mathcal{C}^{\mathrm{op}} \to \mathcal{D}$, set $\mathcal{A} = \mathcal{C} \times \mathcal{C}^{\mathrm{op}} \times \mathcal{C}^{\mathrm{op}}$, and form two new functors $F', G' : \mathcal{A}^{\mathrm{op}} \times \mathcal{A} \to \mathcal{D}$ by taking the composites

$$\begin{aligned} F' = (\mathcal{C}^{\mathrm{op}} \times \mathcal{C} \times \mathcal{C}) \times (\mathcal{C} \times \mathcal{C}^{\mathrm{op}} \times \mathcal{C}^{\mathrm{op}}) &\xrightarrow{\mathrm{proj}} \mathcal{C}^{\mathrm{op}} \times \mathcal{C} \times \mathcal{C} \xrightarrow{F} \mathcal{D} \\ (X', Y', Z'; X, Y, Z) \longmapsto (X', X, Y') &\overset{F}{\mapsto} F(X', X, Y') \\ G' = (\mathcal{C}^{\mathrm{op}} \times \mathcal{C} \times \mathcal{C}) \times (\mathcal{C} \times \mathcal{C}^{\mathrm{op}} \times \mathcal{C}^{\mathrm{op}}) &\xrightarrow{\mathrm{proj}'} \mathcal{C} \times \mathcal{C} \times \mathcal{C}^{\mathrm{op}} \xrightarrow{G} \mathcal{D} \\ (X', Y', Z'; X, Y, Z) \mapsto (Y', Z', Z) &\overset{G}{\mapsto} G(Y', Z', Z). \end{aligned}$$

Now let us put $A' = (X', Y', Z')$ and $A = (X, Y, Z)$, considered as objects in $\mathcal{A}$. An arrow $\varphi : A' \to A$ in $\mathcal{A}$ thus amounts to a triple of arrows $f : X' \to X$, $g : Y \to Y'$, $h : Z \to Z'$, all in $\mathcal{C}$. Following the instructions above, we have $F'(A', A) = F(X', X, Y')$ and $G(A', A) = G(Y', Z', Z)$. Now if we write down a dinaturality hexagon for $\alpha : F' \Rightarrow G'$, we get a diagram of shape

$$\begin{array}{ccccc} F'(A, A') & \xrightarrow{F'(1,\varphi)} & F'(A, A) & \xrightarrow{\alpha_A} & G'(A, A) \\ {\scriptstyle F(\varphi,1)}\downarrow & & & & \downarrow{\scriptstyle G'(\varphi,1)} \\ F'(A', A') & \xrightarrow[\alpha_{A'}]{} & G'(A', A') & \xrightarrow[G'(1,\varphi)]{} & G(A', A) \end{array} \tag{1.20}$$

which translates to a hexagon of shape

$$\begin{array}{ccccc} F(X,X',Y) & \xrightarrow{F(1,f,1)} & F(X,X,Y) & \longrightarrow & G(Y,Z,Z) \\ \Big\downarrow{\scriptstyle F(f,1,g)} & & & & \Big\downarrow{\scriptstyle G(g,h,1)} \\ F(X',X',Y') & \longrightarrow & G(Y',Z',Z') & \xrightarrow[G(1,h,1)]{} & G(Y',Z',Z) \end{array} \tag{1.21}$$

where the unlabelled arrows are the extranatural components. This is the extranaturality hexagon of 1.1.8. □

1.1.3 The Integral Notation for (Co)ends

A suggestive notation to denote the universal object of a (co)end, alternative to the eponymous '$\underline{\text{(co)end}}(F)$', is due to N. Yoneda, who in [Yon60] introduced most of the notions we are dealing with in the setting of Ab-enriched functors $\mathcal{C}^{\text{op}} \times \mathcal{C} \to \text{Ab}$:

Notation 1.1.14. The *integral notation* denotes the end of a functor $F \in \text{Cat}(\mathcal{C}^{\text{op}} \times \mathcal{C}, \mathcal{D})$ as a 'subscripted-integral' $\int_C F(C,C)$, and the coend $\underline{\text{coend}}(F)$ as the 'superscripted-integral' $\int^C F(C,C)$.

From now on we will systematically adopt this notation to denote the universal (co)wedge $\underline{\text{(co)end}}(F)$ or, following a well-established abuse of notation, the object itself; when the domain of F has to be made explicit, we will also employ more pedantic variants of $\int_C F$ and $\int^C F$ like

$$\int_{C\in\mathcal{C}} F(C,C), \qquad \int^{C\in\mathcal{C}} F(C,C). \tag{1.22}$$

Remark 1.1.15. In reading [Yon60], one should be aware that Yoneda employs a reversed notation to denote ends and coends: he calls *integration* what we call a coend, which he denotes as $\int_{C\in\mathcal{C}} F(C,C)$, i.e. in the way we denote an end; and he calls *cointegrations* our ends, which he denotes $\int^*_{C\in\mathcal{C}} F(C,C)$.

No trace of this ambiguity survives in the current literature, so we will not mention the Yoneda convention again.

Functoriality of (co)ends acquires a particularly suggestive flavour when written in integral notation. The dream of every freshman learning calculus is that the integral of a product of functions is just the product of the integrals of the two functions. In category theory this is true, provided the integral of a function is the map induced between two (co)ends, and that product is composition of arrows.

Notation 1.1.16. The unique arrow $\underline{\text{end}}(\eta)$ induced by a natural transformation $\eta : F \Rightarrow G$ between $F, G \in \text{Cat}(\mathcal{C}^{\text{op}} \times \mathcal{C}, \mathcal{D})$ can be written as $\int_C \eta : \int_C F \to \int_C G$, and uniqueness of this induced arrow entails functoriality, i.e. $\int_C(\eta \circ \sigma) = \int_C \eta \circ \int_C \sigma$ and $\int_C \text{id}_F = \text{id}_{\int F}$.

A similar convention holds for coends.

Remark 1.1.17. As [ML98, IX.5] puts it,

> [...] the 'variable of integration' C [in $\int_C F$] appears twice under the integral sign (once contravariant, once covariant) and is 'bound' by the integral sign, in that the result no longer depends on C and so is unchanged if C is replaced by any other letter standing for an object of the category $\mathcal{C}$.

This somehow motivates the integral notation for (co)ends, and yet the analogy between integral calculus and (co)ends seems to be too elusive to justify.

There seems to be no chance of giving a formal explanation of the similarities between integrals and coends, but it is nevertheless very suggestive to employ an informal justification for such an analogy to exist. See Exercise 1.6 or Remark 2.2.6, or even 1.4.5.

1.2 (Co)ends as (Co)limits

A general tenet of elementary category theory is that universal objects (i.e. objects having the property of being initial or terminal in some category) can be equivalently characterised

- U1. as limits (so the existence and uniqueness is simply translated in a category of *diagrams*),
- U2. as adjoints (so the uniqueness follows from uniqueness of adjoints, see A.4.2),
- U3. as the representing object of a certain functor (so the uniqueness follows from the fact that the Yoneda embedding よ$_\mathcal{C}$ is fully faithful, see A.5.4).

The language of (co)ends is no exception: the scope of the following subsection is to characterise the (co)end of a functor $F : \mathcal{C}^{\text{op}} \times \mathcal{C} \to \mathcal{D}$ as a (co)limit over a suitable diagram $\bar{F}$ (obtained from a canonically chosen correspondence $F \mapsto \bar{F}$), and (see A.3.9) consequently as the (co)equaliser of a single pair of arrows.

Remark 1.2.1. Given $F : \mathcal{C}^{\mathrm{op}} \times \mathcal{C} \to \mathcal{D}$ and a wedge $\tau : D \Rightarrow F$, we can build the following commutative diagram

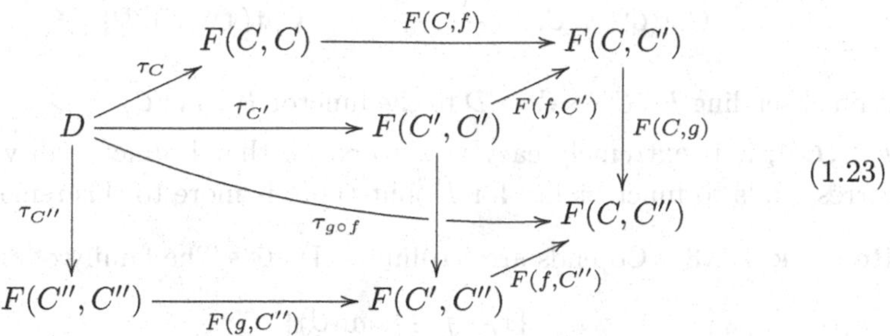

(1.23)

where $C \xrightarrow{f} C' \xrightarrow{g} C''$ are morphisms in $\mathcal{C}$. From this commutativity we deduce the following relations:

$$\begin{aligned}\tau_{g\circ f} &= F(g \circ f, C'') \circ \tau_{C''} = F(C, g \circ f) \circ \tau_C \\ &= F(f, C'') \circ F(g, C'') \circ \tau_{C''} = F(f, C'') \circ \tau_g \\ &= F(C, g) \circ F(C, f) \circ \tau_C = F(C, g) \circ \tau_f,\end{aligned}$$

where τ_f, τ_g are the common values $F(f, C') \circ \tau_{C'} = F(C, f) \circ \tau_C$ and $F(C', g) \circ \tau_{C'} = F(g, C'') \circ \tau_{C''}$ respectively, and $\tau_{g\circ f}$ is the common value $F(C, g) \circ \tau_f = F(f, C'') \circ \tau_g$.

These relations imply that there is a link between (co)wedges and (co)cones, encoded in the following definition.

Definition 1.2.2 (The twisted arrow category of $\mathcal{C}$). For every category $\mathcal{C}$ we define $\mathrm{TW}(\mathcal{C})$, the category of *twisted arrows* in $\mathcal{C}$ as follows.

- The class of objects of $\mathrm{TW}(\mathcal{C})$ is $\hom(\mathcal{C})$, the class (it will in general not be small if $\mathcal{C}$ was not small) of all morphisms of $\mathcal{C}$.
- Given $f : A \to A'$, $g : B \to B'$ a morphism $f \to g$ is given by a pair of arrows $(h : B \to A, k : A' \to B')$, such that the square

$$\begin{array}{ccc} A & \xleftarrow{\;h\;} & B \\ {\scriptstyle f}\downarrow & & \downarrow{\scriptstyle g} \\ A' & \xrightarrow[\;k\;]{} & B' \end{array} \tag{1.24}$$

commutes (asking that the arrow between domains is reversed is *not* a mistake), i.e. that $g = k \circ f \circ h$.

Endowed with the obvious rules for composition and identity, $\text{TW}(\mathcal{C})$ is easily seen to be a category, and now we can find a functor

$$\text{Cat}(\mathcal{C}^{\text{op}} \times \mathcal{C}, \mathcal{D}) \longrightarrow \text{Cat}(\text{TW}(\mathcal{C}), \mathcal{D}) \tag{1.25}$$

defined sending $F : \mathcal{C}^{\text{op}} \times \mathcal{C} \to \mathcal{D}$ to the functor $\bar{F} : \text{TW}(\mathcal{C}) \to \mathcal{D} : \begin{bmatrix} C \\ \downarrow \\ C' \end{bmatrix} \mapsto F(C, C')$; it is extremely easy now to check that bifunctoriality for F corresponds to functoriality for $\bar{F}$, but there is more to this remark.

Remark 1.2.3 ((Co)ends are (co)limits, I). ⦿⦿ The family of arrows

$$\{\tau_f \mid f \in \hom(\mathcal{C})\} \tag{1.26}$$

constructed above is a cone for the functor $\bar{F}$, and conversely any such cone determines a wedge for F, obtained by setting $\{\tau_C = \tau_{\text{id}_C}\}_{C \in \mathcal{C}}$.

A morphism of cones maps to a morphism between the corresponding wedges, and conversely every morphism between wedges induces a morphism between the corresponding cones; these operations are mutually inverse and form an equivalence between the category $\text{Cn}(\bar{F})$ of cones for $\bar{F}$ and the category $\text{Wd}(F)$ of wedges for F. (We leave this to the reader to check and to properly dualise.)

Equivalences of categories obviously preserve initial and terminal objects, thus we have isomorphisms[2]

$$\int_C F(C, C) \cong \lim_{\text{TW}(\mathcal{C})} \bar{F}, \qquad \int^C F(C, C) \cong \operatorname{colim}_{\text{TW}(\mathcal{C}^{\text{op}})^{\text{op}}} \bar{F}. \tag{1.27}$$

Remark 1.2.4. ⦿⦿ According to A.3.9, (co)limits in a category exist as soon as it has (co)products and (co)equalisers. So we would expect a characterisation of (co)ends in terms of these simpler pieces as well; such a characterisation exists, and it turns out to be extremely useful in explicit computations.

It is rather easy to extract from the bare universal property that there must be an isomorphism

$$\textstyle\int_C F(C, C) \cong \text{eq}\left(\prod_{C \in \mathcal{C}} F(C, C) \overset{F^*}{\underset{F_*}{\rightrightarrows}} \prod_{\varphi : C \to C'} F(C, C') \right) \tag{1.28}$$

where the product over morphisms $C \to C'$ can be expressed as a double

[2] Notice that the colimit is taken over the category $\text{TW}^{\text{op}}(\mathcal{C})$, the *opposite* of $\text{TW}(\mathcal{C}^{\text{op}})$: an *object* of $\text{TW}^{\text{op}}(\mathcal{C})$ is an arrow $f : C' \to C$ in $\mathcal{C}^{\text{op}}$, and a *morphism* from $f : C \to C'$ to $g : D \to D'$ is a commutative square (u, v) such that $vgu = f$.

product (over the objects $C, C' \in \mathcal{C}$, and over the arrows f between these two fixed objects), and the arrows F^*, F_* are easily obtained from the arrows whose $(f; C, C')$-components are (respectively) $F(f, C')$ and $F(C, f)$. This is a consequence of the fact that an 'element' in $\prod_C F(C, C)$, regarded as a family $(x_C \mid C \in \mathcal{C})$, equalises both actions of F on arrows at the same time in order to belong to the end $\int_C F(C, C)$. The dual statement of 1.2.4, expressing a coend as a coequaliser,

$$\int^C F(C, C) \cong \text{coeq}\left(\coprod_{C \in \mathcal{C}} F(C, C) \underset{F_*}{\overset{F^*}{\leftleftarrows}} \coprod_{\varphi : C \to C'} F(C', C)\right), \tag{1.29}$$

is left to the reader to formalise as Exercise 1.9.

The following remark is elementary but extremely useful: it asserts that the (co)limit of a functor has the same universal property as the (co)end of the same functor, when it is 'promoted' as mute (see 1.1.9) in its remaining variable.

Remark 1.2.5. Let $F : \mathcal{C} \to \mathcal{D}$ be a functor; we can always regard it as a functor $F' : \mathcal{C}^{\text{op}} \times \mathcal{C} \to \mathcal{D}$ mute (see 1.1.9) in its first variable. This means that we can extend the action of F defining a new functor F' such that $F'(C, C') = FC'$ for every $C, C' \in \mathcal{C}$, and $F'(C, f) = Ff$, $F'(f, C') = \text{id}_{C'}$ for every $f : C \to C'$; from this, and from 1.2.3 above, since all mute functors can be regarded as arising this way, it follows that the (co)end of a functor that is mute in one of its variables coincides with its (co)limit.

Definition 1.2.6. There is an obvious definition of *preservation* of (co)ends from their description as (co)limits, which reduces to the preservation of the particular kind of (co)limit involved in the definition of $\underline{\text{end}}(T)$ and $\underline{\text{coend}}(T)$. Let $F : \mathcal{D} \to \mathcal{E}$ be a functor, and let $T : \mathcal{C}^{\text{op}} \times \mathcal{C} \to \mathcal{D}$ be a functor; we say that

PR1. F *preserves* the end of T if the family of maps

$$F\Big(\int_C T(C, C)\Big) \xrightarrow{F\omega_C} FT(C, C) \tag{1.30}$$

exhibits the universal property of the end of the composed functor $\mathcal{C}^{\text{op}} \times \mathcal{C} \xrightarrow{T} \mathcal{D} \xrightarrow{F} \mathcal{E}$;

PR2. F *reflects* the end of T if, whenever $F\omega_C$ is the end of the composition $F \circ T$, then ω_C exhibits the end of F.

A dual definition gives the concept of preserving and reflecting coends; since ends are limits, it is clear that a functor that preserves all limits preserves all ends, and since limits are ends of mute functors, the converse is also true; dually for coends.

An alternative equivalent way to put this result is the following.

Theorem 1.2.7. Every (co)continuous functor $F : \mathcal{D} \to \mathcal{E}$ preserves the (co)ends that exist in $\mathcal{D}$.

- If $T : \mathcal{C}^{\mathrm{op}} \times \mathcal{C} \to \mathcal{D}$ has an end $\int_C T(C,C)$, and $F : \mathcal{D} \to \mathcal{E}$ commutes with all limits, then
$$F\Big(\textstyle\int_C T(C,C)\Big) \cong \int_C FT(C,C), \tag{1.31}$$
meaning that the image of the terminal wedge of T under F is a terminal wedge for the composite functor $F \circ T : \mathcal{C}^{\mathrm{op}} \times \mathcal{C} \to \mathcal{E}$, and thus the two terminal objects are canonically isomorphic.
- Dually, if $T : \mathcal{C}^{\mathrm{op}} \times \mathcal{C} \to \mathcal{D}$ has a coend $\int^C T(C,C)$ and $F : \mathcal{D} \to \mathcal{E}$ commutes with all colimits, then
$$F\Big(\textstyle\int^C T(C,C)\Big) \cong \int^C FT(C,C), \tag{1.32}$$
meaning that the image of the initial wedge of T under F is an initial cowedge for the composite functor $F \circ T : \mathcal{C}^{\mathrm{op}} \times \mathcal{C} \to \mathcal{E}$, and thus the two initial objects are canonically isomorphic.

Similarly, a functor that reflects all (co)limits reflects all (co)ends.

As a major example of a functor that preserves all ends, we have the hom functors.

Corollary 1.2.8 (The hom functor commutes with integrals)**.** From the fact that the hom bifunctor $\mathcal{C}(\_, =) : \mathcal{C}^{\mathrm{op}} \times \mathcal{C} \to \mathrm{Set}$ is such that
$$\begin{aligned} \mathcal{C}(\operatorname{colim} F, C) &\cong \lim \mathcal{C}(F, C) \\ \mathcal{C}(C, \lim F) &\cong \lim \mathcal{C}(C, F) \end{aligned} \tag{1.33}$$
and from (1.28), (1.29) we deduce that for every $D \in \mathcal{D}$ and every functor $F : \mathcal{C}^{\mathrm{op}} \times \mathcal{C} \to \mathcal{D}$ we have canonical isomorphisms
$$\begin{aligned} \mathcal{D}\Big(\int^C F(C,C), D\Big) &\cong \int_C \mathcal{D}(F(C,C), D) \\ \mathcal{D}\Big(D, \int_C F(C,C)\Big) &\cong \int_C \mathcal{D}(D, F(C,C)). \end{aligned} \tag{1.34}$$

We close the section by recording a notable but somewhat technical result.

Remark 1.2.9 ((Co)ends are (co)limits, II). [ML98, IX.5.1] ☻☻ Define the *subdivision graph* $(\wp\mathcal{C})^\S$ of a category $\mathcal{C}$ as the directed graph having a vertex $C^\S$ for each object $C \in \mathcal{C}$, and a vertex $f^\S$ for each morphism $f : C \to C'$ in $\mathcal{C}$, and edges all the arrows $C^\S \to f^\S$ and $C'^\S \to f^\S$ as $f : C \to C'$ runs over morphisms of $\mathcal{C}$.

The *subdivision category* $\mathcal{C}^\S$ is obtained from $(\wp\mathcal{C})^\S$ formally adding identities and giving to the resulting category the trivial composition law (composition is defined only if one of the arrows is the identity).

Every functor $F : \mathcal{C}^{\mathrm{op}} \times \mathcal{C} \to \mathcal{D}$ induces a functor $F^\S : \mathcal{C}^\S \to \mathcal{D}$, whose limit (provided it exists) is isomorphic to the end of F. More precisely, we have the following result.

Proposition 1.2.10. ☻☻ In the above notation,

- there is a final functor ς from $\mathcal{C}^\S$ to $\mathrm{TW}(\mathcal{C})$; thus, the limit of $\bar{F}$ computed over $\mathrm{TW}(\mathcal{C})$ is the same as the limit of $\bar{F} \circ \varsigma$ computed over $\mathcal{C}^\S$;
- there is a functor $F^\S : \mathcal{C}^\S \to \mathcal{D}$ such that wedges for F correspond bijectively to cones for $F^\S$.

The assignment $F \mapsto F^\S$ sets up an isomorphism of categories between wedges for F and for $F^\S$, and thus the terminal wedge, i.e. the end of F, must correspond to the terminal cone, the limit of $F^\S$. Dually, we have

- there is a cofinal functor ϑ from $\mathcal{C}^\S$ to $\mathrm{TW}(\mathcal{C}^{\mathrm{op}})^{\mathrm{op}}$; thus, the colimit of $\bar{F}$ computed over $\mathrm{TW}(\mathcal{C}^{\mathrm{op}})^{\mathrm{op}}$ is the same as the colimit of $\bar{F} \circ \vartheta$ done over $\mathcal{C}^\S$;
- there is a functor $F^\S : \mathcal{C}^\S \to \mathcal{D}$ such that wedges for F correspond bijectively to cocones for $F^\S$.

This assignment sets up an equivalence of categories between cowedges for F and for $F^\S$, and thus the initial cowedge, i.e. the coend of F, must correspond to the initial cocone, the colimit of $F^\S$.

We just address the case of ends, leaving the dualisation process to the reader. Recall that a functor $F : \mathcal{C} \to \mathcal{D}$ is final if every comma category $(C \downarrow F)$ is non-empty and connected.

To sum up, given a functor $F : \mathcal{C}^{\mathrm{op}} \times \mathcal{C} \to \mathcal{D}$ one can equivalently compute the value of the end $\int_C F$ as

- the terminal wedge $\alpha : \int_C F(C, C) \Rightarrow F$,
- the terminal cone $\alpha^\S : \lim_{\mathcal{C}^\S} F^\S \Rightarrow F^\S$,
- the terminal cone $\bar{\alpha} : \lim_{\mathrm{TW}(\mathcal{C})} \bar{F} \to \bar{F}$.

Proof Let us first define the functor ς. On objects,

$$\varsigma(C^{\S}) = \mathrm{id}_C \qquad \varsigma(f^{\S}) = f \tag{1.35}$$

while on morphisms $f : C \to C'$

$$\varsigma\begin{bmatrix} C^{\S} \\ \downarrow \\ f^{\S} \end{bmatrix} = \begin{array}{ccc} C & = & C \\ \| & & \downarrow \\ C' & \to & C' \end{array} \qquad \varsigma\begin{bmatrix} C'^{\S} \\ \downarrow \\ f^{\S} \end{bmatrix} = \begin{array}{ccc} C' & \leftarrow & C \\ \| & & \downarrow \\ C' & = & C' \end{array} \tag{1.36}$$

This easily shows that ς is surjective on objects, and thus each comma category $(f \downarrow \varsigma)$ is non-empty. The image of ς visibly contains very few morphisms; yet it is still final, because the comma category $(f \downarrow \varsigma)$ is easily seen to be connected.

To prove the second point, we shall show that every functor $F : \mathcal{C}^{\mathrm{op}} \times \mathcal{C} \to \mathcal{D}$ induces a functor $F^{\S} : \mathcal{C}^{\S} \to \mathcal{D}$ from the subdivision category. We can define

- an object function setting $F^{\S}(C^{\S}) = F(C, C)$ and $F^{\S}\begin{bmatrix} C \\ \downarrow \\ C' \end{bmatrix}^{\S} = F(C, C')$,
- a morphism function, setting $F^{\S}\begin{bmatrix} C^{\S} \\ \downarrow \\ f^{\S} \end{bmatrix} = F(C, f)$ and $F^{\S}\begin{bmatrix} C'^{\S} \\ \downarrow \\ f^{\S} \end{bmatrix} = F(f, C')$.

Let now $\alpha : D \Rightarrow F^{\S}$ be a cone for $F^{\S}$; this means that for every morphism of $\mathcal{C}^{\S}$ one among the diagrams

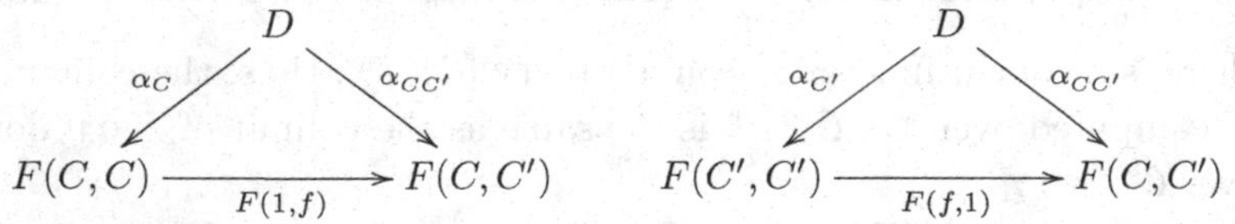

(chosen according to the shape of $f^{\S}$) is commutative. But then the restriction of α to its diagonal component forms a wedge for F with domain D, because the square

$$\begin{array}{ccc} D & \xrightarrow{\alpha_C} & F(C,C) \\ {\scriptstyle \alpha_{C'}}\downarrow & & \downarrow{\scriptstyle F(1,f)} \\ F(C',C') & \xrightarrow[F(f,1)]{} & F(C,C') \end{array} \tag{1.37}$$

has $\alpha_{CC'}$ as diagonal. Vice versa, given a wedge $\alpha : D \Rightarrow F$, the square (1.37) defines the components of a cone $\alpha : D \Rightarrow F^{\S}$ by $\alpha_{CC'} = F(1, f) \circ \alpha_{CC} = F(f, 1) \circ \alpha_{C'C'}$. It is evident that this sets up a bijection between $\mathrm{Wd}(F)$ and $\mathrm{Cn}(F^{\S})$, and that this can in fact be promoted to an isomorphism between the two categories. □

From this it ultimately follows that both characterisations of (co)ends as (co)limits, either as diagrams with domain $\mathcal{C}^§$, or with domain $\text{TW}(\mathcal{C})$, lead to the same theory.

1.3 The Fubini Rule

An absolutely central theorem for coend calculus is the 'exchange rule' for integrals known as the *Fubini rule*. Informally, it says that the result of an integration on more than one variable does not depend on the order in which the operation is performed: in mathematical analysis, if $f : X \times Y \to \mathbb{R}$ is a function such that the integral $\int_{X \times Y} |f| d\mu_{X \times Y}$ with respect to a certain measure exists finite, then the three integrals

$$\int_X \left(\int_Y f(x,y)\, dy \right) dx = \int_Y \left(\int_X f(x,y)\, dx \right) dy = \int_{X \times Y} f(x,y)\, d(x,y) \tag{1.38}$$

are equal. In category theory, if a functor F defined on $\mathcal{C}^{\text{op}} \times \mathcal{C} \times \mathcal{E}^{\text{op}} \times \mathcal{E}$ admits a (co)end, then so do the two functors obtained by integrating over $\mathcal{C}$ first and over $\mathcal{E}$ second, or in the opposite order.

Theorem 1.3.1 (Fubini theorem for (co)ends). ☉☉ Given a functor

$$F : \mathcal{C}^{\text{op}} \times \mathcal{C} \times \mathcal{E}^{\text{op}} \times \mathcal{E} \to \mathcal{D}, \tag{1.39}$$

we can form the end $\int_C F(C, C, -, =)$, obtaining a functor $\mathcal{E}^{\text{op}} \times \mathcal{E} \to \mathcal{D}$ whose end is $\int_E \int_C F(C, C, E, E) \in \mathcal{D}$; we can also form the ends $\int_C \int_E F(C, C, E, E) \in \mathcal{D}$ and $\int_{(C,E)} F(C, C, E, E)$ identifying $\mathcal{C}^{\text{op}} \times \mathcal{C} \times \mathcal{E}^{\text{op}} \times \mathcal{E}$ with $(\mathcal{C} \times \mathcal{E})^{\text{op}} \times (\mathcal{C} \times \mathcal{E})$.

Then, there are canonical isomorphisms between the three objects:

$$\int_{(C,E)} F(C,C,E,E) \cong \int_E \int_C F(C,C,E,E) \cong \int_C \int_E F(C,C,E,E). \tag{1.40}$$

Dually, there are canonical isomorphisms between the iterated coends

$$\int^{(C,E)} F(C,C,E,E) \cong \int^E \int^C F(C,C,E,E) \cong \int^C \int^E F(C,C,E,E). \tag{1.41}$$

As it is customary in category theory, this is an existence and uniqueness result: any of the three objects above in (1.40) (respectively, (1.41)) exists if and only if the other two do, and canonical isomorphisms connect all three.

We shall only prove the statement for ends, (1.41) for coends being the exact dual statement.[3]

One possible strategy to prove the Fubini theorem would be to find a suitable canonical map between the end $\int_{(C,E)} F(C,C,E,E)$ and (say) the end $\int_E \int_C F(C,C,E,E)$ (with a similar reasoning, exchanging the rôle of the integration variable, one can produce a canonical map between the end $\int_{(C,E)} F(C,C,E,E)$ and the end $\int_E \int_C F(C,C,E,E)$), and then prove its invertibility. This is of course a viable option, but has little conceptual content, and can result in a long, unenlightening proof.

Instead, we prefer to offer a more elegant argument, especially because the strategy of our proof will be recycled in Section 7.3 when we prove the Fubini rule for ∞-coends (the analogue of coends in $(\infty, 1)$-category theory, and more specifically in the setting of Joyal–Lurie quasicategories [Joy, Lur09]).

Our strategy goes as follows. First we assume that the category $\mathcal{D}$ admits enough limits to ensure that all ends exist. Thus, as stated in 1.1.7, the correspondence $T \mapsto \int_C T$ is a functor. We shall show that $\int_C$ has a left adjoint $H_{\mathcal{C}} : \mathcal{D} \to \mathrm{Cat}(\mathcal{C}^{\mathrm{op}} \times \mathcal{C}, \mathcal{D})$, and that $H_{\mathcal{C}\times\mathcal{E}} \cong H_{\mathcal{C}} \circ H_{\mathcal{E}} \cong H_{\mathcal{E}} \circ H_{\mathcal{C}}$; the Fubini rule then follows from the uniqueness of adjoints (see A.4.2) because every such 'interchange isomorphism' between the left adjoints must induce a similar isomorphism between the right adjoints.

Note that this argument yields for free that the isomorphisms witnessing the Fubini rule above are natural in F, i.e. that $\int_{(C,E)}, \int_C \int_E, \int_E \int_C$ are isomorphic *as functors*.

Let us define the functor $H_{\mathcal{C}} : \mathcal{D} \to \mathrm{Cat}(\mathcal{C}^{\mathrm{op}} \times \mathcal{C}, \mathcal{D})$ as follows: to the object $D \in \mathcal{D}$ we associate the functor $\hom_{\mathcal{C}} \otimes D : (C, C') \mapsto \mathcal{C}(C, C') \otimes D$, where for a set X and an object $D \in \mathcal{D}$ we denote by $X \otimes D$ the coproduct $\coprod_{x \in X} D$ of X copies of D, also called the *copower* or *tensor* of D by X. See 2.2.3 for the whole definition; note that this yields a chain of isomorphisms

$$\mathcal{D}(X \otimes D, D') \cong \mathcal{D}(D, X \pitchfork D') \cong \mathrm{Set}(X, \mathcal{D}(D, D')), \tag{1.42}$$

where $X \pitchfork D$ is defined dually as $\prod_{x \in X} D$. We will freely employ (1.42) when needed.

We shall prove that there is a bijection

$$\mathrm{Cat}(\mathcal{C}^{\mathrm{op}} \times \mathcal{C}, \mathcal{D})(\hom_{\mathcal{C}} \otimes D, F) \cong \mathcal{D}(D, \textstyle\int_C F). \tag{1.43}$$

[3] We leave the task to dualise properly the proof that follows to the willing reader: the coend functor $\int^C : \mathrm{Cat}(\mathcal{C}^{\mathrm{op}} \times \mathcal{C}, \mathcal{D}) \to \mathcal{D}$ will turn out to be a left adjoint, with right adjoint $H^{\mathcal{C}}$, etc.

If $\alpha : \hom_{\mathcal{C}} \otimes D \Rightarrow F$ is a natural transformation it has components

$$\alpha_{CC'} : \mathcal{C}(C, C') \otimes D \to F(C, C');$$

by (1.42), these maps correspond to

$$\tilde{\alpha}_{CC'} : D \to \mathcal{C}(C, C') \pitchfork F(C, C')$$

under the adjunction isomorphisms. It is now easy to show that these mates form a wedge *in the pair* (C, C'), and thus there is an induced morphism

$$\hat{\alpha} : D \to \int_{(C,C')} \mathcal{C}(C, C') \pitchfork F(C, C'). \tag{1.44}$$

Lemma 1.3.2. The latter integral $\int_{(C,C')} \mathcal{C}(C, C') \pitchfork F(C, C')$ is in fact isomorphic to $\int_C F(C, C)$.

Proof of 1.3.2 We shall first find a candidate wedge

$$\omega_{(CC')} : \int_C F(C, C) \to \mathcal{C}(C, C') \pitchfork F(C, C'); \tag{1.45}$$

for the sake of exposition, we treat the case $\mathcal{D} =$ Set: the core of the argument is the same in the general case, using generalised elements.

Thanks to (1.28), we now know that an 'element' of $\int_C F(C, C)$ is a coherent sequence $(a_C \mid C \in \mathcal{C})$ in the product $\prod_C F(C, C)$, such that $F(u, \mathrm{id}_C)(a_{C'}) = F(\mathrm{id}_{C'}, u)(a_C)$ for every $u : C \to C'$ in $\mathcal{C}$. This means that we can map a coherent sequence $(a_C \mid C \in \mathcal{C})$ into (say) a sequence $(F(u, C')(a_{C'}) \mid u : C \to C')$, and this sets up a map

$$\int_C F(C, C) \xrightarrow{\varpi_{(CC')}} \mathcal{C}(C, C') \pitchfork F(C, C') = \prod_{C,C'} \prod_{u:C\to C'} F(C, C'). \tag{1.46}$$

This in turn defines a wedge in the pair (C, C'), thus inducing a unique morphism

$$\textstyle\int_C F(C, C) \xrightarrow{\bar{\varpi}} \int_{(C,C')} \mathcal{C}(C, C') \pitchfork F(C, C') \tag{1.47}$$

between the ends. Now that we have put all notation in place, we leave as an exercise for the reader to show that $\varpi_{(CC')}$ is a terminal wedge, and thus that $\bar{\varpi}$ is in fact an isomorphism. □

Proof of 1.3.1 A natural transformation $\alpha : \hom_{\mathcal{C}} \otimes D \Rightarrow F$ induces an arrow such as in (1.44), and this induces a unique arrow $\bar{\varpi}^{-1} \circ \hat{\alpha} : D \to \int_C F(C, C)$. The correspondence $\alpha \mapsto \bar{\varpi}^{-1} \circ \hat{\alpha}$ sets up the desired adjunction as in (1.43).

Unwinding its definition, we see that the functor $H_{\mathcal{C}} : \mathcal{D} \to \mathrm{Cat}(\mathcal{C}^{\mathrm{op}} \times \mathcal{C}, \mathcal{D})$ has the property that $H_{\mathcal{C}\times\mathcal{E}} \cong H_{\mathcal{C}} \circ H_{\mathcal{E}} \cong H_{\mathcal{E}} \circ H_{\mathcal{C}}$ (it easily follows from a commutativity of the involved coproducts: we invite the reader to fill in the details). This in turn shows the Fubini rule, since every isomorphism between left adjoint functors $L \Rightarrow L'$ induces an isomorphism of right adjoint functors $R' \Rightarrow R$. □

Remark 1.3.3. From the binary case shown above, an easy induction shows that the iterated (co)end of a functor $F : \prod_{i=1}^{n} \mathcal{C}_i^{\mathrm{op}} \times \mathcal{C}_i \to \mathcal{D}$ gives the same result with respect to their integration variables, taken in whatever order. More formally, if $\sigma : \{1, \dots, n\} \to \{1, \dots, n\}$ is any permutation of an n elements set, then there is a canonical isomorphism

$$\int_{C_{\sigma 1}} \cdots \int_{C_{\sigma n}} F(C_1, C_1, \dots, C_n, C_n) \cong \int_{(C_1,\dots,C_n)} F(C_1, C_1, \dots, C_n, C_n), \tag{1.48}$$

and similarly for the coend of F.

1.4 First Instances of (Co)ends

A basic example exploiting all of the machinery introduced so far is the proof that the set of natural transformations between two functors $F, G : \mathcal{C} \to \mathcal{D}$ can be characterised as an end:

Theorem 1.4.1. Given functors $F, G : \mathcal{C} \to \mathcal{D}$ whose domain is a small category, and whose codomain is locally small, we have the canonical isomorphism of sets

$$\mathrm{Cat}(\mathcal{C}, \mathcal{D})(F, G) \cong \int_C \mathcal{D}(FC, GC). \tag{1.49}$$

Proof A wedge $\tau_C : Y \to \mathcal{D}(FC, GC)$ consists of a function $y \mapsto (\tau_{C,y} : FC \to GC \mid C \in \mathcal{C})$ natural in $C \in \mathcal{C}$ (this is simply a rephrasing of the wedge condition). The equation

$$G(f) \circ \tau_{C,y} = \tau_{C',y} \circ F(f), \tag{1.50}$$

valid for any $f : C \to C'$, means that for a fixed $y \in Y$ the arrows τ_C form the components of a natural transformation $F \Rightarrow G$; thus, there exists a unique way to close the diagram

$$\begin{array}{ccc} Y & \xrightarrow{\tau_C} & \mathcal{D}(FC, GC) \\ & \searrow^{h} & \uparrow \\ & & \mathrm{Cat}(\mathcal{C}, \mathcal{D})(F, G) \end{array} \tag{1.51}$$

with a function sending

$$y \mapsto \tau_{-,y} \in \prod_{C \in \mathcal{C}} \mathcal{D}(FC, GC), \tag{1.52}$$

where $\mathrm{Cat}(\mathcal{C}, \mathcal{D})(F, G) \to \mathcal{D}(FC, GC)$ is the wedge sending a natural transformation to its C-component. The diagram commutes for a single $h : Y \to \mathrm{Cat}(\mathcal{C}, \mathcal{D})(F, G)$, and this is precisely the desired universal property for $\mathrm{Cat}(\mathcal{C}, \mathcal{D})(F, G)$ to be $\int_C \mathcal{D}(FC, GC)$. □

Remark 1.4.2. A suggestive way to express naturality as a 'closure' condition is given in [Yon60, 4.1.1], where for an Ab-enriched functor (see A.7.2) $F : \mathcal{C}^{\mathrm{op}} \times \mathcal{C} \to \mathrm{Ab}$ from a complete Ab-category $\mathcal{C}^{\mathrm{op}} \times \mathcal{C}$, one can prove that natural transformations $F \Rightarrow G$ form the kernel $\ker \delta$ of a map

$$\delta : \bigoplus_{Y \in \mathcal{C}} \mathrm{Ab}(FY, GY) \to \bigoplus_{f : X \to Y} \mathrm{Ab}(FX, GY). \tag{1.53}$$

Remark 1.4.3. If we let $F = G = \mathrm{id}_{\mathcal{C}}$ in the isomorphism above, we get that the end of the hom functor $\hom : \mathcal{C}^{\mathrm{op}} \times \mathcal{C} \to \mathrm{Set}$ is the monoid $M(\mathcal{C})$ of natural transformations from the identity functor $\mathrm{id}_{\mathcal{C}}$ to itself. This monoid is of great importance in homological algebra and algebraic geometry, as it constitutes a precious source of information about $\mathcal{C}$ (morally, $M(\mathcal{C})$ can be thought of as the 0th term of a sequence of cohomology groups $H^n(\mathcal{C})$ associated to $\mathcal{C}$). Similarly, the presence of a 'nice' functor $F : \mathcal{C} \to \mathrm{Set}$ (for example, a faithful and conservative one) yields a fairly rich monoid of endomorphisms $M(F) = \int_C \mathrm{Set}(FC, FC)$.

'Reconstruction theory' is the branch of category theory that asserts that under suitable conditions on F, its domain is a category of $M(F)$-modules, i.e. there is an equivalence (or a full embedding) between $\mathcal{C}$ and $\mathrm{Mod}(M(F))$.

The set of dinatural transformations between two functors can also be characterised as an end, where the hom functor has been 'completely symmetrised'. The result was first proved in [Dub70].

Example 1.4.4 (Dinatural transformations as an end). Let $F, G : \mathcal{C}^{\text{op}} \times \mathcal{C} \to \mathcal{D}$ be two functors; define a new functor

$$\mathcal{D}_d(F,G)[\,\_\,,=] : \mathcal{C}^{\text{op}} \times \mathcal{C} \to \text{Set} \tag{1.54}$$

- on objects, sending the pair (A, B) to the set $\mathcal{D}(F(B,A), G(A,B))$;
- on morphisms, sending a pair of arrows $\left[\begin{smallmatrix} A & X \\ f\downarrow & \downarrow g \\ B & Y \end{smallmatrix}\right]$ to the diagonal of the commutative square

$$\begin{array}{ccc} \mathcal{D}_d(F,G)[B,X] & \longrightarrow & \mathcal{D}_d(F,G)[B,Y] \\ \downarrow & & \downarrow \\ \mathcal{D}_d(F,G)[A,X] & \longrightarrow & \mathcal{D}_d(F,G)[A,Y] \end{array} \tag{1.55}$$

whose horizontal and vertical arrows are defined by the action of F, G on morphisms: for example the left vertical arrow is defined as

$$\begin{aligned} \mathcal{D}(F(X,B), G(B,X)) &\to \mathcal{D}(F(X,A), G(A,X)) \\ u &\mapsto G(f,1) \circ u \circ F(1,f). \end{aligned} \tag{1.56}$$

It is a natural question to wonder what the end of $\mathcal{D}_d(F,G)$ is: [Dub70] was the first to observe that such an end is in fact isomorphic to the set of dinatural transformations $\alpha : F \Rightarrow G$.

In order to prove this, observe that if we consider the parallel morphisms

$$\prod_{C\in\mathcal{C}} \mathcal{D}(F(C,C), G(C,C)) \rightrightarrows \prod_{f:C\to C'} \mathcal{D}(F(C',C), G(C,C')) \tag{1.57}$$

induced by the conjoint action of $\mathcal{D}_d(F,G)$ on a morphism as in (1.28), then u, v are defined as respectively sending a sequence $(x_C \mid C \in \mathcal{C})$ to

$$\begin{cases} (x_C \mid C \in \mathcal{C}) & \mapsto (G(1,f) \circ x_A \circ F(f,1) \mid f : A \to B) \\ (x_C \mid C \in \mathcal{C}) & \mapsto (G(f,1) \circ x_B \circ F(1,f) \mid f : A \to B). \end{cases} \tag{1.58}$$

The equaliser of this pair of maps evidently selects the set of 'coherent' sequences (x_C) such that these two actions are equal, i.e. such that the family $x_C : F(C,C) \to G(C,C)$ specifies a dinatural transformation $F \Rightarrow G$ in the sense of (1.8).

Example 1.4.5 (Stokes' theorem is about (co)ends). ☉☉ This remark was first observed by [Cam] and it follows their exposition almost word-for-word. Let $\mathbb{N}$ be the poset of natural numbers with the usual ordering, and let $\text{Mod}(\mathbb{R})$ be the category of real vector spaces.

Fix a manifold X (or some other sort of smooth space). Then we can define the following functors.

- $C : \mathbb{N}^{\mathrm{op}} \to \mathrm{Mod}(\mathbb{R})$ where C_n is the vector space freely generated by smooth maps $Y \to X$ where Y is a compact, n-dimensional, oriented manifold with boundary, and the induced map $\partial : C_{n+1} \to C_n$ is the boundary map. This is a chain complex, since the composition $\partial\partial$ is the constant at zero.
- $\Omega : \mathbb{N} \to \mathrm{Mod}(\mathbb{R})$ is the de Rham complex; $\Omega_n = \Omega_n(X)$ is the space of n-forms on X and the induced map $d : \Omega_n \to \Omega_{n+1}$ is the exterior derivative. This is the usual de Rham complex of X.

Consider the usual tensor product on $\mathrm{Mod}(\mathbb{R})$; taking the object-wise tensor product of C and Ω, we obtain a functor $C \otimes \Omega : \mathbb{N}^{\mathrm{op}} \times \mathbb{N} \to \mathrm{Mod}(\mathbb{R})$, while there is also the constant functor $\mathbb{R} : \mathbb{N}^{\mathrm{op}} \times \mathbb{N} \to \mathrm{Mod}(\mathbb{R})$.

Then Stokes' theorem asserts that we have a cowedge

$$\textstyle\int : C \otimes \Omega \to \mathbb{R}$$

which, given a map $Y \to X$ and a differential form ω on X, pulls the form back to Y and integrates it (returning 0 if it is of the wrong dimension): $\int_n : C_n \otimes \Omega_n \to \mathbb{R}$ is defined by $(\left[\begin{smallmatrix} Y \\ \downarrow \\ X \end{smallmatrix}\right], \omega) \mapsto \int_Y \varphi^*\omega$ if $\varphi : Y \to X$. (The integral $\int_Y$, here, is not a (co)end.)

In fact, the cowedge condition for $\int$ amounts to the commutativity of the square

$$\begin{array}{ccc} C_{n+1} \otimes \Omega_n & \xrightarrow{\partial \otimes \mathrm{id}} & C_n \otimes \Omega_n \\ {\scriptstyle \mathrm{id} \otimes d}\downarrow & & \downarrow{\scriptstyle \int_n} \\ C_{n+1} \otimes \Omega_{n+1} & \xrightarrow[\int_{n+1}]{} & \mathbb{R} \end{array} \tag{1.59}$$

which according to the definition of $\int$ says that

$$\textstyle\int_{\partial Y} \varphi|_{\partial Y}^* \omega = \int_Y \varphi^* d\omega. \tag{1.60}$$

This is precisely Stokes' theorem.

Exercise for the reader: what is the natural induced map $\int^{n \in \mathbb{N}} C_n \otimes \Omega_n \to \mathbb{R}$? (Hint: express the coequaliser defining $\int^{n \in \mathbb{N}} C_n \otimes \Omega_n$ as the degree-zero cohomology of a suitable bicomplex $(C_\bullet \otimes \Omega_\bullet, D)$.)

Exercises

1.1 Prove equations (1.5) and (1.6). Which commutative diagrams do you need to derive them?

1.2 Show with an example that dinatural transformations $\alpha : P \Rightarrow Q, \beta : Q \Rightarrow R$ cannot be composed in general. Nevertheless, there exists a composition rule of a dinatural $\alpha : P \Rightarrow Q$ with a natural $\eta : P' \Rightarrow P$ which is again dinatural $P' \Rightarrow Q$, as well as a composition $P \Rightarrow Q \Rightarrow Q'$. (Hint: the appropriate diagram results as the pasting of a dinaturality hexagon and two naturality squares.)

1.3 What is the end of the constant functor $\Delta_D : \mathcal{C}^{\mathrm{op}} \times \mathcal{C} \to \mathcal{D}$ at the object $D \in \mathcal{D}$? What is its coend?

1.4 ◔◔ Show that extranatural transformations compose according to these rules:

- (Stalactites) Let F, G be functors of the form $\mathcal{C}^{\mathrm{op}} \times \mathcal{C} \to \mathcal{D}$. If $\alpha_{X,Y} : F(X,Y) \to G(X,Y)$ is natural in X, Y and $\beta_X : G(X,X) \to H$ is extranatural in X (for some object H of $\mathcal{D}$), then

$$\beta_X \circ \alpha_{X,X} : F(X,X) \to H$$

is extranatural in X.

- (Stalagmites) Let G, H be functors of the form $\mathcal{C}^{\mathrm{op}} \times \mathcal{C} \to \mathcal{D}$. If $\alpha_X : F \to G(X,X)$ is extranatural in X (for some object F of $\mathcal{D}$) and $\beta_{X,Y} : G(X,Y) \to H(X,Y)$ is natural in X, Y, then

$$\beta_{X,X} \circ \alpha_X : F \to H(X,X)$$

is extranatural in X.

- (Yanking) Let F, H be functors of the form $\mathcal{C} \to \mathcal{D}$, and let $G : \mathcal{C} \times \mathcal{C}^{\mathrm{op}} \times \mathcal{C} \to \mathcal{D}$ be a functor. If $\alpha_{X,Y} : F(y) \to G(X,X,Y)$ is natural in y and extranatural in X, and if $\beta_{X,Y} : G(X,Y,Y) \to H(X)$ is natural in X and extranatural in Y, then

$$\beta_{X,X} \circ \alpha_{X,X} : F(X) \to H(X)$$

is natural in X.

Express these laws as equalities between suitable string diagrams (explaining also the genesis of the names 'stalactite' and 'stalagmite').

1.5 ◔◔ Prove that dinaturality is strictly more general than extranaturality, following the outline argument below.

Let $\Delta[1] = \{0 \to 1\}$ be the 'generic arrow' category, and S, T :

$\Delta[1]^{\text{op}} \times \Delta[1] \to \text{Set}$ the functors respectively defined by

$$\begin{array}{ccc}
\begin{array}{ccc}\{1\} & \xrightarrow{c_1} & \{1,2\} \\ \downarrow & S & \downarrow \sigma \\ \{1\} & \xrightarrow[c_2]{} & \{1,2\}\end{array} &
\begin{array}{ccc}(0,0) & \longrightarrow & (0,1) \\ \downarrow & & \downarrow \\ (1,0) & \longrightarrow & (1,1)\end{array} &
\begin{array}{ccc}\{1\} & \longrightarrow & \{1\} \\ \downarrow & T & \downarrow c_2 \\ \{1\} & \xrightarrow[c_2]{} & \{1,2\}\end{array}
\end{array}$$

where c_i chooses element $i \in \{1,2\}$, and σ permutes the two elements. Show that there exists a dinatural transformation $T \Rightarrow S$, whose components are identities, which is not extranatural when in 1.1.8 we choose $\mathcal{A} = *$ and $\mathcal{B} = \mathcal{C}^{\text{op}} = \Delta[1]$.

1.6 If (X, Ω, μ) is a measure space, the integral of a vector function $\vec{F} : X \to \mathbb{R}^n$ with the property that each $F_i = \pi_i \circ F : X \to \mathbb{R}$ is measurable and has finite integral, is the vector whose entries are $\left(\int_X F_1 d\mu, \cdots, \int_X F_n d\mu\right)$.

Prove that category theory possesses a similar formula, i.e. that if $F : \mathcal{C}^{\text{op}} \times \mathcal{C} \to \prod_{i=1}^n \mathcal{A}_i$ is a functor into a product category, such that

- each $\mathcal{A}_i$ has both an initial and a terminal object,
- each (co)end $\int(\pi_i \circ F)$ exists.

Then the tuple of objects $\left(\int F_1, \ldots, \int F_n\right) \in \prod \mathcal{A}_i$, is the (base of a universal (co)wedge forming the) (co)end of F. Where did you use the assumption that each $\mathcal{A}_i$ has an initial and a terminal object?

1.7 Let $\mathcal{D}$ be a category. Show that the end of a functor $T : \Delta[1]^{\text{op}} \times \Delta[1] \to \mathcal{D}$ is the pullback of the morphisms

$$T(0,0) \xrightarrow{T(0,d_0)} T(0,1) \xleftarrow{T(d_0,1)} T(1,1),$$

i.e. that the following square is a pullback in $\mathcal{D}$:

$$\begin{array}{ccc}
\int_{i\in\Delta[1]} T(i,i) & \longrightarrow & T(0,0) \\
\downarrow \;\lrcorner & & \downarrow \\
T(1,1) & \longrightarrow & T(0,1)
\end{array}$$

Dualise to the coend being a pushout.

1.8 ◉◉ Let G be a topological group, and let $\text{Sub}(G)$ be the set of its subgroups partially ordered by inclusion; let X be a G-space, i.e. a topological space with a continuous action $G \times X \to X$ (the product $G \times X$ has the product topology).

We can define two functors $\text{Sub}(G) \to \text{Spc}$, sending $(H \leq G) \mapsto$

G/H (this is a covariant functor, and G/H has the induced quotient topology as a space; there is no need for H to be normal) and $(H \le G) \mapsto X^H$ (the subset of H-fixed points for the action; this is a contravariant functor).

- Compute the coend
$$\mathfrak{o}_G(X) = \int^{H \le G} X^H \times G/H$$
in the category Spc of spaces, for $G = \mathbb{Z}/2$ endowed with the discrete topology.
- Give a general rule for computing $\mathfrak{o}_G(X)$ when $G = \mathbb{Z}/n\mathbb{Z}$ is cyclic with n elements.
- Let $\mathrm{Orb}(G)$ be the *orbit category* of subgroups of G, whose objects are subgroups but $\hom(H, K)$ contains *G-equivariant* maps $G/H \to G/K$. Let again $X^{(-)}$ and $G/(\_)$ define the same functors as in the previous exercise, now with different action on arrows. Prove that
$$\int^{H \in \mathrm{Orb}(G)} X^H \times G/H \cong X.$$
(This result is called *Elmendorf reconstruction* [Elm83].)
- Let $E|F$ be a field extension, and $\{H \le \mathrm{Gal}(E|F)\}$ the partially ordered set of subgroups of the Galois group of the extension. Compute (in the category of *rings*) the coend
$$\int^{H} E^H \times \mathrm{Gal}(H|F).$$

1.9 Dualise the construction in Section 1.2 to obtain a characterisation for the coend $\int^C F(C, C)$, as the coequaliser of a pair (F^*, F_*) as in
$$\coprod_{C \to C'} F(C', C) \rightrightarrows \coprod_{C \in \mathcal{C}} F(C, C).$$

1.10 Find an alternative proof that natural transformations can be written as an end (see 1.4.1). Use the characterisation of $\int_C \mathcal{D}(FC, GC)$ as an equaliser in 1.2.4: the end is precisely the subset of the product $\prod_{C \in \mathcal{C}} \mathcal{D}(FC, GC)$ of tuples τ_C such that $\{\tau_C : FC \to GC \mid Gf \circ \tau_C = \tau_{C'} \circ Ff,\ \forall f : C \to C'\}$.

1.11 What is the (co)end of the identity functor $1_{\mathcal{C}^{\mathrm{op}} \times \mathcal{C}} : \mathcal{C}^{\mathrm{op}} \times \mathcal{C} \to \mathcal{C}^{\mathrm{op}} \times \mathcal{C}$? Use the bare definition; use the characterisation of (co)ends as (co)limits; feel free to invoke Exercise 1.6.

1.12 A set of objects $\mathcal{S} \subset \mathcal{C}$, regarded as a full subcategory, *finitely generates* a category $\mathcal{C}$ if for each object $X \in \mathcal{C}$ and each arrow $f : S \to C$ from $S \in \mathcal{S}$, there is a factorisation

$$S \xrightarrow{g} \coprod_{i=1}^{n} S_i \xrightarrow{h_C} C$$

where h_C is an epimorphism and $\{S_1, \dots, S_n\} \subset \mathcal{S}$ (n depends on C and f).

Suppose $T : \mathcal{C}^{\mathrm{op}} \times \mathcal{C} \to \mathrm{Set}$ is a functor, finitely continuous in both variables, and $\mathcal{C}$ is finitely generated by $\mathcal{S}$. Show that if we denote by $T|_{\mathcal{S}} : \mathcal{S}^{\mathrm{op}} \times \mathcal{S} \to \mathrm{Set}$ the restriction, we have an isomorphism

$$\int^{C \in \mathcal{S}} T|_{\mathcal{S}}(C, C) \cong \int^{C \in \mathcal{C}} T(C, C)$$

induced by a canonical arrow $\int^{C \in \mathcal{S}} T'(C, C) \to \int^{C \in \mathcal{C}} T(C, C)$.

1.13 Let $F \dashv U : \mathcal{C} \leftrightarrows \mathcal{D}$ be an adjunction, and $G : \mathcal{D}^{\mathrm{op}} \times \mathcal{C} \to \mathcal{E}$ be a functor; then there is an isomorphism

$$\int^{C} G(FC, C) \cong \int^{D} G(D, UD).$$

Show that the converse of this result is true: if the above isomorphism is true for any G and natural therein, then there is an adjunction $F \dashv U$.

1.14 Give a different proof of the Fubini theorem 1.3.1 as follows: define again $H_{\mathcal{C}} : D \mapsto \hom_{\mathcal{C}} \otimes D$, and find suitable unit and counit maps (see A.4.1).

- The unit of the adjunction $\eta : \mathrm{id}_{\mathcal{D}} \Rightarrow \int_C \circ H_{\mathcal{C}}$ is determined as a morphism

$$\eta_D : D \to \int_C \big(\mathcal{C}(C, C) \otimes D\big)$$

(parenthesisation is important: $\int_C \big(\mathcal{C}(C, C) \otimes D\big)$ is *not* isomorphic to $\big(\int_C \mathcal{C}(C, C)\big) \otimes D$) which in turn corresponds to a wedge

$$\eta_{D,(u)} : D \to \coprod_{v : C \to C} D$$

defined on the component $u : C \to C$ by the inclusion i_u in the coproduct.

- The counit of the adjunction $\epsilon : H_{\mathcal{C}} \circ \left(\int_{\mathcal{C}} -\right) \Rightarrow \mathrm{id}_{\mathsf{Cat}(\mathcal{C}^{\mathrm{op}}\times\mathcal{C},\mathcal{D})}$ is determined as a natural transformation having components
$$\epsilon_{CC'} : \coprod_{u:C\to C'} \int_C F(C,C) \to F(C,C')$$
which in turn have components
$$\epsilon^u_{CC'} : \int_C F(C,C) \to F(C,C')$$
by the universal property of a coproduct, determined by sending a coherent family $(a_C \mid C \in \mathcal{C})$ into $F(C,u)(a_C)$ (or equivalently by the wedge condition, $F(u,C')(a_{C'})$).

Show that these two definitions set up the desired adjunction (prove the triangle identities for η and ϵ, as in A.4.5).

Hint: to show that $\epsilon^u_{CC'}$ is indeed natural in C, C', take two arrows $\begin{smallmatrix} C_0 \\ \alpha\downarrow \\ C_1 \end{smallmatrix}$ and $\begin{smallmatrix} C'_0 \\ \beta\downarrow \\ C'_1 \end{smallmatrix}$ and split the naturality square into the pasting of two smaller squares

$$\begin{array}{ccc}
\coprod_{C_1\to C'_0} \int_C F & \xrightarrow{\epsilon^u_{C_1C'_0}} & F(C_1, C'_0) \\
{\scriptstyle -\circ\beta}\downarrow & & \downarrow{\scriptstyle F(1,\beta)} \\
\coprod_{C_1\to C'_1} \int_C F & \xrightarrow{\epsilon^u_{C_1C'_1}} & F(C_1, C'_1) \\
{\scriptstyle \alpha\circ -}\downarrow & & \downarrow{\scriptstyle F(\alpha,1)} \\
\coprod_{C_0\to C'_1} \int_C F & \xrightarrow[\epsilon^u_{C_0C'_1}]{} & F(C_0, C'_1)
\end{array}$$

each of which commutes for evident reasons.

2

Yoneda and Kan

SUMMARY. In this chapter we begin to learn the rules of (co)end calculus. First, we revisit the well-known Yoneda lemma as an isomorphism of (co)ends. The Yoneda lemma is one of the deepest results that can be examined with the technology built so far: it amounts to say that every functor $F : \mathcal{C}^{\text{op}} \to \text{Set}$ can be decomposed as a coend

$$F \cong \int^C FC \times \mathcal{C}(\_, C).$$

This isomorphism has plenty of consequences; it is in fact equivalent to the assertion that every presheaf is a colimit of a certain diagram $D : \mathcal{C} \to [\mathcal{C}^{\text{op}}, \text{Set}]$ of representables, and that there is a canonical choice for such D. Then we move on to investigate *Kan extensions*; the famous tenet that 'everything is a Kan extension' is, here, translated into the more appealing (for us!) statement that every sufficiently nice thing is a (co)end. We then glimpse *formal category theory*; this latter part draws on [Gra80] and other classical sources from the Australian school of category theory.

> The reason why this technique is so fast is that you are not trying to cut them; you are throwing your sword into them.
>
> sōke M. Hatsumi

Throughout this chapter, abstract 2-categories will be written in sans-serif, $\mathsf{A}, \mathsf{K} \ldots$; objects in K will be written in uppercase roman $A, B, C \ldots$; objects of a 0-cell $A \in \mathsf{Cat}$ will be written in roman lowercase $a, b, a', b' \ldots$

This complies with the common notation for papers in 2-dimensional algebra, and the different notation employed elsewhere should not cause any confusion. Note in particular that we make the distinction between the 1-category Cat (categories and functors) and the 2-category Cat (categories, functors and natural transformations).

2.1 The Yoneda Lemma and Kan Extensions

For ease of exposition, we recall the statement of the *Yoneda lemma.*

Lemma (See A.5.3)**.** Let $\text{よ} : \mathcal{C} \to \mathrm{Cat}(\mathcal{C}^{\mathrm{op}}, \mathrm{Set})$ be the functor sending $X \in \mathcal{C}$ to the presheaf associated to X; then for every $F \in \mathrm{Cat}(\mathcal{C}^{\mathrm{op}}, \mathrm{Set})$ there exists a bijection between the set of natural transformations $\text{よ}X \Rightarrow F$ and the set FX. This bijection is moreover natural in the object X.

We carry out the proof in full detail in our A.5.3 to convince even the most sceptical reader that this result is a tautology. Few tautologies are, however, richer in meaning.

Remark 2.1.1 (The Yoneda–Grothendieck philosophy)**.** The Yoneda lemma entails that there is a copy of a category $\mathcal{C}$ in its category of presheaves $\mathrm{Cat}(\mathcal{C}^{\mathrm{op}}, \mathrm{Set})$; indeed, if we consider a representable $F = \text{よ}B$ (see A.5.2) in the isomorphism above, we see that

$$\mathsf{Cat}(\mathcal{C}^{\mathrm{op}}, \mathrm{Set})(\text{よ}A, \text{よ}B) \cong \mathcal{C}(A, B). \tag{2.1}$$

This means precisely that the Yoneda embedding functor $\text{よ}_{\mathcal{C}} : \mathcal{C} \to \mathrm{Cat}(\mathcal{C}^{\mathrm{op}}, \mathrm{Set})$ is fully faithful. For all practical purposes then, every question about $\mathcal{C}$ is a question about the copy of $\mathcal{C}$ inside its category of presheaves; the latter category is, however, always very well-behaved under certain aspects; for example, it is always complete and cocomplete (see A.3.11) even when $\mathcal{C}$ is not.

Thus, every diagram in $\mathcal{C}$, say $D : \mathcal{J} \to \mathcal{C}$, admits a limit *in* $[\mathcal{C}^{\mathrm{op}}, \mathrm{Set}]$, as we can build the limit

$$\lim \text{よ}(D_I) = \lim \hom_{\mathcal{C}}(\_, D_I). \tag{2.2}$$

If this functor is representable, say by an object L, it is easy to see that L realises the universal property of $\lim D_I$ *in* $\mathcal{C}$; in fact, this is a necessary and sufficient condition. Thus, a completeness requirement on $\mathcal{C}$ has a 1-1 translation in terms of a requirement for representability of certain functors.

The idea that properties of $\mathcal{C}$ can (and should) be translated into representability properties, and more extensively that representability conditions play a major role in category theory, algebra and geometry, was first advocated by A. Grothendieck. This perspective makes heavy use of the Yoneda lemma; thus we colloquially refer to it as the *Yoneda–Grothendieck philosophy.*

The next step of this introduction to the chapter is to single out certain universal 2-cells in a 2-category, and the fundamental properties thereof.

Notation 2.1.2 (Extensions and lifts). Let f, g be 1-cells of a 2-category K, respectively $f : A \to B$ (the extendable arrow) and $g : A \to C$ (the extendant arrow). We say that a pair $\langle u, \eta \rangle$ *exhibits* the left extension $\mathrm{lan}_g f$ of f along g if the 1-cell $u : C \to B$ and the 2-cell $\eta : f \Rightarrow ug$ can be arranged in a triangle

$$\begin{array}{ccc} & A & \\ {}^{f}\swarrow & \overset{\eta}{\Rightarrow} & \searrow^{g} \\ B & \xleftarrow[u]{} & C \end{array} \tag{2.3}$$

initial among all such. This means that every pair $\langle v, \alpha \rangle$ of a 1-cell $\begin{bmatrix} C \\ v\downarrow \\ B \end{bmatrix}$ and a 2-cell $\alpha : f \Rightarrow vg$ factors uniquely through η as a composition $\alpha = (\bar{\alpha} * g) \circ \eta$ (the cell $\bar{\alpha} * g$ is the *whiskering* of A.2.3).

Remark 2.1.3. The notion of left extension is subject to dualisation, in a way that is worth recording explicitly: it is useful to have a diagram illustrating at once all these universal constructions. The diagram below has to be parsed as follows: moving horizontally reverses the direction of 1-cells, but not of 2-cells; moving vertically but not horizontally reverses the direction of 2-cells, but not of 1-cells. The universality requirement acquires the same shape, and when we write (for example) $\dfrac{\mathrm{lift}_g f \Rightarrow h}{f \Rightarrow gh}$ we mean that there is a bijection between 2-cells $\mathrm{lift}_g f \Rightarrow h$ and 2-cells $f \Rightarrow gh$.

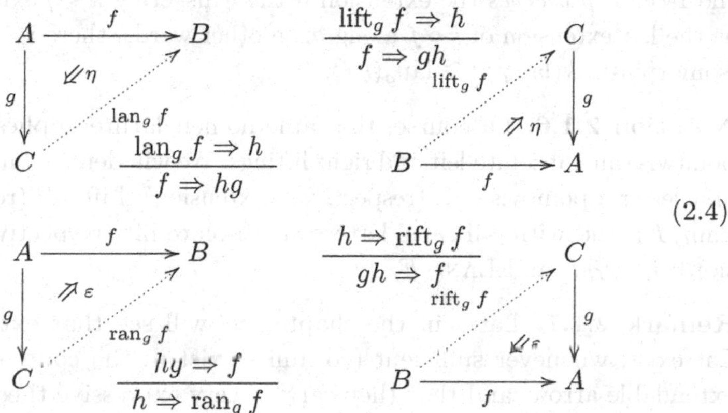

Summing up, a right extension is a left extension in K^{co}, the 2-category where all 2-cells have been reversed; a left lifting is a left extension in

K^{op}, the 2-category where all 1-cells have been reversed; a right lifting is a left extension in $\mathsf{K}^{\mathrm{coop}}$, where both 1- and 2-cells have been reversed.

Definition 2.1.4 (Pointwise extension). We say that a left (respectively, right) extension is *pointwise* if for every object C and $k : X \to C$, the diagram obtained by pasting at g the comma object (see A.2.14) (g/k) (respectively, (k/g)) is again a left extension [Str81, 5.2]; this means that in every diagram of 2-cells

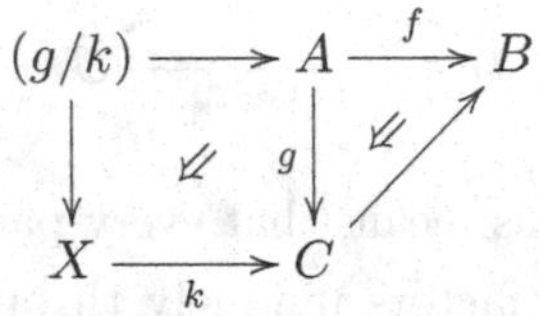

$$\begin{array}{ccccc} (g/k) & \longrightarrow & A & \xrightarrow{f} & B \\ \downarrow & \Swarrow & \downarrow g & \Swarrow \nearrow & \\ X & \xrightarrow[k]{} & C & & \end{array} \tag{2.5}$$

if the left triangle is a left extension and the square is a comma object, then the whole pasting diagram is a left extension; dually (in K^{co}) for a right extension.

Definition 2.1.5 (Absolute extension). We say that a left extension is *absolute* if it is preserved by all 1-cells, in the same sense a functor preserves (co)limits. In the left extension diagram $\langle h, \eta \rangle$

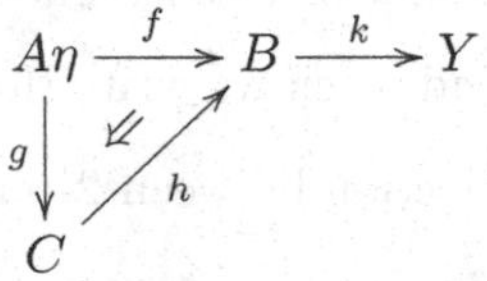

$$\begin{array}{ccccc} A\eta & \xrightarrow{f} & B & \xrightarrow{k} & Y \\ g\downarrow & \Swarrow \nearrow h & & & \\ C & & & & \end{array} \tag{2.6}$$

the 1-cell k *preserves* the extension if the whiskering $k * \eta$ exhibits $k \circ h$ as the left extension of $k \circ f$ along g. In other words, there is a canonical isomorphism $k(\mathrm{lan}_g f) \cong \mathrm{lan}_g(kf)$.

Notation 2.1.6. Of course, the same nomenclature applies to define pointwise and absolute left and right liftings. We will denote with a capital first letter a pointwise lift (respectively, extension): $\mathrm{Lift}_G F$ (respectively, $\mathrm{Lan}_G F$), and with 'all caps' letters an absolute lift (respectively, extension): $\mathrm{LIFT}_G F$ and $\mathrm{LAN}_G F$.

Remark 2.1.7. Later in the chapter we will see that extensions in Cat exist, whenever sufficient (co)limits exist in the codomain of the extendable arrow, and that they carry a very expressive theory (to the point that Mac Lane states that 'everything is a Kan extension' in [ML98]). On the other hand, few lifts (both left and right) exist in Cat.

The deep reason why this is true is that the Yoneda lemma does not hold in the category $\mathsf{Cat}^{\mathrm{op}}$, as the opposite of a functor is not a functor in general.

In fact, one tenet of the present chapter is that left extensions in Cat, where they are called *Kan* extensions, in honour of D. Kan (see 3.1.2, which justifies the notation as lan is a portmanteau for l(-eft) + (K-)an), can be computed as certain explicit (co)limits. If they are left extensions, they are colimits; if they are right extension, they are limits. More precisely, we will see how pointwise Kan extensions can be expressed as (co)ends.

It is not possible to do the same for lifts, due to the peculiar behaviour of the universality requirements in (2.4); the reason is somewhat trivial: the dual of a functor is not a functor in general.

Example 2.1.8. To see that there is no way to compute a lift as a (co)limits, let us consider the following example. Consider the discrete category $\{0,1\}$, and let $\delta_0 \colon \mathcal{C} \to \{0,1\}$, $\delta_1 \colon \mathcal{D} \to \{0,1\}$ be the obvious constant functors (with disjoint images) from any two non-trivial categories $\mathcal{C}, \mathcal{D}$.

Then there are no liftings of δ_0 through δ_1, no matter how complete and cocomplete $\mathcal{C}$ and $\mathcal{D}$ are. In fact, for any functor $F \colon \mathcal{C} \to \mathcal{D}$, there are no natural transformations $\delta_1 \circ F \to \delta_0$ nor $\delta_0 \to \delta_1 \circ F$. The same is true if we replace $\{0,1\}$ with its free cocompletion $\mathrm{Cat}(\{0,1\}^{\mathrm{op}}, \mathrm{Set}) \cong \mathrm{Set}/\{0,1\}$.

The deep reason that we do not have nice formulae for Kan liftings is that $\mathsf{Cat}^{\mathrm{op}}$ lacks an internal concept of (co)end; in other words the internal language of $\mathsf{Cat}^{\mathrm{op}}$ is not expressive enough.

The theory of extensions is much better behaved and it is possible to give a neat characterisation of pointwise extensions.

Proposition 2.1.9. Let $\mathcal{A}, \mathcal{B}, \mathcal{C}$ be categories, f, g be functors as in the triangle below. Then, the following conditions are equivalent.

- The triangle

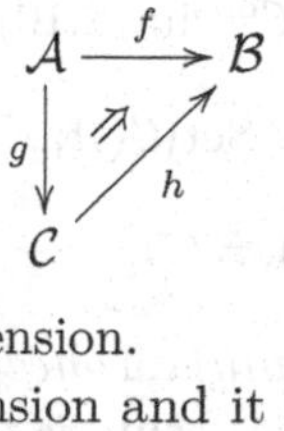

$$\begin{array}{ccc} \mathcal{A} & \xrightarrow{f} & \mathcal{B} \\ {\scriptstyle g}\downarrow & \nearrow\!\!\!\!\nearrow & \nearrow_{h} \\ \mathcal{C} & & \end{array} \tag{2.7}$$

  is a pointwise right Kan extension.
- The triangle is a right extension and it is preserved by every representable $よ^{\vee}(B) = \mathcal{B}(B, \_) : \mathcal{B} \to \mathrm{Set}$, meaning that

$$\mathrm{ran}_g \hom_{\mathcal{B}}(B, f) \cong \hom_{\mathcal{B}}(B, \mathrm{ran}_g f). \tag{2.8}$$

Proof See [Leh14, 5.4]. □

2.2 Yoneda Lemma Using (Co)ends

Tightly linked to the Yoneda lemma is the *density theorem* of A.5.7: every presheaf $F : \mathcal{C}^{\mathrm{op}} \to \mathrm{Set}$ is, canonically, the colimit of a diagram of representables, i.e. functors lying in the image of $よ_{\mathcal{C}} : \mathcal{C} \to \mathsf{Cat}(\mathcal{C}^{\mathrm{op}}, \mathrm{Set})$. We prove this result in full detail in A.5.3, without using coends; however, the scope of the present chapter is to convince the reader that (co)end calculus allows us to concisely and effectively rephrase both the Yoneda lemma and this result.

In Chapter 4, thanks to the machinery of *weighted (co)limits*, the fact that every presheaf is a colimit of representables acquires the more alluring form:

> Every presheaf F results as the weighted colimit of the Yoneda embedding, weighted by F itself.

Proposition 2.2.1 (Ninja Yoneda lemma). For every functor $K : \mathcal{C}^{\mathrm{op}} \to \mathrm{Set}$ and $H : \mathcal{C} \to \mathrm{Set}$, we have the following isomorphisms (natural equivalences of functors):

$$K \cong \int^{C} KC \times \mathcal{C}(\_, C) \tag{2.9}$$

$$K \cong \int_{C} \mathrm{Set}(\mathcal{C}(C, \_), KC) \tag{2.10}$$

$$H \cong \int^{C} HC \times \mathcal{C}(C, \_) \tag{2.11}$$

$$H \cong \int_{C} \mathrm{Set}(\mathcal{C}(\_, C), HC), \tag{2.12}$$

where the functor $\mathcal{C} \times \mathcal{C}^{\mathrm{op}} \times \mathcal{C}^{\mathrm{op}}$ in (2.10) is defined by

$$\lambda AXC.\mathrm{Set}(\mathcal{C}(A, X), KC),$$

and similarly for $\mathrm{Set}(\mathcal{C}(\_, C), HC)$.

Remark 2.2.2. The name *ninja Yoneda lemma* is chosen to instill in the reader the sense that this result *is* the Yoneda lemma in disguise; the name comes from [Lei] where T. Leinster offers the same argument we are about to give as a proof of 2.2.1:

> Th[e above one is] often called the *Density Formula*, or (by Australian ninja category theorists) simply the Yoneda lemma (but Australian ninja category theorists call *everything* the Yoneda lemma...).

Note that the isomorphisms (2.10) and (2.12) follow directly from Theorem 1.4.1, expressing a set of natural transformations as an end. We prove only isomorphism (2.9), as the others can be easily obtained by dualisation.

We put a particular emphasis on giving a detailed proof once, as it is the first argument that truly exploits (co)end *calculus*; it is a paradigmatic example of the style of proofs we are using from now on.

Proof Consider the chain of isomorphisms

$$\begin{aligned}\mathrm{Set}\Big(\int^{C\in\mathcal{C}} KC\times\mathcal{C}(X,C),Y\Big) &\cong \int_{C\in\mathcal{C}}\mathrm{Set}(KC\times\mathcal{C}(X,C),Y)\\ &\cong \int_{C\in\mathcal{C}}\mathrm{Set}(\mathcal{C}(X,C),\mathrm{Set}(KC,Y))\\ &\cong [\mathcal{C},\mathrm{Set}](\mathcal{C}(X,\_),\mathrm{Set}(K\_,Y))\\ &\cong \mathrm{Set}(KX,Y).\end{aligned}$$

The first step is motivated by the coend-preservation property of the hom functor 1.2.8.

The second follows from the fact that Set is a cartesian closed category, where

$$\mathrm{Set}(A\times B,C)\cong\mathrm{Set}(A,\mathrm{Set}(B,C)) \tag{2.13}$$

for all three sets A, B, C, naturally in all arguments, and where $C^B = \mathrm{Set}(B,C)$ is the set of all functions $B\to C$.

The final step exploits Theorem 1.4.1 and the classical Yoneda lemma that says that the set of natural transformations $よ^{\vee}(C)\Rightarrow F$ is in bijection with the set FC for every presheaf $F:\mathcal{C}\to\mathrm{Set}$. Here, $F = \mathrm{Set}(K\_,Y)$.

Every step of this chain of isomorphisms is natural in the object Y; since the Yoneda embedding $よ_{\mathcal{C}}:\mathcal{C}^{\mathrm{op}}\to[\mathcal{C},\mathrm{Set}]$ is fully faithful, the isomorphism of functors

$$\mathrm{Set}\Big(\int^{C} KC\times\mathcal{C}(X,C),Y\Big)\cong\mathrm{Set}(KX,Y) \tag{2.14}$$

ensures in turn that there exists an isomorphism between the represented objects $\int^C KC\times\mathcal{C}(X,C)\cong KX$.

This is moreover natural in the object X. Accepting the validity of

(2.9) and re-doing each step backwards, we can recover the Yoneda lemma in the form expressed in A.5.3. □

From now on we will make frequent use of the notion of *tensor* and *cotensor* in an enriched category; the definitions for a generic $\mathcal{V}$-category can be found for example in [Bor94b, Ch. 6], and in particular its Definition 6.5.1: we have already introduced this definition in the proof of our 1.3.1.

Definition 2.2.3 (Tensor and cotensor in a $\mathcal{V}$-category)**.** Let $\mathcal{C}$ be a $\mathcal{V}$-enriched category (see [Bor94b, 6.2.1] or our A.7.1), then the *tensor* $\otimes : \mathcal{V} \times \mathcal{C} \to \mathcal{C}$ (when it exists) is a functor $(V, C) \mapsto V \otimes C$ such that there is the isomorphism

$$\mathcal{C}(V \otimes C, C') \cong \mathcal{V}(V, \mathcal{C}(C, C')), \tag{2.15}$$

natural in all components; dually, the *cotensor* in an enriched category $\mathcal{C}$ (when it exists) is a functor $(V, C) \mapsto V \pitchfork C$ (contravariant in V) such that there is the isomorphism

$$\mathcal{C}(C', V \pitchfork C) \cong \mathcal{V}(V, \mathcal{C}(C', C)), \tag{2.16}$$

natural in all components.

Example 2.2.4. Every (co)complete, locally small category $\mathcal{C}$ is naturally Set-(co)tensored by choosing $V \pitchfork C \cong \prod_{v \in V} C$ and $V \otimes C \cong \coprod_{v \in V} C$. We have employed this construction in the proof of 1.3.1.

Remark 2.2.5. The tensor, hom and cotensor functors are the prototype of what is called a 'two variable adjunction' (see [Gra80, §1.1]); given the hom-objects of a $\mathcal{V}$-category $\mathcal{C}$, the tensor

$$\_ \otimes = : \mathcal{V} \times \mathcal{C} \to \mathcal{C} \tag{2.17}$$

and the cotensor

$$\_ \pitchfork = : \mathcal{V}^{\text{op}} \times \mathcal{C} \to \mathcal{C} \tag{2.18}$$

can be characterised as adjoint functors to the hom functors, saturated in their first or second component. The usual (co)continuity properties of the (co)tensor functors are implicitly derived from this characterisation.

Remark 2.2.6 (The Yoneda embedding is a Dirac delta)**.** ⚇ In functional analysis, the Dirac delta appears in the following convenient abuse of notation:

$$\int_{-\infty}^{\infty} f(x)\delta(x-y)dx = f(y) \tag{2.19}$$

(the integral sign is not a (co)end). Here $\delta(x-y) := \delta_y(x)$ is the *y-centred delta distribution*, and $f : \mathbb{R} \to \mathbb{R}$ is a continuous, compactly supported function on $\mathbb{R}$.

It is really tempting to draw a parallel between this relation and the ninja Yoneda lemma, conveying the intuition that representable functors on an object $C \in \mathcal{C}$ play the rôle of C-centred delta distributions.

If the relation above is written as $\langle f, \delta_y \rangle = f(y)$, interpreting integration as an inner product between functions (or more precisely as a universal pairing between a space and its dual), then the ninja Yoneda lemma says the same thing for categories: each presheaf $F : \mathcal{C}^{\mathrm{op}} \to \mathrm{Set}$, can be paired with a distribution concentrated on the point C in an 'inner product' $\langle よ_C, F \rangle = \int^X よ_C(X) \times FC$; the latter object is now isomorphic to FC, in the same way that the integral of a smooth function f against a y-centred delta equals $f(y)$.

$$\begin{array}{|c|c|c|c|c|c|}
\int_{x \in \mathbb{R}} & f(x) & \circ & \delta(x-y)dx & = & f(y) \\
\int^X & FX & \times & \mathcal{C}(Y,X) & \cong & FY
\end{array}$$

The analogy between the pairing of a function and a delta distribution, and the ninja Yoneda lemma.

2.3 Kan Extensions Using (Co)ends

In the following series of remarks, $G : \mathcal{C} \to \mathcal{E}$ and $F : \mathcal{C} \to \mathcal{D}$ are functors.

For the sake of exposition we assume that $\mathrm{Lan}_G F$ exists for all such F; we shall see in the future that a sufficient condition for this to be true is that $\mathcal{D}$ is a cocomplete category (see A.3.11), thus we assume that all colimits exist in $\mathcal{D}$.

Remark 2.3.1. The correspondence $F \mapsto \mathrm{Lan}_G F$ is a functor

$$\mathsf{Cat}(\mathcal{C}, \mathcal{D}) \to \mathsf{Cat}(\mathcal{E}, \mathcal{D}); \tag{2.20}$$

to every natural transformation $\alpha : F \Rightarrow F'$ is associated a natural transformation $\mathrm{Lan}_G\,\alpha : \mathrm{Lan}_G F \Rightarrow \mathrm{Lan}_G F'$ defined as the unique 2-cell ξ such that

$$\begin{array}{ccc} & \overset{F}{\curvearrowright}\ \Downarrow\alpha & \\ \mathcal{C} & \longrightarrow & \mathcal{D} \\ {\scriptstyle G}\downarrow & \swarrow\eta' \ \nearrow_{\mathrm{Lan}_G F'} & \\ \mathcal{E} & & \end{array} \quad = \quad \begin{array}{ccc} \mathcal{C} & \xrightarrow{F} & \mathcal{D} \\ {\scriptstyle G}\downarrow & \swarrow\eta \nearrow \ \searrow\xi & \uparrow \\ \mathcal{E} & \xrightarrow[\mathrm{Lan}_G F']{} & \end{array} \tag{2.21}$$

Remark 2.3.2. The correspondence $G \mapsto \mathrm{Lan}_G$ is a functor

$$\mathsf{Cat}(\mathcal{C},\mathcal{E})^{\mathrm{op}} \to \mathsf{Cat}(\mathsf{Cat}(\mathcal{C},\mathcal{D}),\mathsf{Cat}(\mathcal{E},\mathcal{D})); \tag{2.22}$$

this means that to every natural transformation $\alpha : G \Rightarrow G'$ is associated a natural transformation $\mathrm{Lan}_\alpha : \mathrm{Lan}_{G'} \Rightarrow \mathrm{Lan}_G$, defined as the mate of the composite 2-cell

$$F \overset{\eta_F}{\Rightarrow} \mathrm{Lan}_G F \circ G \overset{\mathrm{Lan}_G F * \alpha}{\Rightarrow} \mathrm{Lan}_G F \circ G'. \tag{2.23}$$

In order to show that the correspondence $G \mapsto \mathrm{Lan}_G$ is a pseudofunctor, we have to find suitable coherence isomorphisms $\mathrm{Lan}_{GG'} \cong \mathrm{Lan}_G \circ \mathrm{Lan}_{G'}$ and $\mathrm{Lan}_{\mathrm{id}} \cong \mathrm{id}$; this is the content of Exercise 2.5.

Remark 2.3.3. The functor Lan_G is (the unique up to isomorphism) adjoint to the *inverse image* functor $G^* = \_ \circ G$; this follows directly from the characterisation of $\mathrm{Lan}_G F$ as the functor $\mathcal{E} \to \mathcal{D}$ such that

$$\mathsf{Cat}(\mathcal{E},\mathcal{D})(\mathrm{Lan}_G F, H) \cong \mathsf{Cat}(\mathcal{C},\mathcal{D})(F, HG). \tag{2.24}$$

Remark 2.3.4. Dually, the correspondence $G \mapsto \mathrm{Ran}_G$ is a functor

$$\mathsf{Cat}(\mathcal{C},\mathcal{E})^{\mathrm{op}} \to \mathsf{Cat}(\mathsf{Cat}(\mathcal{C},\mathcal{D}),\mathsf{Cat}(\mathcal{E},\mathcal{D})) \tag{2.25}$$

that acts as right adjoint to the precomposition functor $\_ \circ G$, and it is functorial, contravariant in the extendant component (see Notation 2.1.2). Given $\alpha : G \Rightarrow G'$ the components of Ran_α are the mates of

$$\mathrm{Ran}_{G'} F \circ G \overset{\mathrm{Ran}_{G'} * \alpha}{\Rightarrow} \mathrm{Ran}_{G'} F \circ G' \overset{\epsilon}{\Rightarrow} F. \tag{2.26}$$

Remark 2.3.5. From the universal property of the functor $\mathrm{Lan}_G\,\_$, we can derive the unit and the counit of the adjunctions $\mathrm{Lan}_G \underset{\eta^L}{\overset{\epsilon^L}{\dashv}} G^*$ and $G^* \underset{\eta^R}{\overset{\epsilon^R}{\dashv}} \mathrm{Ran}_G$. We leave this as Exercise 2.2 for the reader to spell out explicitly.

Proposition 2.3.6. Let again $G : \mathcal{C} \to \mathcal{E}$ be a functor and $F : \mathcal{C} \to \mathcal{D}$ be a functor whose domain is a cocomplete category. Since both the (co)tensors (see 2.2.3) and the (co)ends involved in the equations below exist, then the left/right Kan extensions of $G : \mathcal{C} \to \mathcal{E}$ along $F : \mathcal{C} \to \mathcal{D}$ exist and there are isomorphisms (natural in F and G)

$$\mathrm{Lan}_F G \cong \int^C \mathcal{D}(FC, \_) \otimes GC \qquad\qquad \mathrm{Ran}_F G \cong \int_C \mathcal{D}(\_, FC) \pitchfork GC. \tag{2.27}$$

These Kan extensions are pointwise in the sense of 2.1.4.

Since this is our first instance of derivation in (co)end calculus, we apply a certain pedantry to the explanation, duly recording the results that allow each step. Such verbosity will be soon abandoned, to allow the reader to profit from instructive meditation following the chains of isomorphisms.

Proof The proof consists of a string of canonical isomorphisms, exploiting simple remarks and the results established so far. The same argument is offered in [ML98, X.4.1-2].

$$\begin{aligned}
\mathsf{Cat}(\mathcal{D}, \mathcal{E})\Big(\int^C \mathcal{D}(FC, _) \otimes GC, H\Big) &\cong \int_X \mathcal{D}\Big(\int^C \mathcal{D}(FC, X) \otimes GC, HX\Big) \\
1.2.8 &\cong \int_{CX} \mathcal{D}(\mathcal{D}(FC, X) \otimes GC, HX) \\
(2.15) &\cong \int_{CX} \mathsf{Set}(\mathcal{D}(FC, X), \mathcal{E}(GC, HX)) \\
1.4.1 &\cong \int_C [\mathcal{D}(FC, _), \mathcal{E}(GC, H-)] \\
(2.12) &\cong \int_C \mathcal{E}(GC, HFC) \cong \mathsf{Cat}(\mathcal{C}, \mathcal{E})(G, HF).
\end{aligned}$$

The case of $\mathrm{Ran}_F G$ is analogous and we leave it to the reader. □

Corollary 2.3.7. Let $D : \mathcal{A} \to \mathcal{B}$ be a functor.

CC1. If D is a left adjoint, then D preserves all pointwise left Kan extensions that exist in $\mathcal{A}$.

CC2. If D is a right adjoint, then D preserves all pointwise right Kan extensions that exist in $\mathcal{A}$.

Proof Left adjoints commute with tensors, i.e. $D(X \otimes A) \cong X \otimes DA$ for any $(X, A) \in \mathsf{Set} \times \mathcal{C}$, and with colimits (see A.4.6). The result follows, and can easily be dualised. □

Example 2.3.8. Let $T : \mathcal{C} \to \mathcal{C}$ be a monad (see A.6.3) on $\mathcal{C}$; the *Kleisli category* $\mathrm{Kl}(T)$ of T is defined as having the same objects as $\mathcal{C}$ and morphisms given by $\mathrm{Kl}(T)(A, B) := \mathcal{C}(A, TB)$.

Given any functor $F : \mathcal{A} \to \mathcal{C}$, the right Kan extension $T_F = \mathrm{Ran}_F F$, when it exists, is a monad on $\mathcal{C}$, which we call the *codensity monad* of F; using the end expression for $\mathrm{Ran}_F F$ we get that T_F is defined on objects as

$$T_F(C) \cong \int_A \mathcal{C}(C, FA) \pitchfork FA. \tag{2.28}$$

(See 2.2.3 for the $\pitchfork$ operation.) Hom-sets in the Kleisli category $\mathrm{Kl}(T_F)$ can be characterised as

$$\mathrm{Kl}(T_F)(C, C') \cong \int_A \mathrm{Set}(\mathcal{C}(C', FA), \mathcal{C}(C, FA)). \tag{2.29}$$

The multiplication and unit of T_F can be found using its universal property:

- the multiplication is obtained as the mate of
$$\mathrm{Ran}_F F \circ \mathrm{Ran}_F F \circ F \xrightarrow{\mathrm{Ran}_F * \epsilon * F} \mathrm{Ran}_F F \circ F \xrightarrow{\epsilon_F} F \tag{2.30}$$
under the adjunction isomorphism $[\mathcal{C}, \mathcal{C}](H, \mathrm{Ran}_F F) \cong [\mathcal{A}, \mathcal{C}](HF, F)$;
- the unit $\eta : \mathrm{id} \Rightarrow \mathrm{Ran}_F F$ is obtained as the component of the unit of the adjunction $F^* \dashv \mathrm{Ran}_F$ at the identity.

The difficult part is to show that the two maps $\mu : T_F \circ T_F \Rightarrow T_F$ and $\eta : \mathrm{id}_{\mathcal{C}} \Rightarrow T_F$ indeed form a monad; we relegate this proof to Exercise 2.9; the interested reader should try to translate into coend calculus the proof in [Dub70, pp. 67-71].

Remark 2.3.9. There is a dual theory of *density comonads*: given a functor $F : \mathcal{A} \to \mathcal{C}$, the left Kan extension of F along itself, when it exists, has the structure of a comonad S^F (see A.6.17), where

- the comultiplication is obtained as the mate of
$$F \xrightarrow{\eta_F} \mathrm{Lan}_F F \circ F \xrightarrow{\mathrm{Lan}_F F * \eta_F * F} \mathrm{Lan}_F F \circ \mathrm{Lan}_F F \circ F \tag{2.31}$$
under the adjunction isomorphism;
- the counit $\sigma : \mathrm{Lan}_F F \Rightarrow \mathrm{id}$ is obtained as the component of the counit of the adjunction $\mathrm{Lan}_F \dashv F^*$ at the identity.

The reader should find a similar expression for the hom-sets of the coKleisli category $\mathrm{coKl}(S^F)$ (and see Exercise 2.10).

Example 2.3.10 (Stalks of a sheaf). [GAV72, 6.8 and §7.1] Let (X, τ) be a topological space, and let $i_p\colon \{p\} \hookrightarrow X$ be the inclusion of a singleton into X. From this, we get an induced functor

$$\begin{aligned} \mathcal{O}(i_p) : \mathcal{O}(X) &\longrightarrow \mathcal{O}(\{p\}) \\ U &\longmapsto i_p^{-1}(U). \end{aligned} \tag{2.32}$$

Considering now left Kan extensions along the opposite of $\mathcal{O}(i_p)$,

$$\begin{array}{ccc} & & \mathcal{O}(\{p\})^{\mathrm{op}} \\ & \nearrow^{\mathcal{O}(i_p)^{\mathrm{op}}} & \downarrow \mathrm{Lan}_{\mathcal{O}(i_p)^{\mathrm{op}}}\mathcal{F} \\ \mathcal{O}(X)^{\mathrm{op}} & \xrightarrow[\mathcal{F}]{} & \mathrm{Set}, \end{array} \tag{2.33}$$

we obtain a functor $\mathrm{Lan}_{\mathcal{O}(i_p)^{\mathrm{op}}}\colon \mathrm{Cat}(\tau^{\mathrm{op}}, \mathrm{Set}) \to \mathrm{Cat}(\{p\}, \mathrm{Set}) = \mathrm{Set}$, whose image at $\mathcal{F}$ is written $\lceil \mathcal{F}_p \rceil$.[1]

The restriction of this functor to the category of sheaves on X can be identified with the *stalk* functor $(-)_p$: we have $\mathcal{O}(\{p\}) = \{\varnothing \le \{p\}\}$ and computing the images of $\varnothing$ and $\{p\}$ under $\lceil \mathcal{F}_p \rceil$ via the colimit formula for left Kan extensions gives

$$\begin{aligned} \lceil \mathcal{F}_p \rceil(\{p\}) &\cong \operatorname{colim}\Big(\big(\mathcal{O}(\lceil p \rceil) \downarrow \underline{\{p\}}\big)^{\mathrm{op}} \twoheadrightarrow \mathcal{O}(X)^{\mathrm{op}} \xrightarrow{\mathcal{F}} \mathrm{Set}\Big) \\ &\cong \operatorname{colim}_{U \ni p}(\mathcal{F}(U)) \\ &\cong \mathcal{F}_p \\ \lceil \mathcal{F}_p \rceil(\varnothing) &\cong \operatorname{colim}\Big((\mathcal{O}(\lceil p \rceil) \downarrow \underline{\varnothing})^{\mathrm{op}} \twoheadrightarrow \mathcal{O}(X)^{\mathrm{op}} \xrightarrow{\mathcal{F}} \mathrm{Set}\Big) \\ &\cong \operatorname{colim}_{U \hookrightarrow \varnothing}(\mathcal{F}(U)) \\ &\cong \mathcal{F}(\varnothing). \end{aligned}$$

Example 2.3.11 (Analytic functors). A functor $F : \mathrm{Set} \to \mathrm{Set}$ is said to be *analytic* if it results from the left Kan extension of a functor $f : \mathcal{B}(\mathbb{N}) \to \mathrm{Set}$ (the 'species' of F) along the functor $j : \mathcal{B}(\mathbb{N}) \to \mathrm{Set}$; $\mathcal{B}(\mathbb{N})$ is the category having objects natural numbers and such that $\mathcal{B}(\mathbb{N})(m, n)$ are the bijective functions $\{1, \dots, m\} \to \{1, \dots, n\}$ (so this set is empty if $n \neq m$).

In other words, $\mathcal{B}(\mathbb{N})$ is the groupoid arising as the disjoint union of all symmetric groups $\coprod_{n \ge 0} \mathrm{Sym}(n)$.

[1] T. de Oliveira Santos suggested this proof.

Representing the left Kan extension $\mathrm{Lan}_j f$ as a coend we have

$$F(T) \cong \int^n T^n \times f(n); \tag{2.34}$$

a functor is 'analytic' if it can be expressed as a *Taylor series*, and the coend is in a suitable sense that Taylor series). The theory of analytic functors, besides having an intrinsic interest, is used to categorify many phenomena in classical combinatorics. See for example the pioneering paper by A. Joyal [Joy86], but also [AV08, GJ17].

2.3.1 Tannaka Duality Using Coends

In the following remark and example, V is a *finite dimensional* vector space over a field K.

Example 2.3.12. Let $V^\vee$ denote the dual vector space of linear maps $V \to K$. Then there is a canonical isomorphism

$$\int^V V^\vee \otimes_K V \cong K. \tag{2.35}$$

The fastest way to see this is to notice that

$$\int^V \hom(V, \_) \otimes V \cong \mathrm{Lan}_{\mathrm{id}}(\mathrm{id}) \cong \mathrm{id}_{\mathrm{Mod}(K)} \tag{2.36}$$

(compare this argument with any proof trying to explicitly evaluate the coend from its bare definition).

Remark 2.3.13. The universal cowedge $\hom(V, V) \xrightarrow{\alpha_V} K$ sends an endomorphism $f : V \to V$ to its *trace* $\tau(f) \in K$ (which in this way acquires a universal property).

The above argument holds in fact in fair generality, adapting to the case where V is an object of a compact closed monoidal category, and it is linked to the theory of *Tannaka reconstruction.*

Let G be an internal group in a suitable category of spaces (it can be a Lie group or an affine group, i.e. a group in the category of algebraic varieties or schemes). The category $\mathrm{Rep}(G)$ of its finite dimensional representations $\varrho : G \to \mathrm{Mod}(K)$ carries many important properties: it has a monoidal structure and an involution $(\_)^\vee$ turning it into a *rigid* monoidal category (see [Sel10] for a glimpse of the vast zoo of monoidal categories).

Tannaka theory tries to find sufficient conditions on a nice monoidal category $\mathcal{A}$ to ensure that it is equivalent to Rep(G) for some space G, and retrieves sufficient information to *reconstruct* such G (or equivalently by Gel'fand duality, its algebra of functions) from $\mathcal{A}$. Such an algebra of functions is built out of $\mathcal{A}$ alone in a canonical fashion, so long as it comes equipped with a nice functor $\mathcal{A} \to \text{Mod}(K)$ for some ring K.

Remark 2.3.14. ☻☻ For the purposes of this remark, unveiling the coend-y nature of the argument in [Sch13, 10.2.2], we set the following conditions.

- We fix a ground ring R, and we let K be a commutative R-algebra and $\mathcal{A}$ an additive autonomous symmetric monoidal R-linear category; this means that every object in $\mathcal{A}$ is *dualisable.*
- We let $w : \mathcal{A} \to \text{Mod}(K)$ be an R-linear functor that is strong monoidal; since $\mathcal{A}$ is autonomous, this entails that the essential image of w is contained in the subcategory of dualisable (i.e. finitely generated projective) B-modules.
- We also assume that w is comonadic.

Under these assumptions, we can consider the *density comonad* of w, i.e. the left Kan extension of w along itself (of course if w is comonadic, and has a right adjoint r, then $\text{Lan}_w w \cong w \circ r$ because $\text{Lan}_w \cong \_ \circ r$). According to 2.3.6, the left Kan extension in question can be computed as the coend

$$M \mapsto \int^{A \in \mathcal{A}} \text{Mod}(K)(wA, M) \otimes wA. \tag{2.37}$$

We now claim that the object $H = \text{Lan}_w w(K)$ carries a natural Hopf algebra structure in Mod(K): indeed H results as

$$\text{Lan}_w w(K) \cong \int^A \text{Mod}(K)(wA, K) \otimes wA = \int^A (wA)^\vee \otimes wA \tag{2.38}$$

(this is exactly how the object H is introduced in [Bak07, Sch13, Ulb90]). Moreover, $\text{Lan}_w w$ is strong monoidal by doctrinal adjunction [Kel74] and thus $H = \text{Lan}_w w(K)$ must carry a bialgebra structure.

One form of the *Tannaka reconstruction theorem* now asserts that $\mathcal{A}$ is monoidally equivalent to the category of representations of the space $\text{Spec}(H)$.

Here, we prove a slightly less sophisticated version of the theorem using coends. The proof has various steps; we just sketch its backbone to let the reader appreciate how, although there is still a lot of (non-formal) work left to do, the use of coends makes the essential idea crystal clear. We denote mod(K) the category of finite dimensional K-modules, as opposed to Mod(K) (all K-modules).

Theorem 2.3.15. Let K be a ring, $F : \mathcal{A} \to \text{mod}(K)$ a K-linear, faithful, strong monoidal functor with domain a K-linear rigid monoidal category. The codomain mod(K) is the category of finitely generated projective K-modules.

Then there is a bialgebra $B \in \text{Mod}(K)$ (now, not necessarily finitely generated) such that $\mathcal{A}$ is monoidally equivalent to the category of B-modules Mod(B).

Proof First, consider the codensity monad $\text{Ran}_F F : \text{Mod}(K) \to \text{Mod}(K)$, and the density comonad $\text{Lan}_F F$ of F; the image of K under the monad is the object $\text{Ran}_F F(K)$ and corresponds to the module

$$\int_A \hom_K(\hom_K(k, FA), FA) = \int_A \hom_K(FA, FA), \tag{2.39}$$

i.e. to the monoid of endotransformations $\mathsf{Cat}(\mathcal{A}, \text{Mod}(K))(F, F)$ (see 1.4.1; the monoid operation here is vertical composition, which is of course bilinear). We claim this is the algebra B we are looking for. Indeed, the object $\text{Lan}_F F(K)$ is another K-module, and not very far from B:

$$\begin{aligned}
\text{Lan}_F F(k) &\cong \int^A \hom_K(FA, k) \otimes FA \\
&\cong \int^A (FA)^* \otimes FA \\
&\cong \int^A (FA \otimes FA^*)^* \\
&\cong \Big(\int_A FA \otimes FA^*\Big)^* = B^*.
\end{aligned}$$

Now, 'every object of $\mathcal{A}$ is a B-module' in the following sense: the universal wedge of the end $\int_A \hom_K(FA, FA)$ is given by maps

$$\epsilon_A : B \to \hom_K(FA, FA) = \text{End}(FA) \tag{2.40}$$

and this is a ring map giving FA the structure of a B-module; a morphism

$f : A \to A'$ in $\mathcal{A}$ now fits in the commutative square

$$\begin{array}{ccc} B & \longrightarrow & \hom_K(FA, FA) \\ \downarrow & & \downarrow{\scriptstyle Ff_*} \\ \hom_K(FA', FA') & \xrightarrow[Ff^*]{} & \hom_K(FA, FA') \end{array} \tag{2.41}$$

which means that $Ff(b.x) = b.Ff(x)$ for every $b \in B$, where $b.\_ = \epsilon(b)$. We have defined a homomorphism of B-modules.

This is enough to define a functor $\tilde{F} : \mathcal{A} \to \mathrm{Mod}(B)$ (just corestrict F) in such a way that it is an equivalence of categories; it is indeed full and strictly surjective on objects, and strong monoidal and faithful by the initial assumption.

Moreover, the multiplication of B given by vertical composition of natural transformation is compatible with the comultiplication on B^*, and more explicitly (if we use the shorthand $[F, F] := \mathsf{Cat}(\mathcal{A}, \mathrm{Mod}(K))(F, F)$) we have that

$$\frac{([F, F] \otimes [F, F] \to [F, F])^*}{[F, F]^* \otimes [F, F]^* \leftarrow [F, F]^*} \tag{2.42}$$

(and $B \cong B^*$ because considered as a B-module it is of course 1-dimensional). □

2.4 A Yoneda Structure on Cat

2.4.1 Formal Category Theory: A Crash Course

The language of category theory is built upon a certain number of fundamental notions: among these we find the universal characterisation of (co)limits, the definitions of adjunctions, (pointwise) Kan extensions, and the theory of monads.

It is often possible to 'axiomatise' these definitions, pretending that they refer to the 1- and 2-cells of a generic 2-category other than Cat.

In some sense, category theory arises when the way in which abstract patterns interact becomes itself an object of study, and when it is generalised to several different contexts. In a few words, the aim of *formal category theory* is to provide a framework in which this process of conceptualisation can be outlined mathematically. Quoting [Gra80],

> The purpose of category theory is to try to describe certain general

> aspects of the structure of mathematics. Since category theory is also part of mathematics, this categorical type of description should apply to it as well as to other parts of mathematics.

The basic idea is that the category of small categories, Cat, can be promoted to a 2-category Cat with 'formal' properties in the same way that Set is a category with 'formal' properties (leading to the definition of a topos). The aim of formal category theory is to outline these properties, and the assumptions needed to ensure that a certain 2-category behaves like Cat for all practical purposes.

Unfortunately, being too naïve when performing this process does not always give the 'right' answer (because it does not always build an object with the right universal property).

This is ultimately due to the fact that the theory 'behaves differently' in various ways when moving to the setting of $\mathcal{V}$-enriched categories (which is the adjacent step of abstraction from Cat, the 2-category Set-Cat of Set-enriched categories), and some of these differences prevent $\mathcal{V}$-categories from being as expressive as one would have liked them to be (a paradigmatic example of this minor expressiveness is the lack of a *Grothendieck construction* for generic $\mathcal{V}$-presheaves: the way in which the Grothendieck construction of A.5.9 ultimately pertains to formal category theory has been addressed in the early literature on formal category theory (see [Str74, Str80, SW78]).

Formal category theory can be thought as a way to encode the same amount of information carried by the 'concrete' 2-category Cat in other contexts: even though it is always possible to do some constructions by mimicking definitions from Cat (adjunctions and adjoint equivalences, extensions by universal 2-cells, etc.), things get a little hairy when we want to provide the theory with an analogue of the Yoneda lemma.

In the 2-category Cat, we can use the above mentioned Grothendieck construction to 'revert' set-valued functors on an object $\mathcal{B}$ into arrows 'over' $\mathcal{B}$.[2] In the 2-category Cat the comma object of $B : 1 \to \mathcal{B}$ to $\mathrm{id}_{\mathcal{B}} : \mathcal{B} \to \mathcal{B}$ together with its projection $B/\mathcal{B} \to \mathcal{B}$ is a good stand in for the covariant functor represented by B (more generally, *discrete left fibrations* over $\mathcal{B}$ stand in for general functors $\mathcal{B} \to$ Set).

In the 2-category $\mathcal{V}$-Cat, we care about *$\mathcal{V}$-valued* $\mathcal{V}$-functors and we would like to perform the same construction. But for an object B in a

[2] We basically glue together a bunch of fibres $\coprod_B \mathcal{E}_B$ projecting onto $\mathcal{B}$, in the same way we build the *étale space* of a presheaf $F : \mathcal{B}^{\mathrm{op}} \to$ Set; the reader might have noticed that this is secretly the same construction that gives the *étale space* of a presheaf on a topological space (see 3.2.11).

$\mathcal{V}$-enriched category $\mathcal{B}$, the comma category $B/\mathcal{B}$ is more naturally an *internal* category (whose object of objects is $\coprod_{X\in\mathcal{B}}\mathcal{B}(B,X)$, see [Bor94a, Ch. 8]) rather than an *enriched* one (whose objects are morphisms $p : B \to X$ in the underlying category of $\mathcal{B}$). The skewness between these two presentations of category, one 'inside' a universe, and the other 'with respect to' a universe, generates all sorts of subtleties and problems.

Now, we are left with the question:

> Which additional structure on a 2-category K allows us to recognise arrows of K playing the same rôle of discrete (op)fibrations in Cat, thus providing a meaningful notion of (fibrational) Yoneda lemma internal to K (see Section A.5.3)?

The axioms of a *Yoneda structure* provide a possible answer to this question.

Our aim here is to present them not for an arbitrary 2-category, as they appear in [SW78], but for the 2-category Cat. We prove the validity of each axiom as it appears in [SW78]; to do this, we will extensively use the (co)end calculus we know.

To start, we establish the following notation.

YD1. K is a 2-category, fixed once and for all.

YD2. $\mathsf{Ads}(A,B) \subseteq \mathsf{K}(A,B)$ is a full subcategory of 'admissible' 1-cells, which is moreover a *right ideal*, meaning that the composition map restricted to admissible 1-cells restricts as a family of maps

$$c_{XAB}|_{\mathsf{Ads}} : \mathsf{Ads}(A,B) \times \mathsf{K}(X,A) \to \mathsf{Ads}(X,B). \tag{2.43}$$

This means that if $\left[\begin{smallmatrix} X \\ f\downarrow \\ A \end{smallmatrix}\right]$ and $\left[\begin{smallmatrix} A \\ g\downarrow \\ B \end{smallmatrix}\right]$ are 1-cells and g is admissible, then $g \circ f$ is again admissible. We call an *object* A admissible if $\mathrm{id}_A \in \mathsf{Ads}(A,A)$.

Now we assume that the following structure can be found on K:

YS1. for each admissible object $A \in \mathsf{K}$ we can find an admissible 1-cell $よ_A : A \to \boldsymbol{P}A$, called a *Yoneda arrow*;[3]

YS2. for each $f : A \to B$ admissible 1-cell with admissible domain, we

[3] It is desirable for the correspondence $A \mapsto \boldsymbol{P}A$ to be a functor; this will be an axiom. For the moment, the assignment $\boldsymbol{P}A$ is just any object depending on A.

can find a 2-cell

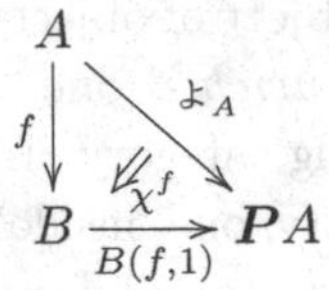

$$\begin{array}{ccc} A & & \\ {\scriptstyle f}\downarrow & \searrow^{\text{よ}_A} & \\ & \Swarrow \chi^f & \\ B & \xrightarrow[B(f,1)]{} & \boldsymbol{P}A \end{array} \tag{2.44}$$

We say that a 2-category K *has a Yoneda structure* if it is endowed with the data above, and if the following axioms are satisfied.

Axiom 2.4.1. In (2.44), the pair $\langle B(f,1), \chi^f \rangle$ exhibits lan_f よ$_A$.

Proof The proof that this axiom holds in Cat consists of the following derivation in coend calculus: let $f : \mathcal{A} \to \mathcal{B}$ be a functor, then

$$\begin{aligned} \mathrm{lan}_f \text{よ}_{\mathcal{A}}(B) &\cong \int^A \mathcal{B}(fA, B) \times \text{よ}_{\mathcal{A}}(A) \\ &\cong \int^A \mathcal{B}(fA, B) \times \mathcal{A}(\_, A) \\ 2.2.1 &\cong \mathcal{B}(f\_, B). \end{aligned}$$

This concludes the proof. □

Axiom 2.4.2. In (2.44), the pair $\langle f, \chi^f \rangle$ exhibits the absolute left lifting $\mathrm{LIFT}_{B(f,1)}$ よ$_A$.

Proof The validity of this axiom in Cat is again a game of coend calculus: if we call $N_f = \mathrm{lan}_f \text{よ}_{\mathcal{A}} = \mathcal{B}(f,1)$ for short, we have $\mathrm{lift}_{N_f} \dashv N_{f,*}$, where $N_{f,*}\colon g \mapsto N_f \circ g$ is the 'direct image' (or *postcomposition*) functor; then we have an isomorphism between sets of 2-cells

$$\begin{aligned} \mathsf{Cat}(\mathcal{A}, \boldsymbol{P}\mathcal{A})(\text{よ}_{\mathcal{A}}, N_f \circ g) &\cong \int_A [\mathcal{A}^{\mathrm{op}}, \mathrm{Set}](\text{よ}_{\mathcal{A}}A, N_f \circ g(A)) \\ &\cong \int_A [\mathcal{A}^{\mathrm{op}}, \mathrm{Set}](\text{よ}_{\mathcal{A}}A, \mathcal{B}(f\_, gA)) \\ &\cong \int_A \mathcal{B}(fA, gA) \\ &\cong \mathsf{Cat}(\mathcal{A}, \mathcal{B})(f, g). \end{aligned}$$

We leave to the reader the proof that this lifting is preserved by every functor; a short equivalent formulation of this axiom is that there is a relative adjunction (see Exercise 2.11) $f \xrightarrow{[\text{よ}_A]}\!\!| \; B(f,1)$ with *relative unit* the Yoneda map よ$_A$. □

Axiom 2.4.3. Given a pair of composable 1-cells $A \xrightarrow{f} B \xrightarrow{g} C$, the pasting of 2-cells

$$\begin{array}{ccc} A & \xrightarrow{\text{よ}_A} & \boldsymbol{P}A \\ {\scriptstyle f}\downarrow & \swArrow \chi^{\text{よ}_B f} & \nearrow {\scriptstyle \boldsymbol{P}f} \\ B & \xrightarrow{\text{よ}_B} & \boldsymbol{P}X \\ {\scriptstyle g}\downarrow & \swArrow \chi^{g} \nearrow {\scriptstyle C(g,1)} & \\ C & & \end{array} \tag{2.45}$$

exhibits the extension $\operatorname{lan}_{gf} \text{よ}_A = C(gf, 1)$, and the pair $\langle \mathrm{id}_{\boldsymbol{P}A}, \mathrm{id}_{\text{よ}_A} \rangle$ exhibits $\operatorname{lan}_{\text{よ}_A} \text{よ}_A$.

Remark 2.4.4. The hidden meaning of this axiom is that $\boldsymbol{P}$ is a pseudofunctor $\mathsf{K}^{\mathrm{coop}} \to \mathsf{K}$ (whose domain is the sub-class of admissible objects).

Let us make this evident: given a pair of composable 1-cells $A \xrightarrow{f} B \xrightarrow{g} C$, the universal property of χ^{gf} entails that there is a unique 2-cell θ^{gf} filling the diagram

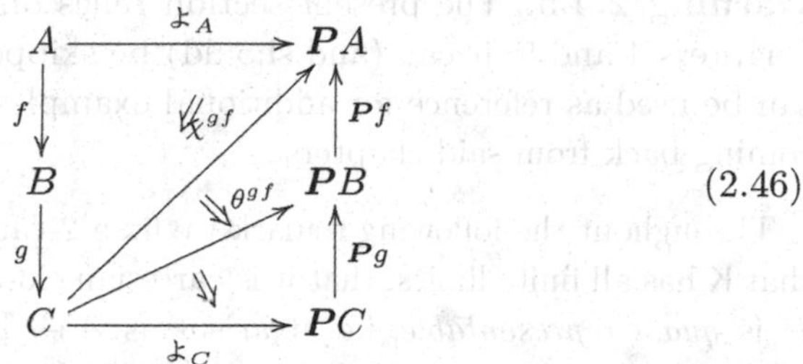

Axiom 2.4.3 is equivalent to the requirement that this arrow is invertible (exercise: draw the correct diagram), and this yields that the above diagram has the same universal property as the square

$$\begin{array}{ccc} A & \xrightarrow{\text{よ}_A} & \boldsymbol{P}A \\ {\scriptstyle gf}\downarrow & \Downarrow & \uparrow{\scriptstyle \boldsymbol{P}(gf)} \\ C & \xrightarrow[\text{よ}_C]{} & \boldsymbol{P}C. \end{array} \tag{2.47}$$

This in turn entails that there is a unique, and invertible, 2-cell $\boldsymbol{P}(gf) \Rightarrow \boldsymbol{P}f \circ \boldsymbol{P}g$. This is of course half of the structure of a pseudofunctor on $\boldsymbol{P}$; the remaining structure is given by the requirement that $\langle \mathrm{id}_{\boldsymbol{P}A}, \mathrm{id}_{\text{よ}_A} \rangle$ exhibits $\operatorname{lan}_{\text{よ}_A} \text{よ}_A$, because this provides a unique, invertible 2-cell

$\mathrm{id}_{\boldsymbol{P}A} \Rightarrow \boldsymbol{P}(\mathrm{id}_A)$ for every admissible object A. This result has a variety of different interpretations: it follows from the Yoneda lemma as stated in A.5.7, and from the coend expression for the density comonad, or by a direct check of the universal property in question.

All the remarks in 2.4.3 are evidently true in Cat, since the (pseudo) functoriality of the correspondence $\mathcal{A} \mapsto [\mathcal{A}^{\mathrm{op}}, \mathrm{Set}]$ can be proved directly without great effort. Nevertheless, Axiom 2.4.3 is still telling us something about a 'reduction rule' for composition of Kan extensions: indeed, it is possible to prove that (in the same notation of Axiom 2.4.3) there is a (canonical!) isomorphism

$$\theta_{gf} : \mathrm{lan}_{gf}\, よ_{\mathcal{A}} \cong \mathrm{lan}_{よ_{\mathcal{B}} f}\, よ_{\mathcal{A}} \circ \mathrm{lan}_g\, よ_{\mathcal{B}}. \tag{2.48}$$

The proof of this statement is another game of coends, using 2.3.6: try to do it as an exercise. (Hint: there will be many coends involved in the proof; choose an appropriate notation to keep it clear.)

2.4.2 Addendum: (Co)ends Inside a Yoneda Structure

Warning 2.4.5. The present section relies on material presented in Chapters 4 and 5; it can (and should) be skipped at first reading, but can be used as reference for additional examples of weighted (co)limits coming back from said chapters.

Throughout the following remark we fix a 2-category K and we assume that K has all finite limits, that it is cartesian closed, and that the functor $\boldsymbol{P}$ is *quasi representable*, i.e. it arises as $A \mapsto [A^\vee, \Omega]$ for some object $\Omega \in \mathsf{K}$ and a *duality involution* $(\_)^\vee$ on K (see [Shu16]). It is possible to prove that there is an isomorphism $\Omega \cong \boldsymbol{P}1$, where 1 is the terminal object of K.

The purpose of this section is to establish how the Yoneda structure having such a $\boldsymbol{P}$ as presheaf construction possesses a formal analogue of (co)end calculus.

Remark 2.4.6. ☻☻ In a quasi representable Yoneda structure $\boldsymbol{P} : \mathsf{Ads} \to \mathsf{K}$, Yoneda embeddings are maps in K of the form $A \to [A^\vee, \boldsymbol{P}1]$; if the product $A^\vee \times A$ is admissible in the Yoneda structure generated by $\boldsymbol{P}$ we can consider the admissible maps

$$\mathfrak{h}_A : A^\vee \times A \to \boldsymbol{P}1 \tag{2.49}$$

as the mates of the Yoneda embeddings $よ_A$. These maps play the rôle of *internal homs*, so that admissible objects can be thought as canonically $\boldsymbol{P}1$-*enriched.* In order to define (co)ends we shall write them as certain weighted (co)limits using the maps $\mathfrak{h}_A$ as weights (see 4.1.13.WC3, (co)ends are precisely hom-weighted (co)limits).[4]

Let us stick for a moment to the case where the ambient 2-category is Cat. By the definition of weighted colimit in a Yoneda structure [SW78, §4], the presence of an isomorphism

$$\int^X \mathcal{A}(A,X) \times \mathcal{A}(X,A') \cong \mathcal{A}(A,A') \tag{2.50}$$

valid for a category $\mathcal{A}$ and objects $A, A' \in A$ means that the left lifting of $よ_{\mathcal{A}^{\text{op}}\times\mathcal{A}}$ along $\boldsymbol{P}1(\mathfrak{h}_\mathcal{A}, 1)$, is $\mathfrak{h}_\mathcal{A}$, and such a lifting is absolute. This yields a relative adjunction $\mathfrak{h}_\mathcal{A} \xrightarrow{[よ_{\mathcal{A}^{\text{op}}\times\mathcal{A}}]} \boldsymbol{P}1(\mathfrak{h}_\mathcal{A}, 1)$; this final requirement means that the triangle

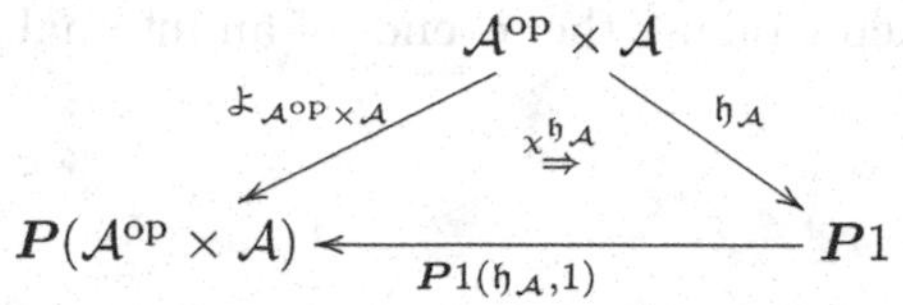

$$\begin{array}{ccc} & \mathcal{A}^{\text{op}} \times \mathcal{A} & \\ {}^{よ_{\mathcal{A}^{\text{op}}\times\mathcal{A}}}\swarrow & \overset{\chi^{\mathfrak{h}_\mathcal{A}}}{\Rightarrow} & \searrow^{\mathfrak{h}_\mathcal{A}} \\ \boldsymbol{P}(\mathcal{A}^{\text{op}} \times \mathcal{A}) & \xleftarrow[\boldsymbol{P}1(\mathfrak{h}_\mathcal{A},1)]{} & \boldsymbol{P}1 \end{array} \tag{2.51}$$

is an absolute left lifting; this is precisely Axiom 2.4.2 applied to $\mathfrak{h}_\mathcal{A}$.

It is worth investigating how far this formalisation of (co)end calculus can go:

Remark 2.4.7 ((Co)ends in a Yoneda structure). ☻☻ The 1-cell $\boldsymbol{P}1(\mathfrak{h}_A, 1)$ admits a left adjoint $\text{Lan}_{よ_{A^\vee\times A}} \mathfrak{h}_A$; computing this Kan extension in Cat we get

$$\begin{aligned} \text{Lan}_{よ_{\mathcal{A}^{\text{op}}\times\mathcal{A}}} \mathfrak{h}_\mathcal{A}(F) &\cong \int^{(A,A')} [\mathcal{A}^{\text{op}} \times \mathcal{A}, \text{Set}](よ_{\mathcal{A}^{\text{op}}\times\mathcal{A}}(A,A'), F) \times \mathfrak{h}_\mathcal{A}(A,A') \\ &\cong \int^{(A,A')} F(A,A') \times \mathcal{A}(A,A') \\ &\cong \int^a F(A,A). \end{aligned}$$

Thus, the adjunction $\text{Lan}_{よ_{A^\vee\times A}} \mathfrak{h}_A \dashv \boldsymbol{P}1(\mathfrak{h}_A, 1)$ is an abstract analogue of the adjunction $\int^A : [\mathcal{A}^{\text{op}} \times \mathcal{A}, \text{Set}] \leftrightarrows \text{Set} : \hom \pitchfork \_$, where the left adjoint $\int^A$ takes the coend of a functor $F : A^{\text{op}} \times A \to \text{Set}$, and the

[4] It is worth recalling that for every small category $A \in \text{Cat}$ the composition maps are given by a family of functions $c_{abc} : A(a,b) \times A(b,c) \to A(a,c)$ such that $c_{a,\_,c}$ is a cowedge, and in fact an initial one.

right adjoint is the hom-weighted limit functor $\hom \pitchfork X : (A, A') \mapsto X^{\hom(A,A')}$, the same that helped prove the Fubini theorem in 1.3.1 (see 4.1.13.WC3 for the precise result).

It is now straightforward to proceed: Axiom 2.4.2 entails that in a Yoneda structure an admissible object A for which $A \times A^{\mathrm{op}}$ is still admissible (this translates into the property that the domain of a relative monad $\boldsymbol{P} : \mathsf{A} \to \mathsf{K}$ is closed under product and duality involution), thus there are absolute left liftings

$$\begin{array}{ccc} & A^{\mathrm{op}} \times A & \\ {}^{\text{よ}_{A^{\mathrm{op}} \times A}} \swarrow & \Rightarrow & \searrow {}^{\mathfrak{h}_A} \\ \boldsymbol{P}(A^{\mathrm{op}} \times A) & \xleftarrow[\boldsymbol{P}1(\mathfrak{h}_A, 1)]{} & \boldsymbol{P}1 \end{array} \tag{2.52}$$

in which $\boldsymbol{P}1(\mathfrak{h}_A, 1)$ has a left adjoint, precisely $\mathrm{Lan}_{\text{よ}_{A^\vee \times A}} \mathfrak{h}_A$. Such a left adjoint is the functor taking the 'coend' of an internal endoprofunctor $F : A^\vee \times A \to \boldsymbol{P}1$.

2.5 Addendum: Relative Monads

Notation 2.5.1. Throughout the present section, categories of functors $\mathcal{X} \to \mathcal{Y}$ are often denoted as $[\mathcal{X}, \mathcal{Y}]$; this complies with the notation for cartesian closed categories and serves to avoid cumbersome accumulation of symbols when considering iterated functor categories such as

$$\mathrm{Cat}(\mathrm{Cat}(\mathrm{Cat}(A^{\mathrm{op}}, \mathrm{Set})_s^{\mathrm{op}}, \mathrm{Set})_s^{\mathrm{op}}, \mathrm{Set}). \tag{2.53}$$

Context always allows us to determine where the hom-category $[\mathcal{X}, \mathcal{Y}]$ lies.

Another notational simplification is the following: every functor $J : \mathcal{X} \to \mathcal{Y}$ induces a left extension functor $\mathrm{Lan}_J = J_! : [\mathcal{X}, \mathcal{Y}] \to [\mathcal{Y}, \mathcal{Y}]$. This notation draws from algebraic geometry and it is chosen because we have to iterate several left extensions and to iteratively apply the functors $J_! \dashv J^*$ and compose their unit and counit maps.

In this section we address a deeper study of the presheaf construction: we shall prove the following.

- Although not an endofunctor, the presheaf construction functor $\boldsymbol{P}$ of the Yoneda structure on Cat is a *relative monad* on the category of functors $\mathsf{cat} \to \mathrm{Cat}$. This means that it is an internal monoid in the skew-monoidal category of those functors; in simple terms, a skew-monoidal category $(\mathcal{M}, \lhd)$ is like a monoidal category, but the associator $\alpha_{XYZ} : (X \lhd Y) \lhd Z \to X \lhd (Y \lhd Z)$ and left and right unitors are not invertible maps. Thus, aggregates of objects

$$X_0 \lhd X_1 \lhd \cdots \lhd X_n \tag{2.54}$$

are not well defined without specifying their parenthesisation. We introduce the notion in full generality, and we prove that, under suitable cocompleteness assumptions on $\mathcal{Y}$, the category $\text{Cat}(\mathcal{X}, \mathcal{Y})$ becomes skew-monoidal.

- We then prove that $\boldsymbol{P}$ is a particularly well-behaved relative monad, of a kind that is called a *yosegi box* in [DLL19]. In simple terms, the action of $\boldsymbol{P}$ on 1-cells is such that

  YB1. every $\boldsymbol{P}_! f$ fits into an adjunction $\boldsymbol{P}_! f \dashv \boldsymbol{P}^* f$;

  YB2. $\boldsymbol{P}_!$ is a relative monad with respect to the inclusion $j : \mathsf{cat} \subset \mathsf{Cat}$;

  YB3. the monad $\boldsymbol{P}_!$ is lax idempotent.

  As shown in [DLL19], it turns out that these three properties characterise the presheaf construction of a Yoneda structure uniquely.

The upshot of the present section is thus the following: provided the left extension along a given functor $J : \mathcal{X} \to \mathcal{Y}$ exists, the functor category $[\mathcal{X}, \mathcal{Y}]$ becomes *skew-monoidal* [Szl12] with respect to the functor $\lhd : [\mathcal{X}, \mathcal{Y}] \times [\mathcal{X}, \mathcal{Y}] \to [\mathcal{X}, \mathcal{Y}]$ sending F, G to $J_! F \circ G$.

Here we offer a characterisation of relative monads as $\lhd$-monoids: we can provide a formal analogue for a similar statement to the one given for $\mathsf{K} = \mathsf{Cat}$ in [ACU10]; while an extremely useful reference, the proof in [ACU10] apparently only works on Cat, whereas the argument appearing here can be easily adapted to an abstract 2-category.

Although a relatively elementary argument, spelling out a complete proof of this fact turns out to be a rather tedious task.

Definition 2.5.2. Let $J : \mathcal{X} \to \mathcal{Y}$ be a functor such that the left extension along J exists, defining a functor $J_! : [\mathcal{X}, \mathcal{Y}] \to [\mathcal{Y}, \mathcal{Y}]$. Then the category $[\mathcal{X}, \mathcal{Y}]$ becomes a *left skew-monoidal category* under the skew multiplication defined by

$$(F, G) \mapsto F \lhd G = J_! F \circ G, \tag{2.55}$$

and there are natural maps of *association*, *left* and *right unit*

$$\begin{array}{rcl} (F \lhd G) \lhd H & \xrightarrow{\gamma_{FGH}} & F \lhd (G \lhd H) \\ J \lhd F & \xrightarrow{\lambda_F} & F \\ F & \xrightarrow{\varrho_F} & F \lhd J \end{array} \tag{2.56}$$

such that the following diagrams are commutative.

SK1. *Skew associativity*:

$$\begin{array}{ccc} ((F \lhd G) \lhd H) \lhd K & \xrightarrow{\gamma_{FG,H,K}} & (F \lhd G) \lhd (H \lhd K) \\ \downarrow{\scriptstyle \gamma_{F,G,H} \lhd K} & & \\ (F \lhd (G \lhd H)) \lhd K & & \downarrow{\scriptstyle \gamma_{F,G,HK}} \\ \downarrow{\scriptstyle \gamma_{F,GH,K}} & & \\ F \lhd ((G \lhd H) \lhd K) & \xrightarrow[F \lhd \gamma_{G,H,K}]{} & F \lhd (G \lhd (H \lhd K)) \end{array} \tag{2.57}$$

SK2. *Skew left and right unit:*

$$\begin{array}{ccc} (F \lhd G) \lhd J & \xrightarrow{\gamma_{F,G,J}} & F \lhd (G \lhd J) \\ \uparrow{\scriptstyle \varrho_F \lhd G} & \nearrow{\scriptstyle F \lhd \varrho_G} \;\; \text{2R} & \\ F \lhd G & & F \lhd G \\ & \nearrow{\scriptstyle \lambda_F \lhd G} \;\; \text{2L} & \uparrow{\scriptstyle \lambda_{F \lhd G}} \\ (J \lhd F) \lhd G & \xrightarrow[\gamma_{J,F,G}]{} & J \lhd (F \lhd G) \end{array} \tag{2.58}$$

SK3. *Zig-zag identity*:

$$\begin{array}{ccccc} & & J \lhd J & & \\ & \nearrow{\scriptstyle \varrho_J} & & \searrow{\scriptstyle \lambda_J} & \\ J & & = & & J \end{array} \tag{2.59}$$

SK4. *Interpolated zig-zag identity*:

$$\begin{array}{ccc} (F \lhd J) \lhd G & \xrightarrow{\gamma_{F,J,G}} & F \lhd (J \lhd G) \\ \uparrow{\scriptstyle \varrho_F \lhd G} & & \downarrow{\scriptstyle F \lhd \lambda_G} \\ F \lhd G & = & F \lhd G \end{array} \tag{2.60}$$

The natural maps γ, λ, ϱ are defined as follows in this specific case.

S1. Given $F, G, H \in [\mathcal{X}, \mathcal{Y}]$, the cell $\gamma_{F,G,H}$ is defined by $\tilde{\gamma}_{F,G} * H$, where $\tilde{\gamma}_{F,G}$ is obtained as the mate of $J_!F * \eta_G$ under the adjunction $J_! \frac{\varepsilon}{\eta}\!\!\dashv J^*$: the arrow

$$J_!F \circ G \xrightarrow{J_!F*\eta_G} J_!F \circ J^*J_!G = J^*(J_!F \circ J_!G) \tag{2.61}$$

mates to a map $J_!(J_!F \circ G) \longrightarrow J_!F \circ J_!G$.

S2. Given $F \in [\mathcal{X}, \mathcal{Y}]$, the cell $\lambda_F : J \triangleleft F \Rightarrow F$ is defined by the whiskering $\sigma * F$, where $\sigma = \sigma_{\mathrm{id}_{\mathcal{Y}}}$ is the counit of the density comonad of J.

S3. Given $G \in [\mathcal{X}, \mathcal{Y}]$, the cell $\varrho_G : G \to G \triangleleft J$ is the G-component η_G of the unit of the adjunction $J_! \dashv J^*$.

Remark 2.5.3. A complete proof of 2.5.2 appears as Theorem 3.1 in [ACU10]; the main argument, however, relies heavily relying on the assumption that $\mathsf{K} = \mathsf{Cat}$. It was our desire to produce a formal proof *ex novo*, while explicitly recording some useful equations satisfied by the skew-monoidal structure in study; as a rule of thumb, the structure maps of this skew-monoidal structure are entirely induced by the $J_! \frac{\eta}{\varepsilon}\!\!\dashv J^*$ adjunction and from (the equations satisfied by) the unit and counit thereof.

Until the end of the section, we adopt the following notation, and employ the following equations wherever needed.

E1. We denote ${}^\bullet\varpi : J_!U \to V$ the *mate* of $\varpi : U \to J^*V$ under the adjunction $J_! \frac{\eta}{\varepsilon}\!\!\dashv J^*$; similarly, we denote by $\chi^\bullet : U \to J^*V$ the mate of $\chi : J_!U \to V$. In this notation, the bijection

$$[\mathcal{X}, \mathcal{Y}](U, J^*V) \cong [\mathcal{Y}, \mathcal{Y}](J_!U, V) \tag{2.62}$$

reads as $({}^\bullet\varpi)^\bullet = \varpi$ and ${}^\bullet(\chi^\bullet) = \chi$.

E2. The first zig-zag identity for the adjunction $J_! \frac{\eta}{\varepsilon}\!\!\dashv J^*$ is $(\varepsilon_B * J) \circ \eta_{BJ} = \mathrm{id}_{BJ}$; in particular, if $B = \mathrm{id}_{\mathcal{Y}}$ we have $(\sigma * J) \circ \eta_J = \mathrm{id}_J$.

E3. The second zig-zag identity for the adjunction $J_! \frac{\eta}{\varepsilon}\!\!\dashv J^*$ is $\varepsilon_{J_!F} \circ J_!(\eta_F) = \mathrm{id}_{J_!F}$.

E4. An irreducible expansion for the associator, all its components explicitly written, is

$$\gamma_{FGH} = \big(\varepsilon_{J_!F \circ J_!G} \circ J_!(J_!F * \eta_G)\big) * H. \tag{2.63}$$

E5. The density comonad of J, whose comultiplication ν and counit σ satisfy the comonad axioms, is defined by the maps

$$\sigma = \varepsilon_{\mathrm{id}_B} : J_!(J) \to 1$$

$$\nu : J_!(J) \xrightarrow{J_!(\eta_J)} J_!(J_!(J) \circ J) \xrightarrow{\tilde{\gamma}_{JJ}} J_!(J) \circ J_!(J). \tag{2.64}$$

In particular, the comultiplication and the associator of the skew-monoidal structure determine each other.

Proof First of all, Axiom SK3 is one of the two triangle identities of the adjunction $J_! \dashv J^*$. It remains to prove the other coherence conditions.

SK1. It is easy to see that one can prove commutativity before precomposing with K; one has then to prove the commutativity of the diagram

$$\begin{array}{ccc}
J_!(J_!(J_!F \circ G) \circ H) & \xrightarrow{\tilde{\gamma}_{F \triangleleft G, H}} & J_!(J_!F \circ G) \circ J_!H \\
{\scriptstyle J_!(\tilde{\gamma}_{FG} * H)}\downarrow & \text{C1} & \\
J_!(J_!F \circ (J_!G \circ H)) & & \downarrow{\scriptstyle \tilde{\gamma}_{FG} * J_!H} \\
{\scriptstyle \tilde{\gamma}_{F, G \triangleleft H}}\downarrow & \text{C2} \quad \dashrightarrow & \\
J_!F \circ J_!(J_!G \circ H) & \xrightarrow[J_!F * \tilde{\gamma}_{GH}]{} & J_!F \circ J_!G \circ J_!H
\end{array}$$

Note that the dashed arrow exists: it is the mate ${}^\bullet\alpha$ of $\alpha = (J_!F \circ J_!G) * \eta_H$. The plan is to prove that the two subdiagrams into which this arrow splits the whole diagram commute separately. In order to do this, we start from the square diagram C1: recall that E2 entails that one of the squares

$$\begin{array}{ccc}
J_!A & \xrightarrow{{}^\bullet a} & X \\
{\scriptstyle J_!f}\downarrow & & \downarrow{\scriptstyle g} \\
J_!B & \xrightarrow[{}^\bullet b]{} & Y
\end{array}
\qquad
\begin{array}{ccc}
A & \xrightarrow{a} & J^*X \\
{\scriptstyle f}\downarrow & & \downarrow{\scriptstyle J^*g} \\
J_!B & \xrightarrow[b]{} & J^*Y
\end{array} \tag{2.65}$$

commutes if and only if the other does. Hence, if we denote $f = \tilde{\gamma}_{FG} * H, g = \tilde{\gamma}_{FG} * J_!H, a = (J_!F \circ G) * \eta_H, b = \alpha$, then C1 commutes if and only if the square

$$\begin{array}{ccc}
J_!F \circ G \circ H & \xrightarrow{(J_!F \circ G) * \eta_H} & J_!(J_!F \circ G) \circ J_!H \circ J \\
{\scriptstyle {}^\bullet(J_!F * \eta_G) * H}\downarrow & & \downarrow{\scriptstyle {}^\bullet(J_!F * \eta_G) * (J_!H \circ J)} \\
J_!F \circ J_!G \circ H & \xrightarrow[(J_!F \circ J_!G) * \eta_H]{} & J_!F \circ J_!G \circ J_!H \circ J
\end{array}$$

commutes. It does, since their common value at the diagonal is simply the horizontal composition ${}^\bullet(J_!F * \eta_G) \boxminus \eta_H$.

A similar argument shows that C2 commutes. We have to establish the commutativity of

$$\begin{array}{ccc} J_!(J_!F \circ J_!G \circ H) & \xrightarrow{\bullet(J_!F * J_!G * \eta_H)} & J_!F \circ J_!G \circ J_!H \\ & \searrow^{\bullet(J_!F * \eta_{G \triangleleft H})} \quad \nearrow_{J_!F * \bullet(J_!G * \eta_H)} & \\ & J_!F \circ J_!(J_!G \circ H) & \end{array}$$

which can be 'straightened' to the left diagram below:

$$\begin{array}{ccc} A & \xrightarrow{\bullet(J_!F * \eta_{G \triangleleft H})} & A \\ \| & & \downarrow{\scriptstyle J_!F * \bullet(J_!G * \eta_H)} \\ A & \xrightarrow[\bullet(J_!F * J_!G * \eta_H)]{} & A \end{array} \qquad \begin{array}{ccc} A & \xrightarrow{J_!F * \eta_{G \triangleleft H}} & A \\ \| & & \downarrow{\scriptstyle J_!F * \bullet(J_!G * \eta_H) * J} \\ A & \xrightarrow[J_!F * J_!G * \eta_H]{} & A \end{array}$$

But the left diagram commutes if and only if the right one does; and the latter commutativity follows from the definition of $\tilde{\gamma}$.

SK2. A separate argument works for diagrams 2R and 2L: unwinding the definitions, the commutativity of 2R amounts to the commutativity of

$$\begin{array}{ccc} J_!(J_!F \circ G) \circ J & \xrightarrow{\gamma_{FGJ}} & J_!F \circ J_!G \circ J \\ {\scriptstyle \eta_{J_!F \circ G}} \downarrow & & \| \\ J_!F \circ G & \xrightarrow{J_!F * \eta_G} & J_!F \circ J_!G \circ J. \end{array} \tag{2.66}$$

If we denote for short $J_!F * \eta_G = \varpi$, this commutativity is equivalent to the fact that $J^*({}^\bullet\varpi) \circ \eta_{J_!F \circ G} = \varpi$, which is true since the left hand side of this equation is $({}^\bullet\varpi)^\bullet$. For axiom 2L, the commutativity of

$$\begin{array}{ccc} J_!(J) \circ F & \longrightarrow & J^* J_!F \\ \| & & \uparrow{\scriptstyle \varepsilon * J_!F} \\ J_!(J) \circ F & \xrightarrow[J_!F * \eta_F]{} & J_!(J) \circ J^* J_!F \end{array} \tag{2.67}$$

follows from the fact that the upper horizontal row coincides with the horizontal composition $\sigma \boxminus \eta_F = \eta_F \circ (\sigma * F)$; this means that 2L is true if and only if the square

$$\begin{array}{ccc} F & \xrightarrow{\eta_F} & J^* J_!F \\ {\scriptstyle \sigma * F} \uparrow & & \uparrow{\scriptstyle \varepsilon * J_!F} \\ J_!J \circ F & \xrightarrow[J_!F * \eta_F]{} & J_!J \circ J^* J_!F \end{array} \tag{2.68}$$

commutes; this is obvious by the naturality property of η.

SKM4) Unwinding the definition, Axiom SK4 asks the diagram

$$\begin{array}{ccc} J_!F \circ J_!(J) & \xrightarrow{J_!F*\sigma} & J_!F \\ {\scriptstyle \gamma_{FJ}}\uparrow & & \| \\ J_!(J_!F \circ J) & \xleftarrow[J_!(\eta_F)]{} & J_!F \end{array} \tag{2.69}$$

to commute. In order to see that it does, we observe that the chain of equivalences

$$\begin{aligned} \underline{(J_!F * \sigma) \circ \varepsilon_{J_!F \circ J_!J}} \circ J_!(J_!F * \eta_J) &= \varepsilon_{J_!F} \circ \underline{J_!(J_!F * \sigma * J) \circ J_!(J_!F * \eta_J)} \\ &= \varepsilon_{J_!F} \circ \underline{J_!\big(J_!F * ((\sigma * J) \circ \eta_J)\big)} \\ &= \varepsilon_{J_!F} \circ \mathrm{id}_{J_!F \circ J} \end{aligned}$$

holds, where the last step is motivated by E2. So, the composition $(J_!F * \sigma) \circ \tilde{\gamma}_{FJ}$ equals $\varepsilon_{J_!F}$; this, together with E3, concludes the proof. □

Remark 2.5.4 (A nice J gives a nice structure). In favourable cases the skew-monoidal structure simplifies until it collapses to a straight monoidal structure:

- if J is fully faithful, the unit $\eta_F : F \to J_!F \circ J$ is invertible for every F, so $F \cong F \triangleleft J$;
- if J is dense, the density comonad σ is isomorphic to the identity functor, so $\lambda_F : F \cong J \triangleleft F$ is an isomorphism;
- if each extension $J_!F$ is absolute, then each $\tilde{\gamma}_{F,G}$ is invertible.

Remark 2.5.5. The bifunctoriality of composition has been implicitly employed in the above proof; we spell it out explicitly for future reference.

In order to prove such bifunctoriality, we have to show the commutativity of the square

$$\begin{array}{ccc} F \triangleleft G & \xrightarrow{F \triangleleft g} & F \triangleleft K \\ {\scriptstyle f \triangleleft G}\downarrow & & \downarrow{\scriptstyle f \triangleleft K} \\ H \triangleleft G & \xrightarrow[H \triangleleft g]{} & H \triangleleft K \end{array} \tag{2.70}$$

given $f : F \to H$ and $g : G \to K$. In fact, unwinding the definition it is easy to realise that this diagram commutes since its diagonal coincides with the horizontal composition $J_!f \boxminus g$.

Remark 2.5.6 (A note on $\triangleleft$-whiskering). This raises the question of how the whiskering of a 1-cell F with a 2-cell α works, on the left and on the right; given the shape of the $\triangleleft$-skew-monoidal structure, it turns out that

$$\begin{aligned} \alpha \triangleleft F &:= J_! \alpha * F \\ F \triangleleft \alpha &:= J_! F * \alpha. \end{aligned} \tag{2.71}$$

Definition 2.5.7 (Monads need not be endofunctors, but are always skew-monoids). We will be interested in the notion of a *J-relative monad*, or simply a relative monad when J is understood from the context; J-relative monads are defined as internal monoids in the skew-monoidal category $([\mathcal{X}, \mathcal{Y}], \triangleleft)$, thus they come equipped with a *unit* $\eta : J \Rightarrow T$ and a multiplication $\mu : T \triangleleft T \Rightarrow T$. However, since the monoidal structure $\triangleleft$ satisfies pretty asymmetrical coherence conditions, the commutativity conditions satisfied by a relative monad must be altered accordingly, in particular as follows.

RM1. The unit axiom amounts to the commutativity of

$$\begin{array}{ccccc} J \triangleleft T & \xrightarrow{\eta \triangleleft T} & T \triangleleft T & \xleftarrow{T \triangleleft \eta} & T \triangleleft J \\ & {\scriptstyle\lambda}\searrow & \downarrow{\scriptstyle\mu} & \nearrow{\scriptstyle\varrho} & \\ & & T & & \end{array} \tag{2.72}$$

where the right triangle 'commutes' in the sense that the composition $\mu \circ (T \triangleleft \eta) \circ \varrho$ makes the identity of T.

RM2. The multiplication μ is 'skew associative':

$$\begin{array}{ccccc} (T \triangleleft T) \triangleleft T & \xrightarrow{\gamma} & T \triangleleft (T \triangleleft T) & \xrightarrow{T \triangleleft \mu} & T \triangleleft T \\ {\scriptstyle\mu \triangleleft T}\downarrow & & & & \downarrow{\scriptstyle\mu} \\ T \triangleleft T & & \xrightarrow[\mu]{} & & T \end{array} \tag{2.73}$$

We are particularly interested in *lax idempotent* monads, i.e. those that satisfy one of the following equivalent properties.

Definition 2.5.8. Let $\boldsymbol{T} : \mathsf{A} \to \mathsf{B}$ be a 2-monad between 2-categories; we say that $\boldsymbol{T}$ is *lax idempotent* if one of the following equivalent conditions is satisfied:

L1. for every pair of $\boldsymbol{T}$-algebras $(a, A), (b, B)$ and morphism $f : A \to B$, the square

$$\begin{array}{ccc} \boldsymbol{T}A & \xrightarrow{\boldsymbol{T}f} & \boldsymbol{T}B \\ {\scriptstyle a}\downarrow & \swarrow & \downarrow{\scriptstyle b} \\ A & \xrightarrow[f]{} & B \end{array}$$

is filled by a unique 2-cell $\bar{f} : b \circ \boldsymbol{T}f \Rightarrow f \circ a$ which is a lax morphism of algebras;

L2. there exists an adjunction $a \dashv \eta_A$ with invertible counit;

L3. there exists an adjunction $\mu \dashv \eta * \boldsymbol{T}$ with invertible counit;

L4. there is a modification $\Delta : \boldsymbol{T} * \eta \Rightarrow \eta * \boldsymbol{T}$ such that $\Delta * \eta = 1$ and $\mu * \Delta = 1$.

The conditions for colax algebras are of course obtained by replacing lax algebra morphisms with *colax* ones:

CX1. for every pair of $\boldsymbol{T}$-algebras a, b and morphism $f : A \to B$, the square

$$\begin{array}{ccc} \boldsymbol{T}A & \xrightarrow{\boldsymbol{T}f} & \boldsymbol{T}B \\ {\scriptstyle a}\downarrow & \nearrow & \downarrow{\scriptstyle b} \\ A & \xrightarrow[f]{} & B \end{array}$$

is filled by a unique 2-cell $\bar{f} : f \circ a \Rightarrow b \circ \boldsymbol{T}f$ which is a colax morphism of algebras;

CX2. there exists an adjunction $\eta_A \dashv a$ with invertible unit;

CX3. there exists an adjunction $\eta * \boldsymbol{T} \dashv \mu$ with invertible unit;

CX4. there is a modification $\Upsilon : \eta * \boldsymbol{T} \Rightarrow \boldsymbol{T} * \eta$ such that $\Upsilon * \eta = 1$ and $\mu * \Upsilon = 1$.

The goal of the next section is to prove that the presheaf construction satisfies the axioms of a lax idempotent monad.

2.5.1 Relative Monads and Presheaves

Warning 2.5.9. Throughout the section we will often blur the distinction between two choices of notation: if the typeface for a 2-category is A, objects, i.e. 0-cells in A, are denoted with Roman letters $A, B, \dots, X, Y$, 1-cell are lowercase Roman $f, g, \dots$

This applies in particular to the case of Cat and its objects, categories $A, B, C, \dots$. We invite the reader to keep in mind this small clash of notation when comparing this section with the rest of the book; a similar choice has been made without further mention discussing particular shapes of 2-limits in Chapter 4 (see 4.2.7 and 4.2.8).

In the present section we study the pair $(\mathsf{Cat}, \boldsymbol{P})$, where $\boldsymbol{P} : \mathsf{cat} \to \mathsf{Cat}$ is the presheaf construction sending $A \mapsto [A^{\mathrm{op}}, \mathrm{Set}]$; this is defined as having domain the category cat of small categories, and codomain the locally small ones. The embedding of cat into Cat will always be denoted as $j : \mathsf{cat} \subset \mathsf{Cat}$. This functor is an inclusion at the level of all cells.

We fix a notation that can be easily generalised to the case of a pair $(\mathsf{K}, \boldsymbol{P})$, where $\boldsymbol{P} : \mathsf{A} \to \mathsf{K}$ is the presheaf construction of a Yoneda structure on K.

Notation 2.5.10 (Presheaves)**.** We consider the functor $\boldsymbol{P} : A \mapsto [A^{\mathrm{op}}, \mathrm{Set}]$ as a *covariant* correspondence on functors and natural transformations; more formally, $\boldsymbol{P}$ acts as a correspondence $\mathsf{cat} \to \mathsf{Cat}$ sending functors $f : A \to B$ to *adjoint pairs* $\boldsymbol{P}_! f : \boldsymbol{P}A \leftrightarrows \boldsymbol{P}B : \boldsymbol{P}^* f$, and its action on 2-cells is determined by our desire to privilege the left adjoint, inducing a 2-cell $\alpha_! : \boldsymbol{P}_! f \Rightarrow \boldsymbol{P}_! g$ for each $\alpha : f \Rightarrow g$. Given $f : A \to B$, the functor $\boldsymbol{P}^* f := \boldsymbol{P}B(よ_B \circ f, 1)$ acts as pre-composition with f, whereas $\boldsymbol{P}_! f$ is the operation of left extension along f. The situation is conveniently depicted in the diagram

$$\begin{array}{ccc} A & \xrightarrow{\ f\ } & B \\ {\scriptstyle よ_A}\downarrow & & \downarrow{\scriptstyle よ_B} \\ \boldsymbol{P}A & \underset{\boldsymbol{P}_! f}{\overset{\boldsymbol{P}^* f}{\leftrightarrows}} & \boldsymbol{P}B \end{array} \tag{2.74}$$

Such a diagram is filled by an isomorphism when it is closed by the 1-cell $\boldsymbol{P}_! f$ and by the cell $\chi^{よ_B \circ f}$ when it is closed by $\boldsymbol{P}^* f$.

Definition 2.5.11 (Small functor and small presheaf)**.** Let X be a category; we call a functor $F : X^{\mathrm{op}} \to \mathrm{Set}$ a *small presheaf* if it results from a *small* colimit of representables. Equivalently, F is small if there exists a small subcategory $i : A \subset X$ and a legitimate presheaf $\bar{F} : A^{\mathrm{op}} \to \mathrm{Set}$ of which the functor F is the left Kan extension along i.

Remark 2.5.12. The same definition applies, of course, to a functor $F : X \to Y$ between two large categories.

Notation 2.5.13 (Small presheaves)**.** It turns out that the category of small presheaves on X is legitimate in the same universe as X (while the category of all functors $X^{\mathrm{op}} \to \mathrm{Set}$ is not). Given a locally small category there is a Yoneda embedding $X \to [X^{\mathrm{op}}, \mathrm{Set}]_s$ having the universal property of free cocompletion of X (see [AR20, §3]). We explicitly record how the functor $[\_^{\mathrm{op}}, \mathrm{Set}]_s : \mathsf{Cat} \to \mathsf{Cat}$ acts on 1- and 2-cells: the universal property of $[\_^{\mathrm{op}}, \mathrm{Set}]_s$ proved in [AR20] implies that the dotted arrow exists in the square

$$\begin{array}{ccc} A & \xrightarrow{f} & B \\ {\scriptstyle よ_A}\downarrow & & \downarrow{\scriptstyle よ_B} \\ [A^{\mathrm{op}}, \mathrm{Set}]_s & \dashrightarrow & [B^{\mathrm{op}}, \mathrm{Set}]_s \end{array} \tag{2.75}$$

(it is the Yoneda extension of $よ_B \circ f$). The action of $[\_^{\mathrm{op}}, \mathrm{Set}]_s$ on 2-cells is uniquely determined as a consequence of this definition.

Lemma 2.5.14. Let $j : \mathsf{cat} \subset \mathsf{Cat}$ be the embedding of small categories in locally small ones. For every large category X there is a natural isomorphism $[X^{\mathrm{op}}, \mathrm{Set}]_s \cong \mathrm{Lan}_j\, \boldsymbol{P}(X)$ (a convenient shorthand is to denote the left extension of $\boldsymbol{P}$ along j as $j_!\boldsymbol{P}$; this is compatible with the notation in 2.5 and we will adopt it without further mention). In particular, there is a canonical isomorphism between $\boldsymbol{P} \triangleleft \boldsymbol{P} = j_!\boldsymbol{P} \circ \boldsymbol{P}$ and the functor $[(\boldsymbol{P}\,\_)^{\mathrm{op}}, \mathrm{Set}]_s$, where for a large category X, the category $[X^{\mathrm{op}}, \mathrm{Set}]_s$ designates small presheaves $X^{\mathrm{op}} \to \mathrm{Set}$.

Proof To show that the universal property of the Kan extension is fulfilled by $[X^{\mathrm{op}}, \mathrm{Set}]_s$ we employ a density argument: given a functor $H : \mathsf{Cat} \to \mathsf{Cat}$, every natural transformation $\alpha : \boldsymbol{P} \Rightarrow H \circ j$ can be extended to a natural transformation $\bar{\alpha}$ from small presheaves to H, using the fact that each $F \in [X^{\mathrm{op}}, \mathrm{Set}]_s$ can be presented as a small colimit of representables: if $F \cong i_!\tilde{F} \in [X^{\mathrm{op}}, \mathrm{Set}]_s$ for $i : A \to X$, the components of $\bar{\alpha}_X$ are defined as

$$[X^{\mathrm{op}}, \mathrm{Set}]_s \xrightarrow{i^*} [A^{\mathrm{op}}, \mathrm{Set}]_s = [A^{\mathrm{op}}, \mathrm{Set}] \xrightarrow{\alpha_A} HjA = HA \xrightarrow{Hi} HX.$$

□

The following corollary is immediate: it will turn out to be useful in the proof of 2.5.17.

Corollary 2.5.15. There is a canonical isomorphism

$$[X^{\mathrm{op}}, \mathrm{Set}]_s \cong \int^{A \in \mathsf{cat}} [X, A]^{\mathrm{op}} \times [A, \mathrm{Set}] \tag{2.76}$$

(in particular, this specific coend exists even if it is indexed over a non-small category).

Remark 2.5.16. It is reasonable to expect $j_!\boldsymbol{P}$ to be the small-presheaf construction; this construction is the legitimate version of the Yoneda embedding associated to a (possibly large) category X. It is important to stress our desire to exploit the results in [FGHW18], but minding that $\boldsymbol{P}$ has additional structure (their approach qualifies $\boldsymbol{P}$ as a monad, but only in the sense that it is a j-pointed functor, endowed with a unit $\eta : j \to \boldsymbol{P}$ and with a 'Kleisli extension' map, instead of a monad multiplication,- sending each $f : jA \to \boldsymbol{P}B$ to $f^* : \boldsymbol{P}A \to \boldsymbol{P}B$). In view of 2.5.2, now we would like to say that [cat, Cat] is a skew-monoidal category with skew unit j, and composition $(F, G) \mapsto j_!F \circ G$. This would yield 'iterated presheaf constructions' $\boldsymbol{P} \triangleleft \boldsymbol{P}$, $\boldsymbol{P} \triangleleft (\boldsymbol{P} \triangleleft \boldsymbol{P}), (\boldsymbol{P} \triangleleft \boldsymbol{P}) \triangleleft \boldsymbol{P}, \dots$, all seen as functors cat $\to$ Cat. Unfortunately, given a functor F : cat $\to$ Cat it is impossible to ensure that $j_!F$ exists in general (cat is a $\mathfrak{U}^+$-category, and Cat cannot be $\mathfrak{U}^+$-cocomplete), thus, if anything, the skew-monoidal structure of 2.5.2 does not exist globally. Fortunately *some* left extensions, precisely those we need, do exist, so we can still employ the 'local' existence of $j_!\boldsymbol{P}$ and its iterates to work as if it was part of a full monoidal structure.

It also turns out (and this is by no means immediate, see 2.5.18) that the unitors $\lambda_{\boldsymbol{P}} : j \triangleleft \boldsymbol{P} \to \boldsymbol{P}$ and $\varrho_{\boldsymbol{P}} : \boldsymbol{P} \to \boldsymbol{P} \triangleleft j$ and the associator $\gamma_{\boldsymbol{P}} : (\boldsymbol{P} \triangleleft \boldsymbol{P}) \triangleleft \boldsymbol{P} \to \boldsymbol{P} \triangleleft (\boldsymbol{P} \triangleleft \boldsymbol{P})$ are all invertible.

The preliminary results in this section make legitimate the practice of naïvely considering iterated presheaf constructions: once we consider small functors, the category $[X^{\text{op}}, \text{Set}]_s$ lives in the same universe as X, and so do all categories $[[A^{\text{op}}, \text{Set}]^{\text{op}}, \text{Set}]_s$, $[[[A^{\text{op}}, \text{Set}]^{\text{op}}, \text{Set}]_s{}^{\text{op}}, \text{Set}]_s$.

Lemma 2.5.17 ($j_!\boldsymbol{P}$ preserves itself)**.** There is a canonical isomorphism

$$\tilde{\gamma}_{\boldsymbol{PP}} : j_!(j_!\boldsymbol{P} \circ \boldsymbol{P}) \cong j_!\boldsymbol{P} \circ j_!\boldsymbol{P}. \tag{2.77}$$

Proof Relying on the previous lemma, we compute the coend

$$j_!(j_!\boldsymbol{P} \circ \boldsymbol{P}) \cong \int^{A \in \mathsf{cat}} [A, X] \times [(\boldsymbol{P}A)^{\text{op}}, \text{Set}]_s \tag{2.78}$$

which is now isomorphic to $[[X^{\text{op}}, \text{Set}]_s^{\text{op}}, \text{Set}]_s$ in view of the Yoneda reduction and of 2.5.14 (note that the definitions of $\lambda A.\boldsymbol{P}A$ and $\lambda A.j_!\boldsymbol{P}A$ entail that $\lambda A.[\boldsymbol{P}A^{\text{op}}, \text{Set}]$ is covariant in A). □

Lemma 2.5.18. There are canonical isomorphisms $\lambda_{\boldsymbol{P}} : j \triangleleft \boldsymbol{P} \cong \boldsymbol{P}$, $\varrho_{\boldsymbol{P}} : \boldsymbol{P} \cong \boldsymbol{P} \triangleleft j$ and $\gamma_{\boldsymbol{PPP}} : (\boldsymbol{P} \triangleleft \boldsymbol{P}) \triangleleft \boldsymbol{P} \cong \boldsymbol{P} \triangleleft (\boldsymbol{P} \triangleleft \boldsymbol{P})$, determined as in S1–S3.

Proof The isomorphisms come from the unit and associativity constraints of S1–S3; as noted in 2.5.4, ϱ is invertible in every component because j is fully faithful, and similarly λ is invertible in every component if we show that j is a dense functor. Once we have shown this, 2.5.17 above concludes the argument, since from (2.63) the associator $\gamma_{\boldsymbol{PPP}}$ coincides with the composition $\tilde{\gamma}_{\boldsymbol{PP}} * \boldsymbol{P}$.

Now, the functor j is dense, because the full subcategory of Cat on the generic commutative triangle $[2] = \{0 \to 1 \to 2\}$ is dense. So cat, being a full supercategory of a dense category, is dense. □

Remark 2.5.19. We often write $\eta\boldsymbol{P}$, $\boldsymbol{P}\eta$, $\mu\boldsymbol{P}$,... to denote what should be written as $\eta \triangleleft \boldsymbol{P}$, $\boldsymbol{P} \triangleleft \eta$, $\mu \triangleleft \boldsymbol{P}$ .... Keeping in mind 2.5.6, which clarifies the formal definition for $\triangleleft$-whiskerings, there is no chance of confusion: $\eta \triangleleft \boldsymbol{P} = j_!\eta * \boldsymbol{P}$, $\boldsymbol{P} \triangleleft \eta = j_!\boldsymbol{P} * \eta$, and similarly for every other whiskering.

Proposition 2.5.20. The presheaf construction $\boldsymbol{P} = [\_^{\text{op}}, \text{Set}]$ is a lax idempotent j-relative monad.

In simple terms, a relative monad $T : \mathcal{X} \to \mathcal{Y}$ is the formal equivalent of a monad in the skew-monoidal structure $([\mathcal{X}, \mathcal{Y}], \triangleleft)$. On [cat, Cat], such structure exists 'locally' in the components we need, allowing us to work as if $\boldsymbol{P}$ really were a $\triangleleft$-monoid. As already said, [FGHW18] does not assume the existence of a multiplication map $\mu : \boldsymbol{P} \triangleleft \boldsymbol{P} \to \boldsymbol{P}$, replacing it with coherently assigned 'Kleisli extensions' to maps $f : jA \to \boldsymbol{P}B$.

A relative monad is now *lax idempotent* (or a *KZ-doctrine*) if it satisfies the 2-dimensional analogue of the notion of idempotency; in short, when the algebra structure on an object A is unique up to isomorphism as soon as it exists. Following [GL12, 2.2], $\boldsymbol{P}$-algebras are cocomplete categories, thus, when an object is a $\boldsymbol{P}$-algebra it is so in a unique way (this is of course a behaviour of all formal cocompletion monads).

Proposition 2.5.21. $\boldsymbol{P}$ is a j-relative monad, if we denote $j : \text{cat} \subset \text{Cat}$ the obvious inclusion.

Proof As already noted, the relevant left extensions involved in this proof exist; the diagrams we have to check the commutativity of are now the following, once we define the Yoneda embedding $よ_A : A \to \boldsymbol{P}A$ as unit, and $\boldsymbol{P}^* よ_A : \boldsymbol{PP}A \to \boldsymbol{P}A$ as multiplication of the desired monad:

$$\begin{array}{ccc} [[(\boldsymbol{P}A)^{\text{op}}, \text{Set}]_s^{\text{op}}, \text{Set}]_s & \xrightarrow{\mu_{\boldsymbol{P}A}} & [(\boldsymbol{P}A)^{\text{op}}, \text{Set}]_s \\ \Big\downarrow{\scriptstyle \boldsymbol{P}_!\mu_A} & & \Big\downarrow{\scriptstyle \mu_A} \\ [(\boldsymbol{P}A)^{\text{op}}, \text{Set}]_s & \xrightarrow[\mu_A]{} & \boldsymbol{P}A \end{array} \tag{2.79}$$

$$\begin{array}{ccccc} \boldsymbol{P}A & \xrightarrow{\text{よ}_{\boldsymbol{P}A}} & [(\boldsymbol{P}A)^{\text{op}^{\text{op}}}, \text{Set}] & \xleftarrow{\boldsymbol{P}_!\text{よ}_A} & \boldsymbol{P}A \\ & \searrow^{=} & \big\downarrow{\scriptstyle \mu_A} & \swarrow_{=} & \\ & & \boldsymbol{P}A & & \end{array} \tag{2.80}$$

In order to show that they commute, we will exploit the adjunctions $\boldsymbol{P}_!\,\text{よ}_A \dashv \boldsymbol{P}^*\,\text{よ}_A$ and $\boldsymbol{P}_!\mu_A \dashv \boldsymbol{P}^*\mu_A$. The functor $\boldsymbol{P}^*\mu_A$ exists because it must coincide with the Yoneda embedding of $[\boldsymbol{P}A^{\text{op}}, \text{Set}]_s$ into the iterated presheaf category $[[\boldsymbol{P}A^{\text{op}}, \text{Set}]_s{}^{\text{op}}, \text{Set}]_s$; it must act as the Yoneda embedding $Q \mapsto [\boldsymbol{P}A^{\text{op}}, \text{Set}]_s(\_, Q)$, and this evidently lands into the category of small presheaves on $[\boldsymbol{P}A^{\text{op}}, \text{Set}]_s$ when restricted to small functors.

Concerning the unit axiom, the commutativity of the left triangle can be deduced from the chain of isomorphisms

$$\begin{aligned} \mu_A \circ \text{よ}_{\boldsymbol{P}A} &\cong \text{Lan}_{\text{よ}_{\boldsymbol{P}A}\text{よ}_A}(\text{よ}_A) \circ \text{よ}_{\boldsymbol{P}A} \\ &\cong \text{Lan}_{\text{よ}_{\boldsymbol{P}A}}(\text{Lan}_{\text{よ}_A}(\text{よ}_A)) \circ \text{よ}_{\boldsymbol{P}A} \\ (\text{よ}_{\boldsymbol{P}A} \text{ is fully faithful}) &\cong \text{Lan}_{\text{よ}_A}(\text{よ}_A) \\ (\text{よ}_A \text{ is dense}) &\cong \text{id}_{\boldsymbol{P}A}. \end{aligned}$$

The right triangle corresponds to the composition

$$\begin{array}{ccccc} [A^{\text{op}}, \text{Set}] & \longrightarrow & [[A^{\text{op}}, \text{Set}]^{\text{op}}, \text{Set}]_s & \longrightarrow & [A^{\text{op}}, \text{Set}] \\ P & \longmapsto & (Q \mapsto [Q, P]) & \longmapsto & (a \mapsto [\text{よ}_A(a), P] \cong Pa) \end{array}$$

which is again isomorphic to the identity of $\boldsymbol{P}A$ thanks to the Yoneda lemma.

In order to show that the multiplication is associative, we prove that $\boldsymbol{P}_!\mu_A \cong \mu_{\boldsymbol{P}A}$ as a consequence of the fact that there is an adjunction $\mu_{\boldsymbol{P}} \overset{1}{\dashv} \boldsymbol{P}\mu$, having moreover invertible counit. The argument will be fairly explicit, building unit and counit from suitable universal properties of the presheaf construction and from the definitions of $\boldsymbol{P}^*\mu_A$ and $\mu_{\boldsymbol{P}A}$:

- $\boldsymbol{P}^*\mu_A$ sends $\zeta : \boldsymbol{P}A^{\text{op}} \to \text{Set}$ into $[\boldsymbol{P}A^{\text{op}}, \text{Set}]_s(\_, \zeta)$ (it plays the exact rôle of a large Yoneda embedding; this will entail that the counit of the adjunction $\mu_{\boldsymbol{P}} \dashv \boldsymbol{P}^*\mu$ is invertible);
- $\mu_{\boldsymbol{P}A}$ acts by sending $\lambda F.\Theta(F) \in [[\boldsymbol{P}A^{\text{op}}, \text{Set}]_s{}^{\text{op}}, \text{Set}]_s$ to the functor $\lambda a.\Theta(\hom(\_, a))$.

With these definitions, the composition $\mu_{\boldsymbol{P}A} \circ \boldsymbol{P}^*\mu_A$ is in fact isomorphic to the identity of $[\boldsymbol{P}A^{\text{op}}, \text{Set}]_s$. We do not find the unit map: we refrain from showing the zig-zag identities as they follow right away from the explicit description of the (co)unit.

The unit will have as components morphisms $\Theta \Rightarrow \boldsymbol{P}^*\mu_A(\mu_{\boldsymbol{P}A}(\Theta))$ natural in $\Theta \in [[\boldsymbol{P}A^{\mathrm{op}}, \mathrm{Set}]_s{}^{\mathrm{op}}, \mathrm{Set}]_s$. Given one of these components, its codomain can be rewritten as the coend

$$\begin{aligned}\boldsymbol{P}^*\mu_A(\mu_{\boldsymbol{P}A}(\Theta)) &= \lambda\chi.[\boldsymbol{P}A^{\mathrm{op}}, \mathrm{Set}]_s(\chi, \mu_{\boldsymbol{P}A}(\Theta)) \\ &\cong \lambda\chi. \int_{F\in\boldsymbol{P}A} \mathrm{Set}(\chi(F), \mu_{\boldsymbol{P}A}(\Theta)(F)) \\ &\cong \lambda\chi. \int_{F\in\boldsymbol{P}A} \mathrm{Set}(\chi(F), \Theta(\boldsymbol{P}A(\_, F))) \\ &\leftarrow \lambda\chi.\Theta(\chi)\end{aligned}$$

and we obtain the candidate morphism in the last line as follows. Its component at $\chi \in [\boldsymbol{P}A^{\mathrm{op}}, \mathrm{Set}]_s$ must be an arrow

$$\Theta(\chi) \longrightarrow \int_{F\in\boldsymbol{P}A} \mathrm{Set}(\chi(F), \Theta(\boldsymbol{P}A(\_, F))) \tag{2.81}$$

which is induced by a wedge

$$\Theta(\chi) \to \mathrm{Set}(\chi(F), \Theta(\boldsymbol{P}A(\_, F))); \tag{2.82}$$

such a wedge comes from (the mate of) Θ^{ar}, the function of Θ on arrows. The Yoneda lemma now entails that such an action induces a map

$$\chi(F) \cong [\boldsymbol{P}A^{\mathrm{op}}, \mathrm{Set}]_s(\boldsymbol{P}A(\_, F), \chi) \xrightarrow{\Theta^{\mathrm{ar}}} \mathrm{Set}(\Theta(\chi), \Theta(\boldsymbol{P}A(\_, F))) \tag{2.83}$$

that by cartesian closure can be reported to

$$\Theta(\chi) \longrightarrow \mathrm{Set}(\chi(F), \Theta(\boldsymbol{P}A(\_, F))). \tag{2.84}$$

Of course, this is a wedge in F, and we conclude. □

Proposition 2.5.22. The monad $\boldsymbol{P}$ is lax idempotent in the sense of 2.5.8: there exists an adjunction $\mu_A \overset{1}{\dashv} \eta_{\boldsymbol{P}A}$.

Proof The existence of an adjunction $\mu_A \overset{1}{\dashv} \eta_{\boldsymbol{P}A}$ will imply all the equivalent conditions in 2.5.8, which we nevertheless recall in 2.5.23 below for the convenience of the reader. From the definition of these maps, there is a natural candidate for the counit, and this will be invertible as a consequence of the Yoneda lemma. Indeed, the isomorphism $1 \cong \mu_A \circ \eta_{\boldsymbol{P}A}$ corresponds to the map

$$\lambda a.Fa \mapsto \lambda G.\,\hom(G, F) \mapsto \lambda a.\,\hom(A(\_, a), F) \cong Fa. \tag{2.85}$$

The unit is instead given by the action of a certain functor on arrows, in a similar way as above: given $\chi \in [\boldsymbol{P}A^{\mathrm{op}}, \mathrm{Set}]_s$, there is a canonical map

$$\begin{aligned}\chi(F) &\to \boldsymbol{P}A(F, \lambda a.\chi(\hom(\_, a))) \\ &\cong \int_{a\in A} \mathrm{Set}(Fa, \chi(\hom(\_, a)))\end{aligned}$$

coming from (the mate of) a wedge

$$Fa \cong \boldsymbol{P}A(\hom(\_, a), F) \to \mathrm{Set}(\chi(F), \chi(\hom(\_, a))) \tag{2.86}$$

defined by the action on arrows of χ. □

The following remark, which is a particular case of 2.5.8, characterises $\boldsymbol{P}$-algebras as categories whose Yoneda embedding $よ_A$ has a left adjoint α. These are the *cocomplete* categories [GL12, §2]; it is rather easy to see that one of the axioms of $\boldsymbol{P}$-algebras asserts that $\alpha(A(\_, a)) \cong a$, and since α is a left adjoint, it is uniquely determined by sending a colimit of representables into the colimit in A of representing objects (in particular, all such colimits exist).

Remark 2.5.23. It turns out from the general theory of lax idempotent monads that $\boldsymbol{P}$ satisfies the following equivalent conditions:

PL1. for every pair of $\boldsymbol{P}$-algebras a, b and morphism $f : A \to B$, the square

$$\begin{array}{ccc} \boldsymbol{P}A & \xrightarrow{\boldsymbol{P}f} & \boldsymbol{P}B \\ {\scriptstyle a}\downarrow & & \downarrow{\scriptstyle b} \\ A & \xrightarrow[f]{} & B \end{array} \tag{2.87}$$

is filled by a unique 2-cell $b \circ \boldsymbol{P}f \Rightarrow f \circ a$ forming a lax morphism of algebras;

PL2. if $\alpha : \boldsymbol{P}A \to A$ is a $\boldsymbol{P}$-algebra, there is an adjunction $\alpha \dashv \eta_A$ with invertible counit;

PL3. there is an adjunction $\mu_A \dashv \eta_{\boldsymbol{P}A}$ with invertible counit.

In particular, we have shown condition PL3; since PL2 holds, there is a single possible choice up to isomorphism for a $\boldsymbol{P}$-algebra structure on an object A, namely the arrow playing the rôle of left adjoint to the Yoneda embedding.

Exercises

2.1 Show that the mate of (2.23) (with the notation therein) is a natural transformation

$$\mathrm{Lan}_\alpha F : \mathrm{Lan}_{G'} F \Rightarrow \mathrm{Lan}_G F,$$

which is the component at F of a transformation $\mathrm{Lan}_\alpha : \mathrm{Lan}_{G'} \Rightarrow \mathrm{Lan}_G$; in other words show that the square

$$\begin{array}{ccc} \mathrm{Lan}_{G'} F & \xrightarrow{\mathrm{Lan}_{G'}\tau} & \mathrm{Lan}_{G'} H \\ {\scriptstyle \mathrm{Lan}_\alpha F}\Downarrow & & \Downarrow{\scriptstyle \mathrm{Lan}_\alpha H} \\ \mathrm{Lan}_G F & \xrightarrow[\mathrm{Lan}_G \tau]{} & \mathrm{Lan}_G H \end{array}$$

commutes for every natural transformation $\tau : F \Rightarrow H$.

2.2 Write explicitly the unit and the counit of the adjunctions $\mathrm{Lan}_G \overset{\epsilon^L}{\underset{\eta^L}{\dashv}} G^*$ and $G^* \overset{\epsilon^R}{\underset{\eta^R}{\dashv}} \mathrm{Ran}_G$.

2.3 Show that all presheaf categories are cartesian closed, using coends: if $\mathrm{Cat}(\mathcal{C}^{\mathrm{op}}, \mathrm{Set})$ is the category of presheaves on a small category $\mathcal{C}$, then there exists an adjunction

$$\mathsf{Cat}(\mathcal{C}^{\mathrm{op}}, \mathrm{Set})(P \times Q, R) \cong \mathsf{Cat}(\mathcal{C}^{\mathrm{op}}, \mathrm{Set})(P, R^Q)$$

where $P \times Q$ is the pointwise product of functors.

Show that $R^Q(c) = \mathsf{Cat}(\mathcal{C}^{\mathrm{op}}, \mathrm{Set})(よ_c \times Q, R)$ does the job (use the ninja Yoneda lemma and 1.4.1).

2.4 Let K be a 2-category, and $e : X \to X$ an endo-1-cell; show that the following conditions are equivalent:

- e is the identity 1-cell of X;
- for every $f : X \to A$ there is a triangle

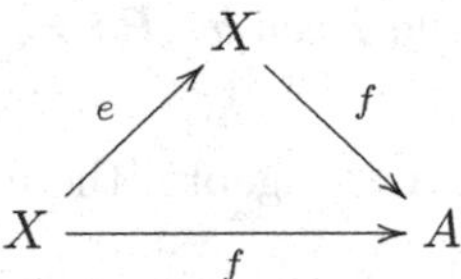

that exhibits f as the right extension of itself along e.

How can this statement be dualised?

2.5 Use equations (2.27) and the ninja Yoneda lemma to show that $\mathrm{Lan}_{\mathrm{id}}$ and $\mathrm{Ran}_{\mathrm{id}}$ are the identity functors, as expected. Use (2.27) and the ninja Yoneda lemma to complete the proof that $F \mapsto \mathrm{Lan}_F$

is a pseudofunctor by showing that for $\mathcal{A} \xrightarrow{F} \mathcal{B}$, $\mathcal{A} \xrightarrow{G} \mathcal{C} \xrightarrow{H} \mathcal{D}$ there is a uniquely determined laxity cell for composition

$$\mathrm{Lan}_H(\mathrm{Lan}_G(F)) \cong \mathrm{Lan}_{HG}(F).$$

(Hint: play with the coend $\mathrm{Lan}_H(\mathrm{Lan}_G(F))D$ until you get

$$\int^{XY} (\mathcal{D}(HX, D) \times \mathcal{C}(GY, X)) \otimes FY;$$

now use the ninja Yoneda lemma plus co-continuity of the tensor, as suggested in Remark 2.2.5.)

2.6 Show that the unit $\eta : F \Rightarrow \mathrm{Lan}_G F \circ G$ is a wedge in the component G; dually, the counit $\epsilon : \mathrm{Lan}_G(H \circ G) \Rightarrow H$ of the same adjunction is a cowedge. Are the two dinatural transformations the end and the coend of the respective functors?

2.7 Let $\mathcal{C}$ be a small compact closed monoidal category (see [Day77]). Show that the functor $Y \mapsto \int^X X^\vee \otimes Y \otimes X$ carries the structure of a monad on the category $\mathcal{C}$.

2.8 Use coend calculus to prove that if a functor G is fully faithful, then so is $\mathrm{Lan}_G(\_)$. Use coend calculus to prove that if a functor $i : \mathcal{C} \to \mathcal{D}$ has small domain and is dense, then the left Kan extension of the Yoneda embedding $よ_\mathcal{C}$ along i is a fully faithful functor $\mathcal{D} \to [\mathcal{C}^{\mathrm{op}}, \mathrm{Set}]$. Chapter 3 will study this kind of phenomena extensively.

2.9 Show that if $T_F = \mathrm{Ran}_F F \in \mathrm{Cat}(\mathcal{D}, \mathcal{D})$ is the codensity monad of a functor $F : \mathcal{C} \to \mathcal{D}$, then the two maps $\mu : T_F \circ T_F \Rightarrow T_F$ and $\eta : \mathrm{id}_\mathcal{C} \Rightarrow T_F$ indeed form a monad, according to A.6.3.

2.10 Use the end expression in (2.29) for the hom-set in the Kleisli category of the codensity monad of $F : \mathcal{A} \to \mathcal{C}$ to define the *Kleisli composition* in $\mathrm{Kl}(T_F)$ by means of the universal property of (2.29). Prove that the Kleisli composition of $f : A \to T_F B$ and $g : B \to T_F C$ is

$$\mu_C \circ Tg \circ f : A \to TC.$$

Do the same for the coKleisli composition in $\mathrm{coKl}(S^F)$, the density comonad of F.

2.11 ◉◉ This exercise draws from a piece of the theory of relative adjunctions: we recall that if $J : \mathcal{A} \to \mathcal{C}$, $F : \mathcal{A} \to \mathcal{B}$, and $G : \mathcal{B} \to \mathcal{C}$ are functors, an adjunction between F and G, relative to J, consists of a family of natural isomorphisms

$$\mathcal{B}(FA, B) \cong \mathcal{C}(JA, GB).$$

If this is the case, we say that G has a *J-relative left adjoint*. This is denoted $F \overset{[J]}{\dashv} G$. Dually (in the same notation), we say that G has a *J-relative right adjoint* F if there exists a natural isomorphism

$$\mathcal{C}(GB, JA) \cong \mathcal{B}(B, FA).$$

This is denoted $G \underset{[J]}{\dashv} F$.

- Show that $F \overset{[J]}{\dashv} G$ if $F \cong \mathrm{lift}_G J$, and that this lifting is absolute; the unit $\eta : J \Rightarrow G \circ \mathrm{lift}_G J = GF$ is the *relative unit* of the adjunction; dually, $G \underset{[J]}{\dashv} F$ if $F \cong \mathrm{rift}_G J$ and this lifting is absolute; the counit $\epsilon : G \circ \mathrm{rift}_G J = GF \Rightarrow J$ is the *relative counit* of the adjunction.
  Note in particular that if $F \overset{[J]}{\dashv} G$ then there is no counit, and if $F \underset{[J]}{\dashv} G$ there is no unit.
- Is this criterion also necessary? Namely, is it true that if $F \overset{[J]}{\dashv} G$ then $F \cong \mathrm{lift}_G J$ is absolute? If not, does this mean that there can be two non-isomorphic G, G' such that $F \overset{[J]}{\dashv} G, G'$? What is the structure, if any, of the class of functors $\{G \mid F \overset{[J]}{\dashv} G\}$?
- Assume that

  $$F \overset{[J_1]}{\underset{G}{\dashv}}{}_1 \quad \text{and } F \underset{[J_2]}{\dashv} G_2$$

  are relative adjunctions; it is then possible to build a 2-cell

  $$J_1 G_2 \overset{\eta * G_2}{\Rightarrow} G_1 F G_2 \overset{G_1 * \epsilon}{\Rightarrow} G_1 J_2$$

  pasting the relative unit of $F \overset{[J_1]}{\dashv} G_1$ with the relative counit of $F \underset{[J_2]}{\dashv} G_2$. Under which conditions is this 2-cell an isomorphism?
- What does the invertibility of the relative unit $\eta : J \Rightarrow GF$ of a relative adjunction $F \overset{[J]}{\dashv} G$ imply for G, F (F is 'relatively fully faithful', what does this mean?). Dual question for the relative counit of $F \underset{[J]}{\dashv} G$.

2.12 Let R be a ring regarded as a one-object category enriched over Ab. A presheaf on R is a right R-module; let $\mathcal{A}$ be a cocomplete Ab-enriched category and $X : R \to \mathcal{A}$ an enriched functor.

- Show that the class of such functors X corresponds to the class of objects of $\mathcal{A}$ with an action of R.
- Compute the left Yoneda extension $X \otimes \_$ of $M : R \to \mathrm{Ab}$ and its right adjoint; why do we use the notation $X \otimes \_$?

2.13 Let $F : \mathcal{C} \to \mathcal{D}$ be a functor, and let $\mathcal{C}^{\triangleright}$ be the category $\mathcal{C}$ with an additional terminal object adjoined (see A.3.5 for the precise notation). There is an obvious embedding functor $\mathcal{C} \hookrightarrow \mathcal{C}^{\triangleright}$. Show that the triangle

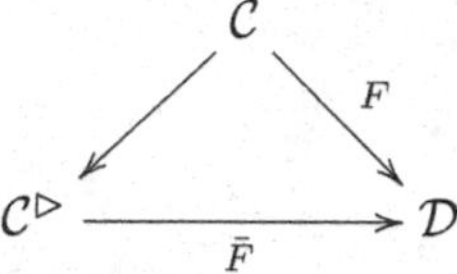

is a right extension; what is the 2-cell $\eta : \bar{F} \Rightarrow F \circ \iota$? Why?

2.14 Prove that in the Yoneda structure on Cat, the following additional axiom holds.

Let $B \underset{g}{\overset{B(f,1)}{\Downarrow \sigma}} \boldsymbol{P}A$ be a 2-cell; if it has the property that the pasting

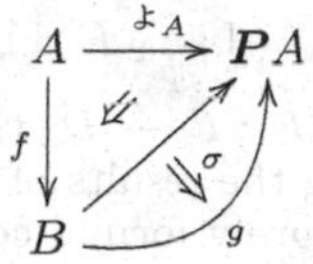

exhibits $\mathrm{lift}_g\, \text{よ}_A$, then σ is invertible.

2.15 Prove that the 2-cell

$$\begin{array}{ccc} A & \xrightarrow{f} & B \\ {\scriptstyle \text{よ}_A}\downarrow & \Rightarrow & \downarrow{\scriptstyle \text{よ}_B} \\ \boldsymbol{P}A & \xleftarrow[\boldsymbol{P}f]{} & \boldsymbol{P}B \end{array}$$

witnesses the fact that there is a *lax dinatural* transformation $\text{よ} : \mathrm{id} \Rightarrow \boldsymbol{P}$. Take the opportunity to define the notion of lax dinaturality; if you do not succeed, wait for Definition 7.1.1.

What is the definition of lax dinaturality in the 2-category Cat? How is the cell $v : \text{よ}_A \Rightarrow \boldsymbol{P}f \circ \text{よ}_B \circ f$ defined in the canonical Yoneda structure on Cat?

3
Nerves and Realisations

SUMMARY. The present chapter studies a single kind of Kan extensions: those where the extendant functor is the Yoneda embedding $よ_C : \mathcal{C} \to [\mathcal{C}^{op}, \text{Set}]$. Far from being too narrow, the resulting theory of *Yoneda extensions* is astoundingly rich and pervasive; the central object of study of the whole chapter are extensions of the form

$$\text{Lan}_よ F \dashv \text{Lan}_F よ$$

induced by a functor $F : \mathcal{C} \to \mathcal{D}$; these are called *nerve-realisation* adjunctions. Exploiting the results of the previous chapter, many theorems turn out to be purely formal consequences of the basic rules of coend calculus.

We illustrate the nature of the nerve-realisation adjunctions with a variety of examples drawing from algebra, topology, geometry, logic, and more category theory.

> Form itself is Void and Void itself is Form.
> Form is not other than Void and Void is not other than Form.
> The same is true of Feelings, Perceptions, Mind, and Consciousness.
>
> Heart sūtra

3.1 The Classical Nerve and Realisation

3.1.1 Overture: the Universal Property of $[\mathcal{C}^{op}, \text{Set}]$

The case in which a Kan extension is done along the Yoneda embedding acquires particular significance in the general theory of Kan extensions: since in a cocomplete category $\mathcal{D}$ all tensors and colimits in (2.27) exist, every functor $F : \mathcal{C} \to \mathcal{D}$ having small domain and cocomplete codomain has an *extension* to the category of presheaves on $\mathcal{C}$, in a diagram

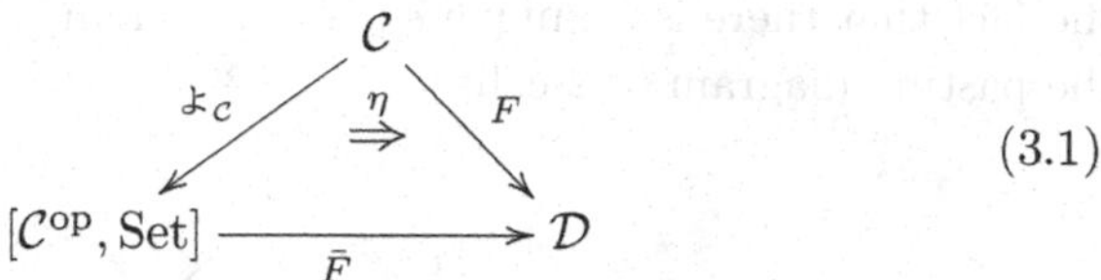

$$\begin{array}{ccc} & \mathcal{C} & \\ {\scriptstyle よ_{\mathcal{C}}}\swarrow & \overset{\eta}{\Rightarrow} & \searrow{\scriptstyle F} \\ [\mathcal{C}^{\mathrm{op}}, \mathrm{Set}] & \xrightarrow[\bar{F}]{} & \mathcal{D} \end{array} \tag{3.1}$$

filled by an invertible 2-cell η. The scope of this introductory section is to outline this as a universal property. The rest of the chapter is devoted to showing how pervasive and expressive the theory of 'Yoneda extensions' is.

Let $\mathcal{C}$ be a small category, $\mathcal{D}$ a cocomplete category; then, precomposition with the Yoneda embedding $よ_{\mathcal{C}} : \mathcal{C} \to [\mathcal{C}^{\mathrm{op}}, \mathrm{Set}]$ determines a functor

$$\mathrm{Cat}([\mathcal{C}^{\mathrm{op}}, \mathrm{Set}], \mathcal{D}) \xrightarrow{- \circ よ_{\mathcal{C}}} \mathrm{Cat}(\mathcal{C}, \mathcal{D}), \tag{3.2}$$

which restricts a functor $G : [\mathcal{C}^{\mathrm{op}}, \mathrm{Set}] \to \mathcal{D}$ to act only on representable functors, identified with objects of $\mathcal{C}$, thanks to the fact that $よ_{\mathcal{C}}$ is fully faithful. We then have the following theorem.

Theorem 3.1.1.

YE1. The universal property of the category $[\mathcal{C}^{\mathrm{op}}, \mathrm{Set}]$ amounts to the existence of a left adjoint $\mathrm{Lan}_{よ_{\mathcal{C}}}$ to precomposition, that has invertible unit (so, the left adjoint is fully faithful).

This means that $\mathrm{Cat}(\mathcal{C}, \mathcal{D})$ is a full subcategory of $\mathrm{Cat}([\mathcal{C}^{\mathrm{op}}, \mathrm{Set}], \mathcal{D})$. Moreover

YI1. the essential image of $\mathrm{Lan}_{よ_{\mathcal{C}}}$ consists of those $F : [\mathcal{C}^{\mathrm{op}}, \mathrm{Set}] \to \mathcal{D}$ that preserve all colimits;

YI2. if $\mathcal{D} = [\mathcal{E}^{\mathrm{op}}, \mathrm{Set}]$, this essential image is equivalent to the subcategory of left adjoints $F : [\mathcal{C}^{\mathrm{op}}, \mathrm{Set}] \to [\mathcal{E}^{\mathrm{op}}, \mathrm{Set}]$.

Proof The first claim asserts that the following are equivalent:

- every $F : \mathcal{C} \to \mathcal{D}$ extends uniquely to $\hat{F} : [\mathcal{C}^{\mathrm{op}}, \mathrm{Set}] \to \mathcal{D}$ (this in turn means that there is such an $\hat{F}$ and that if $\hat{F}', \hat{F}$ both extend F, they are canonically isomorphic);
- there exists an adjunction $\mathrm{Lan}_{よ_{\mathcal{C}}} \underset{\mathrm{id}}{\overset{\epsilon}{\dashv}} - \circ よ_{\mathcal{C}}$ that has invertible unit.

Proving these claims amounts more or less to playing with definitions, and to noticing that since $\hat{F}$ is unique up to isomorphism, the correspondence $F \mapsto \hat{F}$ is a functor determined up to isomorphism. Given the universal property of $\hat{F} = \mathrm{Lan}_{\text{よ}} F$, the counit of the adjunction is determined by the fact that there is a unique $\epsilon : \widehat{G \circ \text{よ}} = \mathrm{Lan}_{\text{よ}}(G\text{よ}) \to G$ induced by the pasting diagram of 2-cells

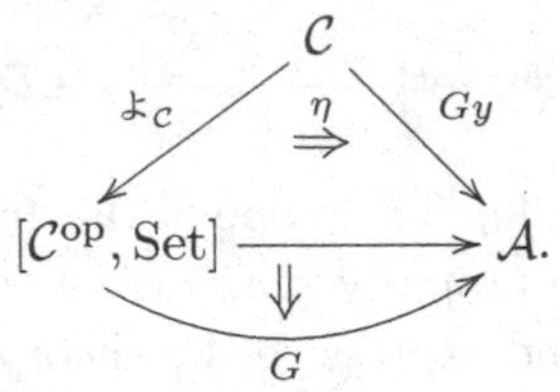

$$\begin{array}{ccc} & \mathcal{C} & \\ {\scriptstyle \text{よ}_{\mathcal{C}}} \swarrow & \overset{\eta}{\Rightarrow} & \searrow {\scriptstyle Gy} \\ [\mathcal{C}^{\mathrm{op}}, \mathrm{Set}] & \xrightarrow{} & \mathcal{A}. \\ & \Downarrow & \\ & \underset{G}{\smile} & \end{array} \tag{3.3}$$

Half of the triangle identities is simply the fact that the above triangle factors the identity of $G \circ \text{よ}_{\mathcal{C}}$. The unit of the adjunction is invertible, as a consequence of a general fact: if the extendant arrow y is a fully faithful functor, then there is a canonical isomorphism $\mathrm{Lan}_y(F) \circ y \cong F$. (In case $y = \text{よ}$ is the Yoneda embedding, there is an easy coend proof for this: find it using the ninja Yoneda lemma.)

Now, we shall prove that the counit $\epsilon_G : \mathrm{Lan}_{\text{よ}_{\mathcal{C}}}(G \circ \text{よ}_{\mathcal{C}}) \Rightarrow G$ of the above adjunction is invertible if and only if the functor G is cocontinuous.

Let G be cocontinuous; then since the counit is obtained as the canonical map

$$\begin{aligned} \mathrm{Lan}_{\text{よ}}(G\text{よ})(P) &\cong \int^X PX \times G(\hom(\_, X)) \\ &\to G\left(\int^X PX \times \hom(\_, X)\right) \\ &\cong GP, \end{aligned}$$

it is easy to see that G is cocontinuous if this is an isomorphism. Conversely, if the counit gives an isomorphism $G \cong \mathrm{Lan}_{\text{よ}}(G\text{よ})$ then

$$\begin{aligned} G(\mathrm{colim}_J P_j) &\cong \int^X \mathrm{colim}_J P_j X \times G(\hom(\_, X)) \\ &\cong \mathrm{colim}_J \int^X P_j X \times G(\hom(\_, X)) \\ &\cong \mathrm{colim}_J G(P_j) \end{aligned}$$

(this is in fact a consequence of a more general result: *every* functor that is of the form $\mathrm{Lan}_{\text{よ}} F$ is cocontinuous; prove this more general fact using the commutation of coends and colimits).

To complete the proof, we just observe that a cocontinuous functor

$L : [\mathcal{C}^{\text{op}}, \text{Set}] \to [\mathcal{E}^{\text{op}}, \text{Set}]$ between presheaf categories has a right adjoint given by the coend

$$\int^{C \in \mathcal{C}} [\mathcal{E}^{\text{op}}, \text{Set}](LC, X) \otimes C, \tag{3.4}$$

where $\otimes$ is a tensor in the sense of 2.2.3. □

3.1.2 Realisations of Simplicial Sets

We start this section by running two examples in parallel. This will make evident how they are particular instances of the same general construction. Consider the category $\mathbf{\Delta}$ of finite non-empty ordinals and monotone functions, and let us consider the Yoneda embedding $よ_{\mathbf{\Delta}} : \mathbf{\Delta} \to [\mathbf{\Delta}^{\text{op}}, \text{Set}]$. The presheaf category over $\mathbf{\Delta}$ is usually called the category sSet of *simplicial sets.* If Spc now denotes a nice category of topological spaces[1] we can define two functors $\rho : \mathbf{\Delta} \to \text{Spc}$ and $i : \mathbf{\Delta} \to \text{Cat}$ which 'represent' every object $[n] \in \mathbf{\Delta}$ either as a topological space or as a small category.

- The category $i[n]$ is obtained by regarding a totally ordered set $\{0 \leq 1 \leq \cdots \leq n\}$ as a category (every poset is a category, in the usual sense; a morphism $x \to y$ between the elements of a poset is the judgment that $x \leq y$).
- The topological space $\rho[n]$ is defined as the *geometric n-simplex* Δ^n embedded in $\mathbb{R}^{n+1}$ as the subset

$$\left\{(x_0, \ldots, x_n) \in \mathbb{R}^{n+1} \mid 0 \leq x_i \leq 1, \textstyle\sum_{i=0}^n x_i = 1\right\} \tag{3.5}$$

  endowed with the subspace topology. (Question for the reader: what is the affine function $\Delta^f : \Delta^m \to \Delta^n$ induced by a monotone function $f : [m] \to [n]$? Draw pictures of $\rho[n]$ for small values of n and define the *face* maps $\Delta^n \to \Delta^{n-1}$ and *degeneracy* maps $\Delta^n \to \Delta^{n+1}$.)

[1] Everyone having gotten their hands a bit dirty with algebraic topology will understand what we mean by 'nice' here; the rest of our readers shall take for granted that the whole category of all topological spaces is not nice, it is, in fact, full of pathological objects we do not want to consider. We instead restrict our argument to a smaller subcategory Spc ⊂ Top, like the category of *compactly generated* spaces or the category of spaces with the homotopy type of a CW-complex.

Looking for the left Kan extension of these two functors along the Yoneda embedding, we are in the situation depicted by the following diagrams:

$$\begin{array}{ccc} \mathbf{\Delta} & \xrightarrow{i} & \mathrm{Cat} \\ {\scriptstyle よ_{\mathbf{\Delta}}}\downarrow & \nearrow_{L} & \\ \mathrm{sSet} & & \end{array} \qquad \begin{array}{ccc} \mathbf{\Delta} & \xrightarrow{\rho} & \mathrm{Spc} \\ {\scriptstyle よ_{\mathbf{\Delta}}}\downarrow & \nearrow_{L'} & \\ \mathrm{sSet} & & \end{array} \tag{3.6}$$

We want to consider the left extensions L, L' of the two functors i, ρ along the Yoneda embedding $よ_{\mathbf{\Delta}} : \mathbf{\Delta} \to \mathrm{sSet}$; according to our 3.1.1 above these extensions are left adjoints.

We denote the adjunctions so determined as

$$\mathrm{Lan}_{よ}\, i \dashv N_i \quad \text{and} \quad \mathrm{Lan}_{よ}\, \rho \dashv N_\rho; \tag{3.7}$$

the two right adjoint functors are called the *nerves* associated to functors i and ρ respectively, and are defined as follows.

- The *categorical nerve* sends a category $\mathcal{C}$ to the simplicial set $N_i(\mathcal{C}) : [n] \mapsto \mathrm{Cat}(i[n], \mathcal{C})$; the set of n-simplices $N_i(\mathcal{C})_n$ coincides with the set of composable tuples of arrows

$$C_0 \xleftarrow{f_1} C_1 \xleftarrow{f_2} C_2 \leftarrow \cdots \leftarrow C_{n-1} \xleftarrow{f_n} C_n. \tag{3.8}$$

  In particular, $N(\mathcal{C})_0$ is the set of objects of $\mathcal{C}$, and $N(\mathcal{C})_1$ its set of morphisms. Thus the category $\mathcal{C}$ can be reconstructed from its nerve.
- The *geometric nerve* sends a topological space to the simplicial set $N_\rho(X) : [n] \mapsto \mathrm{Spc}(\rho[n], X) = \mathrm{Spc}(\Delta^n, X)$ (the *singular complex* of a space X).[2]

The left adjoints to N_ρ and N_i must be thought as 'realisations' of a simplicial set as an object of Spc or Cat.

- The left Kan extension $\mathrm{Lan}_{よ}\, \rho$ is called the *geometric* realisation $|X_\bullet|$ of a simplicial set $X_\bullet$, and it can be characterised as the coend

$$\int^{n \in \mathbf{\Delta}} \Delta^n \times X_n \tag{3.9}$$

  which in turn coincides with a suitable coequaliser in Spc in view of our Section 1.2 and 1.2.4 (and their duals). The product $\Delta^n \times X_n$ is indeed a product of topological spaces when X_n is thought as discrete:

[2] The name is motivated by the fact that if we consider the free abelian group on $N_\rho(X)_n$, the various $C_n = \mathbb{Z} \cdot N_\rho(X)_n = \coprod_{N_\rho(X)_n} \mathbb{Z}$ organise into a chain complex whose differentials are determined as alternating sums of face maps, and whose homology is precisely the singular homology of X.

it is the space $\coprod_{s\in X_n} \Delta^n \subseteq \mathbb{R}^{n+1}$ with the subspace topology on a disjoint union.

The shape of this object is determined in light of our description of the coend as a suitable coequaliser (in 1.2 and 1.2.4 and their duals), i.e. as a quotient space of $\coprod_n \Delta^n \times X_n$ and it agrees with the more classical description presented in almost all books on algebraic topology: the topological space $|X_\bullet|$ is obtained choosing an n-dimensional disk Δ^n for each n-simplex $x \in X_n$ and gluing these disks along the boundaries $\partial\Delta^n$ according to the degeneracy maps of $X_\bullet$.

The resulting space is, almost by definition, a CW-complex, because each standard n-simplex is homeomorphic to a closed disk: this means that $|X_\bullet|$ has the topology induced by a sequential colimit of pushouts of spaces $X_{(0)} \to X_{(1)} \to \cdots$ all obtained starting from a discrete space of 0-simplices $X_{(0)} \cong \Delta^0 \times X_0$ ($\Delta^0 \subset \mathbb{R}$ is a single point).

- The left Kan extension $\mathrm{Lan}_{\text{よ}}\, i$ is the *categorical* realisation $\tau_1(X_\bullet)$ of a simplicial set $X_\bullet$, defined as the coend

$$\int^{n\in\mathbf{\Delta}} i[n] \times X_n. \tag{3.10}$$

The Set-tensor $i[n] \times X_n$ is interpreted as a product in Cat, where the set X_n is thought as a discrete category; faces and degeneracies of X prescribe how the set $\coprod_{s\in X_n}\{0 \to 1 \to \cdots \to n\}$ glues together in the quotient $\left(\coprod_{n\in\mathbf{\Delta}} i[n] \times X_n\right)/\simeq$.

Note that $\tau_1(X_\bullet)$ *realises* $X_\bullet$ in the sense that the set of objects (respectively, of morphisms) of $\tau_1(X_\bullet)$ is precisely the set of 0-simplices (respectively, of 1-simplices) of $X_\bullet$. This is the category whose objects are 0-simplices of $X_\bullet$, arrows are 1-simplices, and where composition is defined by asking that $f, g \in X_1$ compose if there exists a 2-simplex σ having 0th face g and 2nd face f; identities are witnessed by degenerate simplices.

In modern terms (as in [Joy08a, JT07, Lur09]), when $X_\bullet$ is an ∞-category, we call $\tau_1(X_\bullet)$ the *homotopy category* of $X_\bullet$.

Remark 3.1.2. It should now sound reasonable to conjecture that these examples arise as particular instances of a general theorem. This is indeed the case: the general pattern unifying both these constructions, which we will call the 'nerve and realisation paradigm', was first suggested by D.M. Kan's works in algebraic topology, and in particular on the eponymous *Dold–Kan* correspondence. In this perspective we can recover the classical/singular nerve as a particular instance of the paradigm, and many more examples embodied from time to time in different settings; describing this pattern through (co)end calculus is the object of the following section.

3.2 Abstract Realisations and Nerves

The upshot of the present section is that whenever $F : \mathcal{C} \to \mathcal{D}$ has small domain and cocomplete codomain, the universal property of the Yoneda embedding in 3.1.1 determines a functor $\mathrm{Lan}_{\text{よ}} F$; this functor is always a left adjoint (compare this to the axioms of a Yoneda structure in 2.4.1, 2.4.2).

Algebraic topology, representation theory, geometry and logic constitute natural factories for examples of 'nerve and realisations'; more or less everywhere there is an interesting cocomplete category $\mathcal{D}$, there lies an interesting example of nerve-realisation adjunction, induced by a functor $F : \mathcal{C} \to \mathcal{D}$ with small domain.

We now want to lay down the foundations and the terminology allowing us to collect a series of readable and enlightening examples.

Definition 3.2.1 (Nerve and realisation contexts)**.** Any functor $F : \mathcal{C} \to \mathcal{D}$ from a small category $\mathcal{C}$ to a (locally small) *cocomplete* category $\mathcal{D}$ is called a *nerve-realisation context* (an NR *context* for short).

Given an NR context F, we can prove the following result.

Proposition 3.2.2 (Nerve-realisation paradigm)**.** The left Kan extension of F along the Yoneda embedding $\text{よ} : \mathcal{C} \to [\mathcal{C}^{\mathrm{op}}, \mathrm{Set}]$, i.e. the functor

$$L_F = \mathrm{Lan}_{\text{よ}} F : [\mathcal{C}^{\mathrm{op}}, \mathrm{Set}] \to \mathcal{D} \tag{3.11}$$

is a left adjoint, $L_F \dashv N_F$. L_F is called the *$\mathcal{D}$-realisation functor* or the *Yoneda extension* of F, and its right adjoint the *$\mathcal{D}$-coherent nerve*.

Proof The cocomplete category $\mathcal{D}$ is Set-tensored, and hence $\mathrm{Lan}_{\text{よ}} F$ can be written as the coend in (2.27); so the claim follows from the chain of isomorphisms

$$\begin{aligned} \mathcal{D}\big(\mathrm{Lan}_{\text{よ}} F(P), D\big) &\cong \mathcal{D}\Big(\int^{C} [\mathcal{C}^{\mathrm{op}}, \mathrm{Set}](\text{よ}_C, P) \otimes FC, D\Big) \\ &\cong \int_C \mathcal{D}([\mathcal{C}^{\mathrm{op}}, \mathrm{Set}](\text{よ}_C, P) \otimes FC, D) \\ &\cong \int_C \mathrm{Set}\big([\mathcal{C}^{\mathrm{op}}, \mathrm{Set}](\text{よ}_C, P), \mathcal{D}(FC, D)\big) \end{aligned}$$

$$\cong \int_C \text{Set}(PC, \mathcal{D}(FC, D)).$$

If we define $N_F(D)$ to be $C \mapsto \mathcal{D}(FC, D)$, this last set becomes canonically isomorphic to $[\mathcal{C}^{\text{op}}, \text{Set}](P, N_F(D))$. □

It is straightforward to recognise the choice of F leading to the nerves N_ρ and N_i. Also, in light of 3.1.1, the previous result can be rewritten as follows.

Remark 3.2.3. There is an equivalence of categories, induced by the universal property of the Yoneda embedding $よ_\mathcal{C} : \mathcal{C} \to [\mathcal{C}^{\text{op}}, \text{Set}]$,

$$\_ \circ よ_\mathcal{C} : \text{Cat}(\mathcal{C}, \mathcal{D}) \cong \text{Cat}_!(\text{Cat}(\mathcal{C}^{\text{op}}, \text{Set}), \mathcal{D}) \tag{3.12}$$

whenever $\mathcal{C}$ is small and $\mathcal{D}$ is locally small and cocomplete. The left hand side is the category of *cocontinuous* functors $[\mathcal{C}^{\text{op}}, \text{Set}] \to \mathcal{D}$.

Remark 3.2.4. It is well-known to algebraic topologists that the geometric realisation functor $L_\rho = |\_| : \text{sSet} \to \text{Spc}$ commutes with finite products. Coend calculus simplifies this result a lot reducing it to a direct check on representables; it is unfortunately not powerful enough to simplify the discussion any further.

In fact, we can only define a *bijection* between the sets $|\Delta[n] \times \Delta[m]| \cong \Delta^n \times \Delta^m$; after that, a certain amount of dirty work is necessary to show that this bijection is also a homeomorphism with respect to the natural topologies on the two sets. The formal proof that L_ρ commutes with finite products is left as an exercise at the end of the chapter.

3.2.1 Examples of Nerves and Realisations

A natural factory of NR contexts is *homotopical algebra*, as such nerve and realisation functors are often used to build Quillen equivalences between model categories. The existence of such Quillen equivalences is somewhat related to the fact that 'transfer theorems' (in the sense of [Cra95]) for model structures often apply to well-behaved nerve functors: if $\mathcal{D}$ is a model category and $F : \mathcal{C} \to \mathcal{D}$ is an NR functor, oftentimes there is a Quillen adjunction Set $\leftrightarrows \mathcal{D}$, and sometimes a Quillen *equivalence* induced by the right adjoint N_F.

Quillen adjunctions between model categories are certainly not the only examples of NR paradigms. The following list attempts to gather other important examples from different areas of algebra, geometry and logic: for the sake of completeness, we repeat the description of the two above-mentioned examples of the topological and categorical realisations.

Example 3.2.5 (Categorical nerve and realisation). If our NR context is $i : \mathbf{\Delta} \to \text{Cat}$, we obtain the *classical nerve* N_i of a (small) category $\mathcal{C}$, whose left adjoint is the *categorical realisation* (the *fundamental category* $\tau_1 X$ of X described in [Joy08a]). The NR adjunction

$$\tau_1 : \text{sSet} \leftrightarrows \text{Cat} : N_i \tag{3.13}$$

gives a Quillen adjunction between the Joyal model structure on sSet (see [Joy08a]) and the folk model structure on Cat. This adjunction yields a Quillen adjunction between the category Cat and the category of simplicial sets; the fibrant objects in sSet, ∞*-categories*, constitute a model for studying category theory in a homotopy coherent fashion.

Example 3.2.6 (Geometric nerve and realisation). If our NR context is the functor $\rho : \mathbf{\Delta} \to \text{Spc}$, the realisation of a representable $[n]$ is the standard topological simplex, and we obtain the adjunction between the *geometric realisation* functor $X_\bullet \mapsto |X_\bullet|$ of a simplicial set $X_\bullet$ and the *singular complex* of a topological space Y, i.e. the simplicial set $Y_\bullet$ having as set of n-simplices the continuous functions $\Delta^n \to Y$.

If we apply the free abelian group functor $\mathbb{Z}[\_] : \text{Set} \to \text{Ab}$ objectwise to this simplicial set we obtain the simplicial abelian group $\mathbb{Z}Y_\bullet$, which under the Dold–Kan correspondence 3.2.10 gives rise to a (positive degree) chain complex, the *singular complex* of Y. The homology of this chain complex coincides with the singular homology of Y.

Example 3.2.7 (sSet-coherent nerve and realisation). Let $F : \mathbf{\Delta} \to \text{Cat}_{\mathbf{\Delta}}$ be the functor that realises every simplex $[n]$ as a simplicial category having objects the same set $[n] = \{0, 1, \dots, n\}$ and as $\hom(i, j)$ the simplicial set obtained as the nerve of the poset $P(i, j)$ of subsets of the interval $[i, j]$ which contain both i and j.[3]

This sets up an NR context, and if we consider $\text{Lan}_{\text{よ}} F$ we obtain the *(Cordier) simplicially coherent nerve and realisation*, defined as follows.

- The left adjoint sends a simplicial set into the simplicial category

$$\int^n F[n] \times X_n \tag{3.14}$$

obtained in a similar fashion to $\tau_1(X_\bullet)$ by taking simplicial categories

[3] In particular if $i > j$ then $P(i, j)$ is empty and hence its nerve is the constant simplicial set on $\varnothing$.

$F[n]$ as shapes and the simplices $s \in X_n$ as gluing instructions. Of course, colimits of (enriched) categories tend to be wildly complicated, but it is an instructive exercise to try to understand how the functor $X_\bullet \mapsto \|X_\bullet\|$ behaves on simple examples. We direct the reader to [DS11, Rie].

- The right adjoint $N_{\mathbf{\Delta}} : \text{Cat}_{\mathbf{\Delta}} \to \text{sSet}$ sends a simplicial category $\mathcal{C}$ into a simplicial set constructed remembering that $\mathcal{C}$ carries a simplicial structure. Intuitively, simplicial functors $F[n] \to \mathcal{C}$ carry more information than plain set-enriched functors $[n] \to \mathcal{C}$.

This establishes another Quillen adjunction $\text{sSet} \leftrightarrows \text{Cat}_{\mathbf{\Delta}}$ which restricts to an equivalence between quasicategories (fibrant objects in the Joyal model structure on sSet [Joy08b]) and fibrant simplicial categories (with respect to the Bergner model structure on $\text{Cat}_{\mathbf{\Delta}}$ [Ber07]).

Example 3.2.8 (Moerdijk generalised intervals). The construction giving the topological realisation of $\Delta[n]$ extends to the case of any 'interval' in the sense of [Moe95, §III.1], i.e. any ordered topological space J having 'endpoints' $0, 1$. Indeed every such space J defines a generalised topological n-simplex $\Delta^n(J)$ as follows:

$$\Delta^n(J) := \{(x_1, \dots, x_n) \mid x_i \in J, x_0 \le \cdots \le x_n\} \subseteq J^{n+1} \tag{3.15}$$

endowed with the subspace and product topology. These data assemble into an NR context $\Delta^\bullet(J) : \mathbf{\Delta} \to \text{Spc}$ that gives rise to an adjunction $\text{Lan}_{\text{よ}} \Delta^\bullet(J) \dashv N_{\Delta^\bullet(J)}$. Instead of going deep into the technicalities, we direct the reader to [Moe95, §III.1] for more information.

Example 3.2.9 (Toposophic nerve and realisation). The correspondence $D : [n] \mapsto \text{Sh}(\Delta^n)$ defines a *cosimplicial topos*, i.e. a functor from $\mathbf{\Delta}$ to the category of (spatial) toposes, which serves as an NR context. Some geometric properties of this nerve/realisation are studied in [Moe95, §III]. We outline an instance of a problem where this adjunction naturally arises: let $\mathcal{X} = \text{Sh}(X), \mathcal{Y} = \text{Sh}(Y)$ be the categories of sheaves over topological spaces X, Y.

Let $\mathcal{X} \star \mathcal{Y}$ be the *join* (see A.3.13) of the two toposes seen as categories: this blatantly fails to be a topos, but there is a canonical 'replacement' procedure

$$\begin{array}{ccccccc} \text{Cat} \times \text{Cat} & \xrightarrow{\star} & \text{Cat} & \xrightarrow{N_i} & \text{sSet} & \xrightarrow{\text{Lan}_{\text{よ}} D} & \text{Topos} \\ (\mathcal{X}, \mathcal{Y}) & \longmapsto & \mathcal{X} \star \mathcal{Y} & \longmapsto & \text{Cat}(\Delta^\bullet, \mathcal{X} \star \mathcal{Y}) & \longmapsto & \mathcal{X} \odot \mathcal{Y} \end{array} \tag{3.16}$$

that builds a topos out of $\mathcal{X}$ and $\mathcal{Y}$; various questions about this construction are left as Exercise 3.7

Example 3.2.10 (The Dold–Kan correspondence). The Dold–Kan correspondence [Dol58] asserts that there is an equivalence of categories between simplicial abelian groups $[\boldsymbol{\Delta}^{\mathrm{op}}, \mathrm{Ab}]$ and chain complexes $\mathrm{Ch}^+(\mathrm{Ab})$ concentrated in positive degree, and it can be seen as an instance of the NR paradigm.

In this case, the functor $dk : \boldsymbol{\Delta} \to \mathrm{Ch}^+(\mathrm{Ab})$ sending $[n]$ to $\mathbb{Z}^{\Delta[n]}$ (the free abelian group on $\Delta[n]$) and then to the *Moore complex* $M(\mathbb{Z}^{\Delta[n]})$ determined by any simplicial group $A \in [\boldsymbol{\Delta}^{\mathrm{op}}, \mathrm{Ab}]$ as in [GJ09] is the NR context. The resulting adjunction

$$\mathrm{Lan}_{\text{よ}}(dk) = DK : [\boldsymbol{\Delta}^{\mathrm{op}}, \mathrm{Ab}] \leftrightarrows \mathrm{Ch}^+(\mathrm{Ab}) : N_{dk} \tag{3.17}$$

sets up an equivalence of categories.

Example 3.2.11 (Étale spaces as Kan extensions). The present example reworks [Bre97, 1.5]. Let X be a topological space, and $o(X)$ its poset of open subsets; let Spc/X be the slice category of morphisms with codomain X and commutative triangles as morphisms.

There exists a tautological functor

$$\iota : o(X) \to \mathrm{Spc}/X \tag{3.18}$$

sending $U \subseteq X$ to itself, regarded as an object $\left[\begin{smallmatrix} U \\ \downarrow \\ X \end{smallmatrix}\right]$; this works as an NR context, yielding a pair of adjoint functors

$$\mathrm{Lan}_{\text{よ}}\, \iota \dashv N_\iota \tag{3.19}$$

where N_A is defined as taking the (pre)sheaf of sections of $p \in \mathrm{Spc}/X$. The resulting left adjoint is precisely the functor sending a presheaf $F : o(X) \to \mathrm{Set}$ to the disjoint union of stalks $\tilde{F} = \coprod_{x \in X} F_x$, endowed with the final topology that makes continuous all maps of the form $\tilde{s} : U \to \tilde{F}$ sending $x \in X$ to the equivalence class $[s]_x \in F_x$.

In order to see this, let us unpack the definition of the left Kan extension in study: we have to compute the coequaliser

$$\coprod_{V \subseteq U} FU \otimes \left[\begin{smallmatrix} V \\ \downarrow \\ X \end{smallmatrix}\right] \underset{r_{UV}}{\overset{l_{UV}}{\rightrightarrows}} \coprod_{U \in o(X)} FU \otimes \left[\begin{smallmatrix} U \\ \downarrow \\ X \end{smallmatrix}\right] \longrightarrow \mathrm{Lan}_y\, \iota(F) \tag{3.20}$$

where the parallel maps are defined on components by $l_{VU}(s, V) = (s|_V, V)$ and $r_{VU}(s, V) = (s, U)$, and $\otimes$ is the canonical Set-tensor of

Spc/X. Such a coequaliser imposes on $\coprod_{U\in o(X)} FU \otimes \left[\begin{smallmatrix} U \\ \downarrow \\ X\end{smallmatrix}\right]$ a relation identifying (s, U) with all its restrictions $(s|_V, V)$ to smaller open sets.

This means that in the equivalence relation generated by these pairs, a section $(s \in FU)$ is identified with a section $(t \in FV)$ if they coincide on at least an open $W \subseteq U \cap V$. This is very near to the universal property defining the set of all germs $\coprod_{x\in X} F_x$, and in fact it is exactly what is needed to define a natural map $q_U : FU \otimes \left[\begin{smallmatrix} U \\ \downarrow \\ X\end{smallmatrix}\right] \to \coprod_{x\in X} F_x$ for each $U \in o(X)$ (the cocone condition for these q_U entails that they are 'defined on germs of sections', in a precise sense that can be easily spelled out).

In fact, $q_U : FU \otimes \left[\begin{smallmatrix} U \\ \downarrow \\ X\end{smallmatrix}\right] \to \coprod_{x\in X} F_x$ corresponds, by the universal property of the tensor functor on Spc/X, to a natural family of functions

$$\bar{q}_U : FU \to \mathrm{Spc}/X(\iota[U], \textstyle\coprod_x F_x). \tag{3.21}$$

This is to say, every abstract section $s \in FU$ gives rise to a 'true section' $\dot{s} : U \to \coprod_x F_x$.

A routine argument now shows that the family $(q_U \mid U \in o(X))$ is also initial: every other cocone for the same diagram, say $\zeta_U : FU \otimes \left[\begin{smallmatrix} U \\ \downarrow \\ X\end{smallmatrix}\right] \to Z$ for some space Z, must determine a unique map $\bar{\zeta} : \coprod_x F_x \to Z$, whose components are $\bar{\zeta}_x : F_x \to Z$, sending $[s]_x \in F_x$ to $\zeta_u(s)$ for some (the cocone condition makes this choice well defined) $(s, U) \in FU \times \mathcal{U}[x]$ ($\mathcal{U}[x]$ is the filter of neighbourhoods of $x \in X$) having germ $[s]_x$ at x.

This adjunction restricts to an equivalence of categories between the subcategory Sh(X) of sheaves on X and the subcategory Ét(X) of *étale spaces* over X, giving a formal method for proving [MLM92, II.6.2].

Example 3.2.12 (The tensor product of modules as a coend)**.** Any ring R can be regarded as an Ab-enriched category (see A.7.1) having a single object, whose set of endomorphisms is the ring R itself. Once we have noticed this, we obtain natural identifications for the categories of modules over R as covariant and contravariant enriched presheaves on R:

$$\begin{aligned} \mathrm{Mod}_R &\cong \mathrm{Cat}(R^{\mathrm{op}}, \mathrm{Ab}) \\ {}_R\mathrm{Mod} &\cong \mathrm{Cat}(R, \mathrm{Ab}). \end{aligned} \tag{3.22}$$

Given $A \in \mathrm{Mod}_R$, $B \in {}_R\mathrm{Mod}$, we can define a functor $T_{AB} : R^{\mathrm{op}} \times R \to$ Ab which sends the unique object to the tensor product of abelian groups $A \otimes_{\mathbb{Z}} B$. The coend of this functor can be computed as the coequaliser

$$\mathrm{coker}\Big(\textstyle\coprod_{r\in R} A \otimes_{\mathbb{Z}} B \underset{1\otimes r}{\overset{r\otimes 1}{\rightrightarrows}} A \otimes_{\mathbb{Z}} B \Big), \tag{3.23}$$

which quotients the object $A \otimes_{\mathbb{Z}} B$ by the submodule generated by the sums $ra \otimes b - a \otimes rb$.

In other words, there is a canonical isomorphism $\int^{*\in R} T_{AB} \cong A \otimes_R B$.

Remark 3.2.13. The previous construction is in fact part of a richer structure: we can define a bicategory Mod (see A.7.11) having

- 0-cells the rings $R, S, \dots$,
- 1-cells $R \to S$ the modules ${}_R M_S$, regarded as functors $M : R \times S^{\text{op}} \to$ Ab,
- 2-cells $f : {}_R M_S \Rightarrow {}_R N_S$ the module homomorphisms $f : M \to N$.

This bicategory Mod has a fairly rich structure induced by the structure of ${}_R\text{Mod}_S$: for example, bifunctoriality of the tensor product $\otimes$ amounts to the interchange law in the bicategory Mod.

Chapter 5 on profunctors will extensively generalise this point of view, extending it to the case of multi-object categories, enriched over a generic base.

Remark 3.2.14. The previous point of view on tensor products can be generalised further (see [ML98, §IX.6], but more on this has been written in [Yon60, §4]). Given functors $F, G : \mathcal{C}^{\text{op}}, \mathcal{C} \to \mathcal{V}$ having values in a cocomplete (see A.3.11) monoidal category, we can define the *tensor product* of F, G as the coend

$$F \boxtimes_{\mathcal{C}} G := \int^{C} FC \otimes_{\mathcal{V}} GC. \tag{3.24}$$

Chapter 4 generalises this point of view in some ways, regarding this example as an instance of a *weighted colimit* of F with weight G, in the case where $\mathcal{C}$ is a $\mathcal{V}$-enriched category.

To appreciate the next example we need to recall the following.

Proposition 3.2.15. The following properties for a functor $F : \mathcal{C} \to \text{Set}$ are equivalent:

FL1. F commutes with finite limits;
FL2. $\text{Lan}_{\text{よ}} F$ commutes with finite limits;
FL3. the *category of elements* $\mathcal{C} \int F$ of F (see Definition A.5.9 and Proposition 4.1.11) is cofiltered.

Proof We provide a proof using as much coend calculus as we can: first of all, from A.5.7 we know that every presheaf is the colimit of the Yoneda embedding over its own category of elements, thus if F commutes with finite limits, then its category of elements $\mathcal{C} \int F$ has the property that

$$\mathrm{colim}_{\mathcal{C}\int F} \lim_J よ(A)(D_i) \cong \lim_J \mathrm{colim}_{\mathcal{C}\int F} よ(A)(D_i) \tag{3.25}$$

for every object $X \in \mathcal{C}$ and diagram $D : J \to \mathcal{C}$ with finite domain. This is true only if $\mathcal{C} \int F$ is filtered.

Condition FL2 now implies condition FL1 because $\mathrm{Lan}_よ F \circ よ \cong F$ and the left hand side is a composition of finitely continuous functors.

Last, condition FL3 implies condition FL2, because

$$\begin{aligned}
\mathrm{Lan}_よ F(P) &\cong \int^C PC \otimes FC \\
&\cong \int^C PC \otimes \big(\mathrm{colim}_J よ(D_j)C\big) \\
&\cong \int^C \mathrm{colim}_J \big(PC \otimes よ(D_j)C\big) \\
&\cong \mathrm{colim}_J \int^C \big(PC \otimes よ(D_j)C\big) \\
&\cong \mathrm{colim}_J P(D_j).
\end{aligned}$$

Now, assume P results as a finite limit $\lim_{A\in\mathcal{A}} P_A$; the category J is filtered, thus we can complete the step as

$$\begin{aligned}
\mathrm{Lan}_よ F(\lim_{\mathcal{A}} P_A) &\cong \mathrm{colim}_J \lim_{\mathcal{A}} P_A(D_j) \\
&\cong \lim_{\mathcal{A}} \mathrm{colim}_J P_A(D_j) \\
&\cong \lim_{\mathcal{A}} \mathrm{Lan}_よ F(P_A).
\end{aligned}$$

This conclude the proof. □

Example 3.2.16 (Giraud theorem using coends). *Giraud's theorem* asserts that every Grothendieck topos is equivalent to a left exact localisations of a presheaf category $[\mathcal{C}^{\mathrm{op}}, \mathrm{Set}]$ (a Grothendieck topos is a category of *sheaves* $\mathrm{Sh}(\mathcal{C}, \mathrm{j})$ with respect to a *Grothendieck topology* j).

Such a classical 'representation' theorem is deeply intertwined with the theory of locally presentable categories (see [AR94, CV06]), described at the end of [MLM92].

We now try to outline an argument that employs coend calculus and Yoneda extensions; we will realise a Grothendieck topos $\mathcal{E}$ as a full subcategory of $[\mathcal{C}^{\mathrm{op}}, \mathrm{Set}]$ for $\mathcal{C} = \mathcal{E}_{<\omega} \subset \mathcal{E}$ the subcategory of *compact* (or *finitely presentable*) objects of $\mathcal{E}$.[4] A finite limit of finitely presentable

4 Recall that an object $X \in \mathcal{E}$ is *compact* or *finitely presentable* if the functor

objects is again finitely presentable, and thus the inclusion $\iota : \mathcal{E}_{<\omega} \subseteq \mathcal{E}$ preserves finite limits.

The full embedding $\iota : \mathcal{C} \subset \mathcal{E}$ works as an NR context as in 3.2.1, and moreover it is a dense functor; this enables us to use coend calculus to prove the following.

NR1. The ι-nerve N_ι is full and faithful; it will turn out to be the inclusion of sheaves into presheaves $[\mathcal{C}^{\text{op}}, \text{Set}]$.

NR2. $\text{Lan}_{\text{よ}}\, \iota$ is a (left exact, thanks to 3.2.15 and to the fact that ι is left exact) reflection of $[\mathcal{C}^{\text{op}}, \text{Set}]$ onto $\mathcal{E}$.

Hence, the NR adjunction has the following form:

$$\text{Lan}_{\text{よ}}(\iota) : [\mathcal{C}^{\text{op}}, \text{Set}] \rightleftarrows \mathcal{E} : N_\iota. \tag{3.26}$$

As said, since ι is dense, the nerve N_ι is fully faithful (see Exercise 3.5): it only remains to prove that the functor $\text{Lan}_{\text{よ}}(\iota)$ behaves like sheafification. In view of the shape of unit and counit of $\text{Lan}_{\text{よ}}(\iota) \overset{\epsilon}{\underset{\eta}{\dashv}} N_\iota$ (see Exercise 3.3), we have to manipulate the following chain of (iso)morphisms:

$$\begin{aligned}
\text{Lan}_{\text{よ}}(\iota)(P)(C) &\cong \mathcal{E}(\iota C, \text{Lan}_{\text{よ}}\, \iota(P)) \\
&\cong \mathcal{E}\Big(\iota C, \int^A PA \times \iota A\Big) \\
&\leftarrow \int^A \mathcal{E}(\iota C, PA \times \iota A) \\
&\cong \int^A PA \times \mathcal{E}(\iota C, \iota A) \\
&\cong \int^A PA \times \mathcal{C}(C, A) \\
&\cong PC.
\end{aligned}$$

This gives an arrow $PC \to LPC$ which is easily seen to be the component of a natural transformation, and in fact of a reflection (to show that η_P has the correct universal property, it suffices to show that every morphism $P \Rightarrow F$ where F is an object in $\mathcal{E}$ uniquely extends to $LP \Rightarrow F$).

It remains to prove that this functor is left exact. To do this we invoke 3.2.15, since $\mathcal{E}_{<\omega}$ is closed under finite limits. It also remains to characterise *sheaves* as those P such that η_P is invertible, but this is an

hom$(X, _)$ commutes with filtered colimits. This essentially means that X can be presented with finitely many generators and relations in the 'theory' of $\mathcal{E}$. Also, an object is finitely *generated* if it commutes with filtered colimits of *monomorphisms*.

equivalent characterisation of orthogonal classes, addressed in Exercise A.11.

Example 3.2.17 (Simplicial subdivision functor). Once again, let $\mathbf{\Delta}$ be the category of non-empty finite ordinals. The *Kan Ex$^\infty$ functor* is an endofunctor of sSet turning every simplicial set $X_\bullet$ into a *Kan complex*.[5] This construction is of fundamental importance in simplicial homotopy theory, and we now want to revisit the classical argument given by Kan in the modern terms of an NR paradigm on $\mathbf{\Delta}$, following [GJ09].

First of all, we note from [GJ09] that the non-degenerate m-simplices of $\Delta[n]$ are in bijective correspondence with the subsets of $\{0, \dots, n\}$ of cardinality $m+1$; this entails that the set of non-degenerate simplices of $\Delta[n]$ becomes a poset $\mathrm{s}[n]$ ordered by inclusion.

We can then consider the nerve $N_\rho(\mathrm{s}[n]) \in \mathrm{sSet}$ (see Example 3.2.6). This organises into a functor $\mathrm{sd} : \mathbf{\Delta} \to \mathrm{sSet}$, which works as an NR paradigm: using 3.2.2 we obtain the adjunction

$$\mathrm{sSet} \underset{\mathsf{Sd}}{\overset{\mathrm{Ex}}{\underset{\longleftarrow}{\overset{\longrightarrow}{\top}}}} \mathrm{sSet} \tag{3.27}$$

where Ex is the nerve N_{sd} associated to the NR paradigm sd (so a right adjoint to $\mathsf{Sd} = \mathrm{Lan}_{\text{よ}}\,\mathrm{sd}$). The set of m-simplices $\mathrm{Ex}(X)_n$ is $\mathrm{sSet}(\mathrm{sd}(\Delta[n]), X)$.

There is a canonical map $\mathrm{sd}(\Delta[n]) \to \Delta[n]$ which by precomposition and by the Yoneda lemma, induces a map $X_n = \mathrm{sSet}(\Delta[n], X) \xrightarrow{j^*} \mathrm{sSet}(\mathrm{sd}(\Delta[n]), X) = \mathrm{Ex}(X)_n$, natural in $X \in \mathrm{sSet}$. This gives $\mathrm{Ex}(\_)$ the structure of a pointed functor, and in fact a *well-pointed* functor in the sense of [Kel80]; this, finally, means that we can define

$$\mathrm{Ex}^\infty(X) \cong \mathrm{colim}\left(X \xrightarrow{\eta} \mathrm{Ex}(X) \xrightarrow{\eta * \mathrm{Ex}} \mathrm{Ex}^2(X) \xrightarrow{\eta * \mathrm{Ex}^2} \cdots\right) \tag{3.28}$$

as an endofunctor on sSet. The functor Ex^∞ enjoys a great deal of formal properties useful in the study of simplicial homotopy theory (the most important of which is that $\mathrm{Ex}^\infty(X)$ is a Kan complex for each $X \in \mathrm{sSet}$, see [GJ09]). A more intrinsic characterisation of this construction is contained in [EP08], and defines not only $\mathrm{Sd} = \mathrm{Lan}_{\text{よ}}\,\mathrm{sd}$ as a left Kan

[5] A Kan complex is a simplicial set Y such that the functor $\hom(\_, X)$ turns each *horn inclusion* $\Lambda^k[n] \to \Delta[n]$ (see [GJ09]) into an epimorphism.

extension, but also sd. The authors consider the diagram of 2-cells

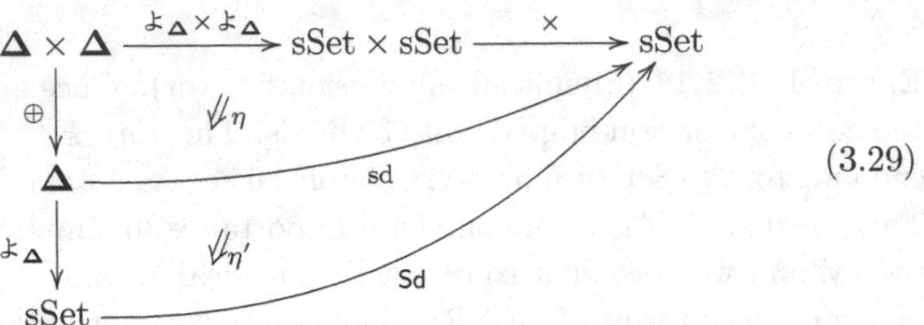

(3.29)

where $\oplus : \mathbf{\Delta}\times\mathbf{\Delta} \to \mathbf{\Delta}$ is the *ordinal sum* defined by $[m]\oplus[n] = [m+n+1]$. Exercise 3.9 expands on this. In the notation of our Chapter 6, the functor sd is the *convolution* $よ_{\mathbf{\Delta}} * よ_{\mathbf{\Delta}}$.

Example 3.2.18 (Isbell duality). Let $\mathcal{V}$ be a (co)complete symmetric monoidal category $\mathcal{V}$ (this will be called a *Bénabou cosmos* in subsequent sections), and $\mathcal{C} \in \mathcal{V}\text{-Cat}$ a $\mathcal{V}$-enriched category (see A.7.1); then, we have an adjunction

$$\mathcal{V}\text{-Cat}(\mathcal{C},\mathcal{V})^{\mathrm{op}} \underset{\mathsf{Spec}}{\overset{\mathsf{O}}{\underset{\longleftarrow}{\overset{\longrightarrow}{\bot}}}} \mathcal{V}\text{-Cat}(\mathcal{C}^{\mathrm{op}},\mathcal{V}). \tag{3.30}$$

This means that we find a bijection of hom-sets

$$\begin{aligned}\mathcal{V}\text{-Cat}(\mathcal{C},\mathcal{V})^{\mathrm{op}}\big(\mathsf{O}(X),Y\big) &= \mathcal{V}\text{-Cat}(\mathcal{C},\mathcal{V})\big(Y,\mathsf{O}(X)\big)\\ &\cong \mathcal{V}\text{-Cat}(\mathcal{C}^{\mathrm{op}},\mathcal{V})(X,\mathsf{Spec}(Y))\end{aligned} \tag{3.31}$$

induced by the functors

$$\mathsf{O} : X \mapsto \Big(C \mapsto \mathcal{V}\text{-Cat}(\mathcal{C},\mathcal{V})\big(X, よ_{\mathcal{C}^{\mathrm{op}}}(C)\big)\Big),$$

$$\mathsf{Spec} : Y \mapsto \Big(C \mapsto \mathcal{V}\text{-Cat}(\mathcal{C},\mathcal{V})^{\mathrm{op}}\big(よ_{\mathcal{C}^{\mathrm{op}}}(C), Y\big)\Big).$$

The adjunction property is a simple derivation in coend calculus:

$$\begin{aligned}\mathcal{V}\text{-Cat}(\mathcal{C},\mathcal{V})\big(Y,\mathsf{O}(X)\big) &= \int_D \mathcal{V}\Big(YD, \int_A \mathcal{V}(XA,\mathcal{C}(A,D))\Big)\\ &\cong \int_{DA} \mathcal{V}(YD,\mathcal{V}(XA,\mathcal{C}(A,D)))\\ &\cong \int_A \mathcal{V}\Big(XA, \int_D \mathcal{V}(YD,\mathcal{C}(A,D))\Big)\\ &= \mathcal{V}\text{-Cat}(\mathcal{C}^{\mathrm{op}},\mathcal{V})\big(X,\mathsf{Spec}(Y)\big).\end{aligned}$$

Exercises

3.1 Use coend calculus to show that the geometric realisation L_ρ commutes with finite products of simplicial sets, assuming that it does commute with finite products *of representables.*

3.2 Compute the J-realisation (see Example 3.2.8) of $X \in \text{sSet}$ in the case where J is the Sierpiński space $\{0 < 1\}$ with topology $\{\varnothing, J, \{1\}\}$. (Hint: the colimit $\int^n \Delta^n(J) \times X_n$ stops after two steps.)

3.3 Let $F : \mathcal{C} \to \mathcal{D}$ be an NR context. Write the unit and counit of the nerve and realisation adjunction $\text{Lan}_{\text{よ}} F \dashv N_F$ explicitly.

3.4 ☉☉ Without using coend calculus, show that the nerve functor N_F is canonically isomorphic to $\text{Lan}_F \text{よ}$, so that there is an adjunction

$$\text{Lan}_{\text{よ}} F \dashv \text{Lan}_F \text{よ}.$$

3.5 Show that the following two conditions are equivalent, for a nerve-realisation context $F : \mathcal{C} \to \mathcal{D}$.

- The nerve $N_F : \mathcal{D} \to \text{Cat}(\mathcal{C}^{\text{op}}, \text{Set})$ is fully faithful.
- The functor F is *dense*, i.e. the density comonad $\text{Lan}_F F$ is the identity functor of $\mathcal{D}$.

3.6 ☉☉ A *tensor-hom cotensor* situation (*THC situation* for short) consists of a triple $(\otimes, \wedge, [\_, =])$ of functors between three categories $\mathcal{S}, \mathcal{A}, \mathcal{B}$, whose covariance type is defined by the adjunctions

$$\mathcal{B}(S \otimes A, B) \cong \mathcal{S}(S, [A, B]) \cong \mathcal{A}(A, S \wedge B).$$

More precisely, if $\otimes : \mathcal{S} \times \mathcal{A} \to \mathcal{B}$, then $\wedge : \mathcal{S}^{\text{op}} \times \mathcal{B} \to \mathcal{A}$, and $[\_, =] : \mathcal{A}^{\text{op}} \times \mathcal{B} \to \mathcal{S}$. The aim of this exercise is to show that given a THC situation $(\otimes, \wedge, [\_, =])$, we can induce a new one $(\boxtimes, \curlywedge, \langle \_, = \rangle)$, on the categories $\mathcal{S}^{I^{\text{op}} \times J}, \mathcal{A}^I, \mathcal{B}^J$, for any $I, J \in \text{Cat}$.

- Define $F \boxtimes G \in \mathcal{B}^J$ starting from $F \in \mathcal{S}^{I^{\text{op}} \times J}$ and $G \in \mathcal{A}^I$ as the coend

$$\int^i F(i, \_) \otimes Gi$$

  and show that there is an adjunction

$$\mathcal{B}^J(F \boxtimes G, H) \cong \mathcal{S}^{I^{\text{op}} \times J}(F, \langle G, H \rangle) \cong \mathcal{A}^I(G, F \curlywedge H)$$

  for suitable functors $\langle \_, = \rangle$ and $\_ \curlywedge =$, developing $\mathcal{B}^J(F \boxtimes G, H) = \dots$ in two ways, with coend calculus.

- Assuming that the relevant structure exists, is it true that composition of the 2-category Cat is the $\otimes$ functor of a THC situation $\otimes : \mathsf{Cat}(\mathcal{B},\mathcal{C}) \times \mathsf{Cat}(\mathcal{A},\mathcal{B}) \to \mathsf{Cat}(\mathcal{A},\mathcal{C})$? What are the parametric adjoints of $\_ \circ F$ and $G \circ \_$?

3.7 ◉◉ Example 3.2.9 can be expanded and studied more deeply.

- Is $\odot$ a monoidal structure on the category Topos of toposes and geometric morphisms?
- Under which conditions on X, Y is $\mathcal{X} \odot \mathcal{Y}$ equivalent to a topos of sheaves on a topological space $X \odot Y$?
- What are the properties of the bifunctor $(X, Y) \mapsto X \odot Y$? Does this operation resemble or extend the topological join?

3.8 ◉◉ In this exercise Spc is a nice category for algebraic topology. Define the category Γ having objects the power-sets of finite sets, and morphisms the functions $f : 2^n \to 2^m$ preserving unions and set-theoretical differences.

(a) Show that there is a functor $\mathbf{\Delta} \hookrightarrow \Gamma$, sending the chain $\{0 < 1 < \cdots < n\}$ in $\mathbf{\Delta}$ to $\{\varnothing \subset \{0\} \subset \cdots \subset \{0, \dots, n\}\}$ in Γ.

(b) The category of presheaves of spaces $A : \Gamma^{\text{op}} \to \mathsf{Spc}$ is called the category of Γ-*spaces*; a Γ-space is said to satisfy the *Segal condition* (or to be Segal) if it turns pushouts in Γ into homotopy pullbacks in Spc. Describe pushouts in Γ. Show that a Γ-space is Segal if and only if the following two conditions are satisfied: (i) $A(0)$ is contractible; (ii) the canonical map $A(n) \to \prod_{i=1}^{n} A(1)$ is a homotopy equivalence in Spc.

(c) Let $X \in \mathsf{Spc}$ and $A : \Gamma^{\text{op}} \to \mathsf{Spc}$; define $X \otimes A$ to be the coend (in Spc)

$$\int^{n \in \Gamma} X^n \times A(n).$$

Show that this defines a bifunctor $\Gamma^{\text{op}} \times \Gamma \to \mathsf{Spc}$. Find a canonical map connecting the tensor product of S^1 and A with the geometric realisation of the simplicial space $\mathbf{\Delta}^{\text{op}} \to \Gamma^{\text{op}} \xrightarrow{A} \mathsf{Spc}$.

(d) Let $\mathcal{C} : \Gamma^{\text{op}} \to \mathsf{Spc}\text{-}\mathsf{Cat}$; let $X \otimes_\Gamma \mathcal{C}$ be the coend (in the category of topological categories)

$$\int^{n \in \Gamma} X^n \times \mathcal{C}(n).$$

Show that $X \otimes_\Gamma (\_) : \mathsf{Spc}\text{-}\mathsf{Cat} \to \mathsf{Cat}$ commutes with finite products, i.e. if $\mathcal{C}, \mathcal{D}$ are topological categories, then

$$X \otimes_\Gamma (\mathcal{C} \times \mathcal{D}) \cong (X \otimes_\Gamma \mathcal{C}) \times (X \otimes_\Gamma \mathcal{D}).$$

3.9 Write suitable coends for the Kan extensions that define sd and Sd in (3.29), and for their right adjoints.

3.10 Generalise the NR paradigm to the setting of *separately cocontinuous* (also called *multi-linear*) functors. Given $F : \mathcal{C}_1 \times \cdots \times \mathcal{C}_n \to \mathcal{D}$, where each $\mathcal{C}_i$ is small and $\mathcal{D}$ is cocomplete, show that there exists an equivalence of categories

$$\mathrm{Cat}(\mathcal{C}_1 \times \cdots \times \mathcal{C}_n, \mathcal{D}) \cong \mathrm{Mult}([\mathcal{C}_1^{\mathrm{op}}, \mathrm{Set}] \times \cdots \times [\mathcal{C}_n^{\mathrm{op}}, \mathrm{Set}], \mathcal{D})$$

where $\mathrm{Mult}(\_, =)$ is the category of all functors that are cocontinuous in each variable once all the others have been fixed (hint: show it 'by induction' composing multiple Kan extensions). Given $\theta \in \mathrm{Cat}(\mathcal{C}_1 \times \cdots \times \mathcal{C}_n, \mathcal{D})$, describe the right adjoint of each $\theta^{(i)} : [\mathcal{C}_i^{\mathrm{op}}, \mathrm{Set}] \to \mathcal{D}$ (it fixes all components on objects $C_j \in \mathcal{C}_j$ for $j \neq i$, and the ith component runs free). All these functors assemble to a 'vector-nerve' $N : \mathcal{D} \to [\mathcal{C}_1^{\mathrm{op}}, \mathrm{Set}] \times \cdots \times [\mathcal{C}_n^{\mathrm{op}}, \mathrm{Set}]$.

3.11 Let $よ_\mathcal{C} : \mathcal{C} \to [\mathcal{C}^{\mathrm{op}}, \mathrm{Set}]$ be the Yoneda embedding, and $\check{よ}_\mathcal{C} : \mathcal{C}^{\mathrm{op}} \to \mathrm{Cat}(\mathcal{C}, \mathrm{Set})$ its contravariant counterpart sending C into the functor $\mathcal{C}(C, \_)$. Show that in Example 3.2.18 we can characterise the functor O from Isbell duality (see 3.2.18) as $\mathrm{Lan}_{よ_\mathcal{C}}(\check{よ}_\mathcal{C})$.

3.12 Regard a combinatorial species $f : \mathcal{B}(\mathbb{N}) \to \mathrm{Set}$ as an NR context. Show that there is an isomorphism

$$\mathrm{Lan}_よ f(N_f(X)) \cong \mathrm{Lan}_j f(X)$$

in the notation of 2.3.11, if $j : \mathcal{B}(\mathbb{N}) \to \mathrm{Set}$ is the natural functor. What can you say about the composition

$$N_f \circ \mathrm{Lan}_j f : \mathrm{Set} \to \mathrm{Set} \to [\mathcal{B}(\mathbb{N})^{\mathrm{op}}, \mathrm{Set}]$$

for the same species f?

3.13 ◉◉ Make the world a better place by providing a deeper account of Isbell duality.

- An object of a category $\mathcal{A}$ is called *Isbell autodual* if it is a fixed point of the comonad $\mathsf{O} \circ \mathsf{Spec}$; is there a general way to characterise all Isbell autodual objects $X \in \mathcal{A}^\urcorner$ as a subcategory? (Hint: start simple. What if $\mathcal{A}$ is a monoid?)
- The *Isbell envelope* of a category $\mathcal{A}$ consists of the category having

  - objects the triples $(A \in \mathcal{A}, \xi : \hom \Rightarrow P \times Q)$ where $P : \mathcal{A}^{\text{op}} \to \text{Set}, Q : \mathcal{A} \to \text{Set}$ and ξ is a natural transformation with components $\xi_A : \hom(A, A') \to PA \times QA'$;
  - morphisms $(\alpha, \beta) : (A, \xi) \to (A', \eta)$ the pairs $\alpha : P \Rightarrow P'$ and $\beta : Q' \Rightarrow Q$ such that $(Q * \alpha) \circ \xi = (\beta * P') \circ \eta$.

  What are the properties of this category? Does it have an initial object? A terminal object? Is there a functor between $\mathrm{I}(\mathcal{A})$ and the category $\mathcal{A}^{\urcorner} \subset \mathcal{A}$ of Isbell autoduals?
- Explicitly determine the unit and counit of the Isbell adjunction in 3.2.18.
- Let $\Gamma \subset \mathsf{cat}$ be a subcategory of cat; let

$$\text{Cat}(\mathcal{A}^{\text{op}}, \text{Set})_\Gamma \subset \text{Cat}(\mathcal{A}^{\text{op}}, \text{Set})$$

  be the subcategory of presheaves commuting with limits of functors having domain in Γ; does the $\mathcal{O}$ functor land in the subcategory $\text{Cat}(\mathcal{A}, \text{Set})_\Gamma^{\text{op}}$? A particular case of this is when $\mathcal{A}$ is a Grothendieck site, and we want to know whether $\mathcal{O}(F)$ is a sheaf if F is.
- What does Isbell duality look like, when $\text{Cat}(\mathcal{A}^{\text{op}}, \text{Set})$ is identified with the category of discrete opfibrations over $\mathcal{A}$, using A.5.14?

4
Weighted (Co)limits

SUMMARY. The present chapter introduces the theory of *weighted (co)limits*. Such universal objects constitute a cornerstone of enriched category theory, which can be easily formulated and understood in terms of (co)end calculus.

After having introduced the main definition of weighted limit and colimit, we show that in the presence of (co)tensors in a $\mathcal{V}$-category $\mathcal{C}$ the limit $\lim^W F$ of $F : \mathcal{J} \to \mathcal{C}$ weighted by a functor $W : \mathcal{J} \to \mathcal{V}$ can be written as an end

$$\int_J WJ \pitchfork FJ$$

and dually the colimit $\operatorname{colim}^W F$ is a coend

$$\int^J WJ \otimes FJ$$

for $W : \mathcal{J}^{\mathrm{op}} \to \mathcal{V}$. This allows us to revisit many of the results we already know (for example, Kan extensions are weighted (co)limits), and to find that many constructions in category theory (comma objects, laxified versions of (co)limits,...) can all be expressed as weighted (co)limits for suitable weights W. It turns out that weighted (co)limits are the correct notion of such universal objects in enriched category theory. Thus, we conclude the chapter discussing the theory of *enriched (co)ends.*

> No, Time, thou shalt not boast that I do change:
> Thy pyramids built up with newer might
> To me are nothing novel, nothing strange.
>
> William Shakespeare, Sonnet CXXIII

We recall the fundamentals of enriched category theory in a section of our Appendix, A.7.1, but the material therein is by no means sufficient to provide a self-contained introduction to the topic. Instead, the reader unfamiliar with the basic notions can (and should) consult classical references such as [EK66a, Kel05a] or [Bor94b, §6.2].

Limits and colimits constitute a cornerstone of elementary category theory, because of their ubiquity in describing universal construction. Nevertheless, the notion soon becomes too strict when one moves to the world of *enriched* categories. The 'conical' shape of a classical (co)limit is not general enough to encompass the fairly rich variety of shapes in which universal objects in $\mathcal{V}$-categories arise.

This feature of the enriched-categorical world can be justified in many ways: naïvely speaking, the notion of *cone* for a functor $F : \mathcal{J} \to \mathcal{C}$ is based on the notion of constant functor. Yet, in many cases, there is no such thing as a 'constant' $\mathcal{V}$-enriched functor $F : \mathcal{J} \to \mathcal{C}$ (we shall notice in 4.3.8 that for the exact same reason $\mathcal{V}$-enriched categories do not support a sensible notion of dinaturality). Weighted (co)limits circumvent this issue by declaring that a limit depends on *two* arguments. The diagram F *of which* we want to compute the (co)limit, and a functor $W \in \mathrm{Cat}(\mathcal{J}, \mathcal{V})$ *along which* we 'weight' the (co)limit. The span $\mathcal{V} \xleftarrow{W} \mathcal{J} \xrightarrow{F} \mathcal{C}$ gives rise to an object of W-shaped F-diagrams, and the terminal such diagram constitutes the W-weighted limit of F (and dually, the initial object is the W-weighted colimit of F).

In the presence of a terminal functor (for example, when $\mathcal{V} = \mathrm{Set}$), *conical* (co)limits arise when we choose $W : \mathcal{J} \to \mathcal{V}$ to be the *terminal* presheaf, sending every object of $\mathcal{J}$ to the singleton of Set (and every morphism to the identity function of the singleton).

The theory of weighted (co)limits is fairly rich and spans several chapters of category theory. We cannot touch but the surface of this intricate topic: the interested reader can consult [Kel89], a presentation of unmatched clarity filled with enlightening examples.

Remark. As a consequence of the vastness of the topic, our approach is a compromise between full generality and usability, for we are more interested in translating the fundamentals about weighted (co)limits into results about suitable (co)ends, than in drawing a fully general theory. The focus is not on proofs; a few arguments, not directly linked to our main topic, are only briefly sketched. The underlying structure of the present chapters draws a lot from [Rie14, II.7]. Quite often, the choice of notation appears to be similar.

Notation. Throughout this chapter, a *weight* is a $\mathcal{V}$-enriched functor $W : \mathcal{J} \to \mathcal{V}$, or more generally $W : \mathcal{I} \times \mathcal{J}^{\mathrm{op}} \to \mathcal{V}$; we call *ordinary* a category which is enriched over Set with its obvious cartesian closed structure. All bases of enrichment $\mathcal{V}$ are *Bénabou cosmoi*, i.e. symmetric monoidal closed, complete and cocomplete categories. These are the 'good places' to do enriched category theory.

Throughout the chapter we make heavy use of the *category of elements* of a weight $W : \mathcal{C}^{\mathrm{op}} \to \mathrm{Set}$; the reader is invited to follow the present section having thoroughly meditated on A.5.11 and A.5.13.

4.1 Weighted Limits and Colimits

Remark 4.1.1 (A sophisticated look at classical (co)limits). Let $F : \mathcal{J} \to \mathcal{C}$ be a functor between small ordinary categories.

- The *limit* $\lim F$ of F can be characterised as the representing object of a suitable presheaf: to define $\lim F$ up to isomorphism we have the natural isomorphism

$$\mathcal{C}(C, \lim F) \cong \mathrm{Cat}(\mathcal{J}, \mathrm{Set})(1, \mathcal{C}(C, F\_)) \tag{4.1}$$

  where 1 is a shorthand to denote the terminal functor $\mathcal{C} \to \mathrm{Set} : C \mapsto 1$ sending every object to the singleton set, and $\mathcal{C}(C, F\_)$ is the functor $\mathcal{J} \to \mathrm{Set}$ sending J to $\mathcal{J}(C, FJ)$.
- Dually, the colimit $\mathrm{colim}\, F$ of $F : \mathcal{J} \to \mathcal{C}$ can be characterised, in the same notation, as the representing object in the natural isomorphism

$$\mathcal{C}(\mathrm{colim}\, F, C) \cong \mathrm{Cat}(\mathcal{J}^{\mathrm{op}}, \mathrm{Set})(1, \mathcal{J}(F\_, C)). \tag{4.2}$$

So, $\mathrm{Cat}(\mathcal{J}^{\mathrm{op}}, \mathrm{Set})(1, \mathcal{C}(F\_, C))$ is a set (of natural transformations), for every C, $\mathcal{J}(F\_, C) : \mathcal{J}^{\mathrm{op}} \to \mathrm{Set}$ is a presheaf, as well as $C \mapsto \mathrm{Cat}(\mathcal{J}^{\mathrm{op}}, \mathrm{Set})(1, \mathcal{J}(F\_, C)) : \mathcal{C} \to \mathrm{Set}$. If each functor of this sort is representable, we say that F *admits a limit*, precisely the representing object for $\mathrm{Cat}(\mathcal{J}, \mathrm{Set})(1, \mathcal{J}(F, \_))$. Clearly, this is nothing but an instance of the Yoneda–Grothendieck philosophy introduced in 2.1.1.

The leading idea behind the definition of weighted (co)limit is to generalise this construction to admit shapes other than the terminal presheaf for the domain functor in $1 \to \mathcal{J}(F\_, C)$. We can package this rough idea in the following definition.

Definition 4.1.2 (Weighted limit and colimit). Given a diagram of functors

$$\mathcal{C} \xleftarrow{F} \mathcal{J} \xrightarrow{W} \mathrm{Set} \tag{4.3}$$

we define the *weighted limit* of F by W (or, equally often, the limit of F, weighted by W) as a representing object for the presheaf sending $C \in \mathcal{C}$ to $\mathrm{Cat}(\mathcal{J}, \mathrm{Set})(W, \mathcal{C}(C, F\_))$.

In other words the weighted limit of F by W is an object $\lim^W F \in \mathcal{C}$ such that the isomorphism

$$\mathcal{C}(C, \lim^W F) \cong \mathrm{Cat}(\mathcal{J}, \mathrm{Set})(W, \mathcal{C}(C, F\_)) \tag{4.4}$$

holds naturally in $C \in \mathcal{C}$. Dually, we define the *colimit* of $F : \mathcal{J} \to \mathcal{C}$ weighted by $W : \mathcal{J}^{\mathrm{op}} \to \mathrm{Set}$ to be an object $\mathrm{colim}^W F \in \mathcal{C}$ such that the natural isomorphism

$$\mathcal{C}(\mathrm{colim}^W F, C) \cong \mathrm{Cat}(\mathcal{J}^{\mathrm{op}}, \mathrm{Set})(W, \mathcal{C}(F\_, C)) \tag{4.5}$$

holds naturally in $C \in \mathcal{C}$.

Notation 4.1.3. A common alternative notation for the object $\lim^W F$ is $\{W, F\}$; a common alternative notation for $\mathrm{colim}^W F$ is $W \otimes F$; this is meant to evoke a THC situation, through the isomorphisms

$$\mathcal{C}(C, \{W, F\}) \cong \mathrm{Cat}(\mathcal{J}, \mathrm{Set})(W, \mathcal{C}(C, F\_)) \tag{4.6}$$

$$\mathcal{C}(W \otimes F, C) \cong \mathrm{Cat}(\mathcal{J}^{\mathrm{op}}, \mathrm{Set})(W, \mathcal{C}(F\_, C)). \tag{4.7}$$

Although this is not properly the same arrangement of functors as in Exercise 3.6, the intuition is fruitful.

Example 4.1.4. Let $[1]$ be the 'generic arrow' category $\{0 \to 1\}$, and let $\ulcorner f \urcorner : [1] \to \mathcal{C}$ be the functor choosing an arrow $f : X \to Y$ in $\mathcal{C}$, and $W : [1] \to \mathrm{Set}$ the functor choosing the arrow $W0 = \{0, 1\} \to \{0\} = W1$; then a natural transformation $W \Rightarrow \mathcal{C}(C, f)$ consists of arrows $W0 \to \mathcal{C}(C, X), W1 \to \mathcal{C}(C, Y)$, namely the choice of two arrows $h, k : C \to X$ such that $fh = fk$. The universal property for $\lim^W f$ implies that this is the *kernel pair* of the arrow f, namely that h, k fill in the pullback

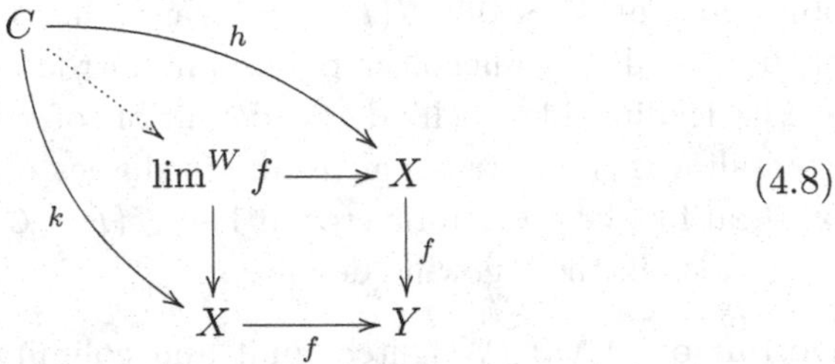

(4.8)

of the arrow f along itself. (The reader is invited to compute the colimit of the same diagram as a preparatory exercise.)

In order to characterise weighted (co)limits as (co)ends, we employ the same notation of 4.1.2.

Proposition 4.1.5 (Weighted limits as ends). When the end below and the Set-cotensor (see 2.2.3) $(X, A) \mapsto X \pitchfork A$ exist, we can express the weighted limit $\lim^W F$ for $F : \mathcal{J} \to \mathcal{C}$ as an end, as follows from the chain of computations

$$\begin{aligned}
\mathrm{Cat}(\mathcal{J}, \mathrm{Set})(W, \mathcal{C}(C, F)) &\cong \int_{J \in \mathcal{J}} \mathrm{Set}(WJ, \mathcal{C}(C, FJ)) \\
&\cong \int_{J \in \mathcal{J}} \mathcal{C}(C, WJ \pitchfork FJ) \\
&\cong \mathcal{C}\Big(C, \int_{J \in \mathcal{J}} WJ \pitchfork FJ\Big).
\end{aligned}$$

The above derivation implies that there is a canonical isomorphism

$$\lim^W F \cong \int_{J \in \mathcal{J}} WJ \pitchfork FJ. \tag{4.9}$$

The reader might have noticed that we did not provide a dual statement to 4.1.5; this dualisation process is left as an easy exercise, spelled out explicitly in Exercise 4.1: the weighted colimit of F by W is a coend, precisely

$$\mathrm{colim}^W F \cong \int^{J \in \mathcal{J}} WJ \otimes FJ. \tag{4.10}$$

Example 4.1.6. Consider the particular case of two parallel functors $W, F : \mathcal{C} \to \mathrm{Set}$; then we can easily see that $\lim^W F$ coincides with the set of natural transformations $W \Rightarrow F$, since the cotensor $WC \pitchfork FC$ is precisely the set $\mathrm{Set}(WC, FC)$. So, the limit of a presheaf F weighted by a parallel presheaf W is the set of natural transformations $W \Rightarrow F$.

Example 4.1.7 (Kan extensions). The ninja Yoneda lemma 2.2.1, rewritten in this notation, says that $\lim^{\mathcal{C}(C,-)} F \cong FC$ (or, in case F is contravariant, $\lim^{\mathcal{C}(-,C)} F \cong FC$).

A short slogan to remember this fact is the following.

> Representably weighted (co)limits are evaluations on the representing object.

(This hides a slightly more general fact: the category of element A.5.9 of W has a terminal object if and only if the weight W is representable, say $\text{よ}A \cong W$; then, $\lim^W F \cong FA$. See also our 2.2.6 about representables playing the role of Dirac delta functions.) It also suggests that *Kan extensions* can be expressed as suitable weighted (co)limits, and more precisely that they can be characterised as those weighted (co)limits where the weight is a representable functor (possibly 'twisted' by a functor $K : \mathcal{C} \to \mathcal{D}$):

$$\mathrm{Ran}_K F_- \cong \int_{C\in\mathcal{C}} \mathcal{D}(-, KC) \pitchfork FC \cong \lim^{\mathcal{D}(-,K=)} F \tag{4.11}$$

$$\mathrm{Lan}_K F_- \cong \int^{C\in\mathcal{C}} \mathcal{D}(KC, -) \otimes FC \cong \mathrm{colim}^{\mathcal{D}(K-,=)} F. \tag{4.12}$$

More precisely, $\mathcal{D}(-, K=) : \mathcal{D}^{\mathrm{op}} \times \mathcal{C} \to \mathrm{Set}$ is a functor, and for every $D \in \mathcal{D}$ the functor $W_D = \mathcal{D}(C, K=) : \mathcal{C} \to \mathrm{Set}$ works as a weight; the weighted limit of F along W_D is the value of $\mathrm{Ran}_K F$ on the object D.

Note that if K is the identity functor, we obtain the various forms of ninja Yoneda lemma (2.9)–(2.12) as special cases.

Example 4.1.8. ☉☉ Ends themselves can be computed as weighted limits: given $T : \mathcal{C}^{\mathrm{op}} \times \mathcal{C} \to \mathcal{D}$ we can take the hom functor $\mathcal{C}(-, =) : \mathcal{C}^{\mathrm{op}} \times \mathcal{C} \to \mathrm{Set}$ as a weight, and if the weighted limit exists, we have the chain of isomorphisms

$$\begin{aligned} \lim^{\mathcal{C}(-,=)} T &\cong \int_{(C,C')} \mathcal{C}(C, C') \pitchfork T(C, C') \\ &\cong \int_C \Big(\int_{C'} \mathcal{C}(C, C') \pitchfork T(C, C') \Big) \\ 2.2.1 &\cong \int_C T(C, C). \end{aligned}$$

Remark 4.1.9. It is particularly instructive to unwind the statement above and directly compute the end of $T : \mathcal{C}^{\mathrm{op}} \times \mathcal{C} \to \mathcal{D}$ as the equaliser of a pair of maps

$$\begin{array}{c} \prod_{C,C'\in\mathcal{C}} \mathcal{C}(C, C') \pitchfork T(C, C') \\ \downarrow\downarrow \\ \prod_{(f,g):(C,C')\to(C'',C''')} \mathcal{C}(C, C') \pitchfork T(C'', C''') \end{array} \tag{4.13}$$

determined by the action of T on arrows. In fact, this is exactly what we did in 1.3.2 to prove the Fubini rule 1.3.1.

The following Remark and Proposition constitute a central observation.

Remark 4.1.10 (The Grothendieck construction absorbs weights). ☉☉ Our 4.1.2 can be extended to the case $F : \mathcal{J} \to \mathcal{C}$ is a $\mathcal{V}$-enriched functor between $\mathcal{V}$-categories, and $W : \mathcal{J} \to \mathcal{V}$ is a $\mathcal{V}$-presheaf; this is the setting where the notion of a weighted limit proves itself to be the correct one over the 'conical' one (where the weight is the terminal presheaf). When $\mathcal{V} = \text{Set}$, indeed, the *Grothendieck construction* sending a presheaf into its category of elements turns out to simplify the theory of Set-weighted limits, reducing every weighted limit to a conical one.

This statement is clarified by the following proposition: recall from A.5.9 the definition of the *category of elements* of a functor $W : \mathcal{J} \to \text{Set}$.

Proposition 4.1.11 (Set-weighted limits are limits)**.** As shown in A.5.13, the category of elements $\mathcal{J} \int W$ comes equipped with a discrete fibration $\Sigma : \mathcal{J} \int W \to \mathcal{J}$; such a fibration is universal, in the sense that for any functor $F : \mathcal{J} \to \mathcal{C}$ one has

$$\lim{}^W F \cong \lim_{\mathcal{J} \int W} \left(\mathcal{J} \textstyle\int W \xrightarrow{\Sigma} \mathcal{J} \xrightarrow{F} \mathcal{C} \right). \tag{4.14}$$

Proof Using 4.1.8, the characterisation of the end $\int_{J \in \mathcal{J}} WJ \pitchfork FJ$ as an equaliser (as in 1.2.4), and the characterisation of Set-cotensors as iterated products, we can see that

$$\begin{aligned}
\int_{J \in \mathcal{J}} WJ \pitchfork FJ &\cong \text{eq}\Big(\prod_{J \in \mathcal{J}} WJ \pitchfork FJ \rightrightarrows \prod_{J \to J'} FWJ \pitchfork J' \Big) \\
&\cong \text{eq}\Big(\prod_{J \in \mathcal{J}} \prod_{x \in WJ} FJ \rightrightarrows \prod_{J \to J'} \prod_{x \in WJ} FJ' \Big) \\
(\star) &\cong \text{eq}\Big(\prod_{(J,x) \in \mathcal{J} \int W} FJ \rightrightarrows \prod_{(J,x) \to (J',x') \in \mathcal{J} \int W} FJ' \Big) \\
&\cong \lim_{(J,x) \in \mathcal{J} \int W} F \circ \Sigma
\end{aligned}$$

(step $(\star)$ is motivated by the fact that, thanks to the discrete fibration property for Σ, every arrow $f : \Sigma(J, x) \to J'$ has a unique lift $(J, x) \to (J', x')$ since $W(f)(x) = x'$). □

Remark 4.1.12. If the weight has the form $W = \mathcal{D}(D, K\_)$ for an object $D \in \mathcal{D}$, and a functor $K : \mathcal{J} \to \mathcal{D}$, then the category of elements $\mathcal{J} \int W$ is precisely the *comma category* $(D \downarrow K)$. Thus the right Kan extension of F along K can be computed as the conical limit of the functor $F \circ \Sigma$, where $\Sigma : (D \downarrow K) \to \mathcal{J}$ is the obvious forgetful functor. We have just rediscovered (4.11) and (4.12).

When *every* weighted limit exists in $\mathcal{C}$, we can prove that the correspondence $(W, F) \mapsto \lim^W F$ is a bifunctor:

$$\lim^{(-)}(=) : \mathrm{Cat}(\mathcal{J}, \mathrm{Set})^{\mathrm{op}} \times \mathrm{Cat}(\mathcal{J}, \mathcal{C}) \to \mathcal{C}. \tag{4.15}$$

From this, we have at once the following.

- Functoriality in the W component of $\lim^W F$ entails that the terminal morphism $W \Rightarrow 1$ induces a *comparison arrow* between the W-weighted limit of any $F : \mathcal{J} \to \mathcal{C}$ and the classical (conical) limit: every weighted limit is a 'fattened up' version of the conical limit, and there is a comparison arrow $\lim F \to \lim^W F$. This intuition has some connection with homotopy theory: it will become clearer in 4.2.5.
  As an example, consider that the conical limit of the functor $f : [1] \to \mathcal{C}$ described in Example 4.1.4 consists of the object $\mathrm{src}(f)$; hence the comparison arrow consists of the unique factorisation of two copies of the identity of $\mathrm{src}(f)$ along the kernel pair of f.
- Using 4.1.8 one can prove that the functor $\lim^{(-)} F$ is *continuous*, i.e. the isomorphism

$$\lim^{\left(\operatorname{colim}_{\mathcal{I}} W_i\right)} F \cong \lim_{\mathcal{I}} \left(\lim^{W_i} F\right), \tag{4.16}$$

  holds for every small diagram of weights $\mathcal{I} \to \mathrm{Cat}(\mathcal{C}, \mathrm{Set}) : i \mapsto W_i$. Indeed, for a generic object $X \in \mathcal{C}$ we have

$$\begin{aligned} \mathcal{C}\Big(X, \lim^{\left(\operatorname{colim}_{\mathcal{I}} W_i\right)} F\Big) &\cong \mathcal{C}\Big(X, \textstyle\int_A \left(\operatorname{colim}_{\mathcal{I}} W_i A\right) \pitchfork FA\Big) \\ &\cong \mathcal{C}\Big(X, \textstyle\int_A \lim_{\mathcal{I}}(W_i A \pitchfork FA)\Big) \\ &\cong \mathcal{C}(X, \lim_{\mathcal{I}} \lim^{W_i} F) \end{aligned}$$

  so the two objects $\lim^{\left(\operatorname{colim}_{\mathcal{I}} W_i\right)} F$ and $\lim_{\mathcal{I}} \left(\lim^{W_i} F\right)$ must be canonically isomorphic.

The above observation will turn out to be useful during our discussion of *simplicially coherent* (co)ends in 7.2.15.

Weighted *colimits* can be obtained as a straightforward dualisation of the above arguments. The boring technicalities are left to the reader to expand in Exercise 4.1; for the record, we state the proper dualisation in a separate remark.

Remark 4.1.13.

WC1. (Weighted colimits as coends) Let

$$\mathcal{C} \xleftarrow{F} \mathcal{J} \xrightarrow{W} \mathrm{Set} \tag{4.17}$$

be two functors; if $\mathcal{C}$ admits the coend below, we can express the weighted colimit $\mathrm{colim}^W F$ as

$$\mathrm{colim}^W F \cong \int^{J\in\mathcal{J}} WJ \otimes FJ \tag{4.18}$$

where we used, as we always do, the Set-tensoring structure of $\mathcal{C}$.

WC2. (Left Kan extensions as weighted colimits) Let $F : \mathcal{J} \to \mathcal{C}$ and $K : \mathcal{J} \to \mathcal{D}$ be functors; then

$$\mathrm{Lan}_K F\_ \cong \int^{J\in\mathcal{J}} \mathcal{D}(KJ, \_) \otimes FJ \cong \mathrm{colim}^{\mathcal{D}(K=,-)} F. \tag{4.19}$$

WC3. (Coends as hom-weighted colimits) The coend of $T : \mathcal{C}^{\mathrm{op}} \times \mathcal{C} \to \mathcal{D}$ can be written as $\mathrm{colim}^{\mathcal{C}(-,=)} T$, regarding $\hom_{\mathcal{C}} : \mathcal{C}^{\mathrm{op}} \times \mathcal{C} \to \mathrm{Set}$ as a weight:

$$\int^{C} T(C,C) \cong \int^{C,C'} \mathcal{C}(C,C') \otimes T(C,C'). \tag{4.20}$$

WC4. If the weight W is Set-valued, the colimit of $F : \mathcal{J} \to \mathcal{C}$ weighted by $W : \mathcal{J}^{\mathrm{op}} \to \mathrm{Set}$ can be written as a conical colimit over $\mathcal{J}^{\mathrm{op}} \int W$ using $\Sigma : \mathcal{J}^{\mathrm{op}} \int W \to \mathcal{J}^{\mathrm{op}}$:

$$\mathrm{colim}^W F \cong \mathrm{colim}_{(J,x)\in(\mathcal{J}^{\mathrm{op}} \int W)^{\mathrm{op}}} (F \circ \Sigma^{\mathrm{op}}). \tag{4.21}$$

WC5. (Functoriality) If the W-colimit of $F : \mathcal{J} \to \mathcal{C}$ always exists, then the correspondence $(W, F) \mapsto \mathrm{colim}^W F$ is a functor, cocontinuous in its first variable:

$$\mathrm{colim}^{(-)}(=) : \mathrm{Cat}(\mathcal{J}^{\mathrm{op}}, \mathrm{Set}) \times \mathrm{Cat}(\mathcal{J}, \mathcal{C}) \to \mathcal{C},$$
$$\mathrm{colim}^{\left(\mathrm{colim}_{\mathcal{I}} W_i\right)} F \cong \mathrm{colim}_{\mathcal{I}} \left(\mathrm{colim}^{W_i} F\right). \tag{4.22}$$

WC6. (Comparison) There is a canonical natural transformation $W \to 1$, inducing a canonical *comparison arrow* from the W-colimit of any $F : \mathcal{J} \to \mathcal{C}$ to the conical colimit.

From the description 4.1.8 and 4.1.13.WC1 above of $\lim^W F$ and $\mathrm{colim}^W F$ as (co)ends, it is immediate that hom functors preserve weighted limits.

Remark 4.1.14. For every $C \in \mathcal{C}$, every functor $F : \mathcal{J} \to \mathcal{C}$ and weight $W : \mathcal{J}^{\mathrm{op}} \to \mathcal{V}$ we have a canonical isomorphism

$$\mathcal{C}\left(\operatorname{colim}^W F, C\right) \cong \lim^W \mathcal{C}(F, C). \tag{4.23}$$

Dually, for every functor $F : \mathcal{J} \to \mathcal{C}$ and weight $W : \mathcal{J} \to \mathcal{V}$ we have a canonical isomorphism

$$\mathcal{C}\left(C, \lim^W F\right) \cong \lim^W \mathcal{C}(C, F). \tag{4.24}$$

As an immediate consequence, every functor that preserves (co)ends preserves weighted (co)limits; for example, a left adjoint must preserve all weighted colimits, and a right adjoint must preserve all weighted limits.

4.2 Examples of Weighted Colimits

Due to their deep connections with enriched category theory, homological algebra and algebraic topology are a natural factory of examples of weighted colimits.

Example 4.2.1 (The cone construction as a weighted colimit). Let K be a ring, and $\mathcal{V} = \mathrm{Ch}(K)$ the category of chain complexes of K-modules. Considering $\mathcal{V}$ as self-enriched, suitably defining the chain complex of maps $C_* \to D_*$, we aim to prove that the *mapping cone* $\mathrm{C}(f) = X_*[1] \oplus Y_*$ of a chain map $f : X_* \to Y_*$ [Wei94, 1.5.1] in $\mathcal{V}$ can be characterised as $\operatorname{colim}^W f$, where $f : [1] \to \mathcal{V}$ is the arrow f, and $W : [1]^{\mathrm{op}} \to \mathcal{V}$ is the functor which chooses the map $S^1(K)_* \to D^2(K)_*$, where $S^n(K)_* = K[n]_*$ is the chain complex with the only non-zero term K concentrated in degree $-n$, and $D^n(K)_*$ is the complex

$$\cdots \longrightarrow 0 \longrightarrow K \mathop{=\!=} K \longrightarrow 0 \longrightarrow \cdots,$$

where the first non-zero term is in degree $-n$. There is an obvious inclusion $S^n_* \hookrightarrow D^{n+1}_*$:

$$\begin{array}{ccccccccc}
\cdots \longrightarrow & 0 & \longrightarrow & 0 & \longrightarrow & K & \longrightarrow & 0 & \longrightarrow \cdots \\
 & & & \downarrow & & \| & & & \\
\cdots \longrightarrow & 0 & \longrightarrow & K & = & K & \longrightarrow & 0 & \longrightarrow \cdots .
\end{array}$$

We now aim to prove that

$$\mathrm{C}(f) \cong \int^i W(i) \otimes f(i). \tag{4.25}$$

In view of (the dual of) Exercise 1.7, it is enough to show that there is a pushout square

$$\begin{array}{ccc} W(1)\otimes f(0) & \longrightarrow & W(1)\otimes f(1) \\ \downarrow & \ulcorner & \vdots \\ W(0)\otimes f(0) & \cdots\cdots\to & \mathrm{C}(f) \end{array}$$

This is a rather simple exercise in universality, given the maps

$$B \xrightarrow{\binom{0}{1}} \mathrm{C}(f) \xleftarrow{\left(\begin{smallmatrix} 0 & 1 \\ f & 0 \end{smallmatrix}\right)} A \oplus A[1]. \tag{4.26}$$

The following example is juicier. In addition to those of A.5.11 there is a fourth characterisation for the category of elements of a presheaf as a suitable coend over Cat.

Example 4.2.2 (The category of elements of a presheaf). ☉☉ The category of elements of a functor $F : \mathcal{C} \to \mathrm{Set}$ introduced in A.5.9 can be characterised as a Cat-weighted colimit: it results as the colimit

$$\mathcal{C} \textstyle\int W \cong \int^{C\in\mathcal{C}} C/\mathcal{C} \times WC \tag{4.27}$$

where WC is a set, regarded as a discrete category; in other words it is isomorphic to the weighted colimit $\operatorname{colim}^S W$, where $S : \mathcal{C}^{\mathrm{op}} \to \mathrm{Cat}$ (S as 'slice') is the functor $C \mapsto C/\mathcal{C}$ (the 'coslice' category of arrows $C \to X$ and commutative triangles under C).

To prove this statement, we verify that $\mathcal{C} \int W$ has the universal property of the coequaliser of the pair

$$\coprod_{f:A\to B} B/\mathcal{C} \times WA \overset{\alpha}{\underset{\beta}{\rightrightarrows}} \coprod_{C\in\mathcal{C}} C/\mathcal{C} \times WC \tag{4.28}$$

where α has components $\alpha_f : B/\mathcal{C} \times WA \xrightarrow{1\times Ff} B/\mathcal{C} \times WB$ sending $\left(\left[\begin{smallmatrix} B \\ \downarrow \\ X \end{smallmatrix}\right], u\right) \mapsto \left(\left[\begin{smallmatrix} B \\ \downarrow \\ X \end{smallmatrix}\right], F(f)u\right)$ and β has components $\beta_f : B/\mathcal{C} \times WA \xrightarrow{f^*\times WA} A/\mathcal{C} \times WA$ sending $\left(\left[\begin{smallmatrix} B \\ \downarrow \\ X \end{smallmatrix}\right], u\right) \mapsto \left(\left[\begin{smallmatrix} A \xrightarrow{f} B \\ \downarrow \\ X \end{smallmatrix}\right], u\right)$. Of course, these maps are composed with the suitable coproduct injections.

It is rather easy now to define a functor

$$\theta : \coprod_{A\in\mathcal{C}} A/\mathcal{C} \times WA \to \mathcal{C} \textstyle\int W \tag{4.29}$$

having components $\theta_A : A/\mathcal{C} \times WA \to \mathcal{C} \int W$ defined by

$$\left(\begin{bmatrix} A \\ f\downarrow \\ B \end{bmatrix}, u \in FA \right) \mapsto (b, F(f)(u) \in Fb), \tag{4.30}$$

which coequalises α and β. This functor θ has the universal property of the coequaliser: given any other $\zeta : \coprod_{A \in C} A/\mathcal{C} \times WA \to \mathcal{K}$ we can define a functor $\overline{\zeta} : \mathcal{C} \int W \to \mathcal{K}$ such that

$$\overline{\zeta}(A, u \in FA) = \zeta(\mathrm{id}_A, u). \tag{4.31}$$

Now notice that every map ζ' that coequalises (α, β) has the property that

$$\zeta' \left(\begin{bmatrix} B \\ \downarrow \\ X \end{bmatrix}, F(f)u \right) = \zeta' \left(\begin{bmatrix} A \xrightarrow{f} B \\ \downarrow \\ X \end{bmatrix}, u \right). \tag{4.32}$$

It is now a routine verification to see that $\overline{\zeta} \circ \theta_A = \zeta_A$, and every other functor with this property must coincide with our $\overline{\zeta}$. This concludes the proof.

Remark 4.2.3 (Again an alternative characterisation of the category of elements)**.** The reader may have noticed that all the above discussion gives a fifth presentation for the category of elements $\mathcal{C} \int W$, as the image of W under the Kan extension $\mathrm{Lan}_{\text{よ}} J$. In the language of Chapter 3, $S : \mathcal{C}^{\mathrm{op}} \to \mathrm{Cat}$ is the NR context of the paradigm

$$\mathcal{C} \int \_ : \mathrm{Cat}(\mathcal{C}, \mathrm{Cat}) \rightleftarrows \mathrm{Cat} : N_S \tag{4.33}$$

where $N_S : \mathrm{Cat} \to \mathrm{Cat}(\mathcal{C}, \mathrm{Cat})$ is the 'nerve' functor sending $\mathcal{D}$ to the functor $C \mapsto \mathrm{Cat}(C/\mathcal{C}, \mathcal{D})$.

Remark 4.2.4. An alternative approach to characterising $\mathcal{C} \int W$ is the following: the category $\mathcal{C} \int W$ is precisely the lax limit of W regarded as a Cat-valued presheaf [Kel89, §4], [Gra80, Str76].

We can express the Bousfield–Kan construction for the homotopy (co)limit functor using (co)end calculus (see 7.2.1 for a crash course on homotopy (co)limits). We summarise some key aspects of the Bousfield–Kan construction in the following series of examples.

Theorem 4.2.5 (The Bousfield–Kan formula for homotopy (co)limits)**.** ⚈⚈ Let $F : \mathcal{J} \to \mathcal{M}$ be a diagram in a model category $\mathcal{M}$; let moreover $\mathcal{M}$ be equipped with a THC situation (see Exercise 3.6) by functors $\_ \pitchfork = : \mathrm{sSet}^{\mathrm{op}} \times \mathcal{M} \to \mathcal{M}$, $[\_, =] : \mathcal{M}^{\mathrm{op}} \times \mathcal{M} \to \mathrm{sSet}$ (so, in particular, $\mathcal{M}$ is sSet-enriched), and $\_ \otimes = : \mathrm{sSet} \times \mathcal{M} \to \mathcal{M}$.

Let us consider the nerve functor of 3.2.5. The functo $N : \mathrm{Cat} \to \mathrm{sSet}$

sends a category $\mathcal{C}$ to the simplicial set of its n-tuples of composable arrows, for each $[n] \in \Delta$.

Then the homotopy limit $\operatorname{hlim} F$ of F can be computed as the end

$$\int_J N(\mathcal{J}/J) \pitchfork FJ, \tag{4.34}$$

and the homotopy colimit $\operatorname{hcolim} F$ of F can be computed as the coend

$$\int^J N(J/\mathcal{J}) \otimes FJ. \tag{4.35}$$

Remark 4.2.6. These two universal objects are weighted (co)limits in an evident way: it is possible to rewrite $\operatorname{hlim} F \cong \lim^{N(J/-)} F$ and $\operatorname{hcolim} F \cong \operatorname{colim}^{N(-/J)} F$.

The idea behind this characterisation is to replace the terminal weight with a homotopy equivalent, but a fibrant one (in the case of limits, cofibrant in the case of colimits).

The Bousfield–Kan formula arises precisely when we replace the terminal weight with a fibrant one: for every object J, both $N(J/\mathcal{J})$ and $N(\mathcal{J}/J)$ are contractible categories, and they are linked to $N(1)$ by a homotopy equivalence induced by the terminal functor.

Then, the categories $N(J/\_), N(\_/J)$ must be thought as proper *replacements* for the (co)limit functor that correct its failure to preserve weak equivalences (see [Str11, Ch. 6] for an extremely hands-on account of the theory of homotopy (co)limits in algebraic topology, and [Hov99] for a standard, easy reference on model categories).

A fairly large class of interesting examples of weighted (co)limits comes from the theory of 2-categories; many 2-dimensional constructions are captured by the above formalism. Moreover, (co)end calculus expresses very concretely the shape of the universal object $\lim^W F$, as well as its 1- and 2-dimensional universal properties. Like in the previous section, we do not seek utter generality, but instead clarity of exposition. Thus, we restrict our attention to the 2-category Cat of categories, strict functors and natural transformations, avoiding the case of pseudofunctors, pseudonatural transformations, etc.

We first study a simple example of Cat-enriched limit, and its dual; the description of other shapes of weighted (co)limits is way more instructive when it is left as an exercise for the reader (see [Kel89]). Recall Warning 2.5.9.

Example 4.2.7 (Inserters in Cat). ⦿⦿ Let $\mathcal{C}$ be a 2-category, and

$f, g : X \rightrightarrows Y$ two parallel 1-cells in $\mathcal{C}$; the *inserter* $I(f,g)$ is a pair (p,λ) where $p : I(f,g) \to X$ is a 1-cell and $\lambda : fp \Rightarrow gp$ is a 2-cell, universal with respect to the property of connecting fp, gp. This means that the pair (p,λ) enjoys the following properties.

IN1. A 1-dimensional universal property: given a diagram

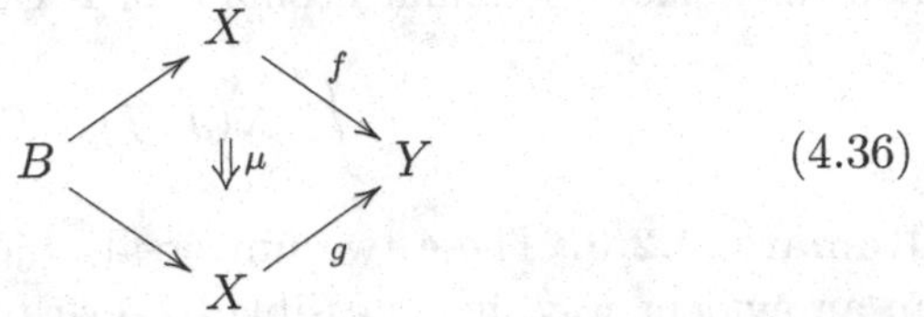

this can be split as the whiskering

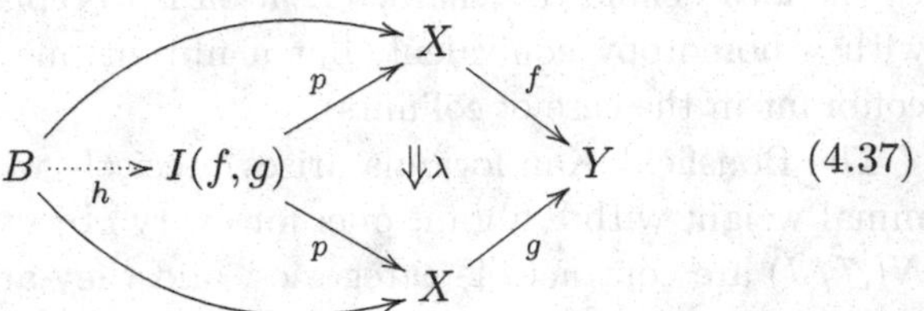

for a unique 1-cell $h : b \to I(f,g)$ in $\mathcal{C}$. This means, again, that $ph = q$ and $\lambda * h = \mu$.

IN2. A 2-dimensional universal property: given parallel 1-cells $h, k : A \to I(f,g)$ and a 2-cell $\beta : ph \Rightarrow pk$ such that

$$\begin{array}{ccc} A & \xrightarrow{h} & I(f,g) \\ {\scriptstyle k}\downarrow & \swarrow\!\!\!\swarrow_{\beta} & \downarrow{\scriptstyle p} \\ I(f,g) & \xrightarrow{p} & B \\ {\scriptstyle p}\downarrow & \swarrow\!\!\!\swarrow_{\lambda} & \downarrow{\scriptstyle f} \\ B & \xrightarrow[g]{} & C \end{array} \quad = \quad \begin{array}{ccc} A & \xrightarrow{k} & I(f,g) \\ {\scriptstyle h}\downarrow & {}^{\beta}\nearrow\!\!\!\nearrow & \downarrow{\scriptstyle p} \\ I(f,g) & \xrightarrow[p]{} & B \\ {\scriptstyle p}\downarrow & {}^{\lambda}\nearrow\!\!\!\nearrow & \downarrow{\scriptstyle g} \\ B & \xrightarrow[f]{} & C \end{array} \tag{4.38}$$

there is a unique $\bar{\beta} : h \Rightarrow k$ such that

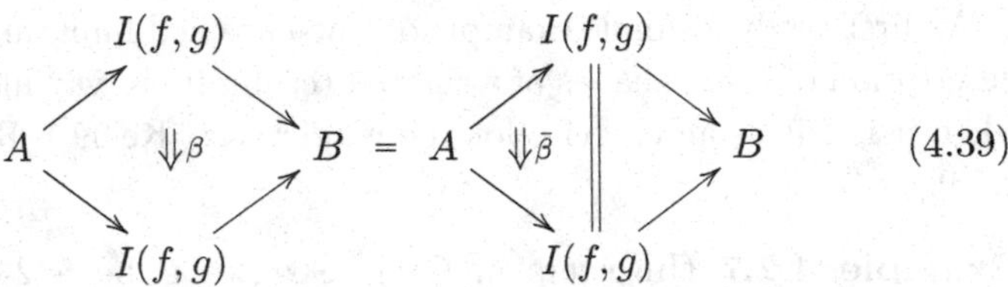

Now, if $\mathcal{J}$ is the category $\{0 \rightrightarrows 1\}$ with two objects and two parallel non-identity arrows, the inserter $I(f,g)$ is the limit of the functor $F : \mathcal{J} \to \mathcal{C}$ choosing the two 1-cells f, g, weighted by the weight $W : \mathcal{J} \to \mathsf{Cat}$ choosing the parallel 'source' and 'target' functors $s, t : [0] \rightrightarrows [1]$.

We shall deduce the shape of the inserter when $\mathcal{C} = \mathsf{Cat}$ is the 2-category of categories. In this case, the end

$$\int_{J \in \mathcal{J}} WJ \pitchfork FJ \tag{4.40}$$

that according to 4.1.2 defines the weighted limit boils down to the object of Cat-natural transformations $W \Rightarrow F$ (see also 1.4.1). It is indeed the case that such a natural transformation is determined as a pair $(b, u : c \to c') \in B \times C^{[1]}$ such that $u : fb \to gb$. In fact, naturality corresponds to the commutativity of the following two squares:

$$\begin{array}{ccc} [0] & \xrightarrow{b} & B \\ {\scriptstyle s}\downarrow & & \downarrow{\scriptstyle f} \\ [1] & \xrightarrow[u]{} & C \end{array} \qquad \begin{array}{ccc} [0] & \xrightarrow{b} & B \\ {\scriptstyle t}\downarrow & & \downarrow{\scriptstyle g} \\ [1] & \xrightarrow[u]{} & C \end{array} \tag{4.41}$$

and this means precisely that u has fb as domain, and gb as codomain.

Let us show that this object, as a subobject $p : (B \times C^{[1]})^\circ \subseteq B \times C^{[1]}$ has the desired universal property. First of all, $(B \times C^{[1]})^\circ$ clearly is a subcategory of the product category $B \times C^{[1]}$. There is an obvious natural transformation $\lambda : fp \Rightarrow gp$ defined on components as $\lambda_{(b,u)} : fb \xrightarrow{u} gb$. We leave to the reader to check that this is indeed the component of a natural transformation.

- IN1. Every $\alpha : fq \Rightarrow gq$ with components $\alpha_a : fqa \to gqa$ is such that $(qa, \alpha_a) \in (B \times C^{[1]})^\circ$, and the map $h : A \to (B \times C^{[1]})^\circ$ that sends a into (qa, α_a) is a functor because α was natural.
- IN2. A similar argument shows that every $\beta : ph \Rightarrow pk$ satisfying (4.38) factors through $(B \times C^{[1]})^\circ$.

Altogether, these two properties show that there is a unique isomorphism between $(B \times C^{[1]})^\circ$ and $I(f,g)$. This concludes the proof.

Example 4.2.8 (Comma objects). Let $\mathcal{C}$ be a 2-category, and f, g a cospan of 1-cells like

$$\begin{array}{ccc} & & C \\ & & \downarrow g \\ B & \xrightarrow{f} & X \end{array} \tag{4.42}$$

This can be regarded as the image of a functor $F : \Lambda^2_2 \to \mathcal{C}$, where Λ^2_1 is the 'generic cospan' $\{0 \to 2 \leftarrow 1\}$. Let us consider the weight $W : \Lambda^2_2 \to \text{Cat}$ whose image is

$$W\begin{bmatrix} & 1 \\ & \downarrow \\ 0 \to & 2 \end{bmatrix} = \begin{array}{ccc} & & [0] \\ & & \downarrow d_1 \\ [0] & \xrightarrow{d_0} & [1] \end{array} \tag{4.43}$$

where $d_i : \{i\} \to \{0 \to 1\}$ chooses the object i. Let us prove that the limit of F weighted by W is the *comma object* of f, g; evidently, in the special case of $\mathcal{C} = \text{Cat}$, the limit $\lim^W F$ is the *comma category* of A.2.14.

Let us fix an object A of $\mathcal{C}$; a natural transformation $W \Rightarrow \mathcal{C}(A, F\_)$ consists of the following data:

C1. a 1-cell $u : A \to B$;
C2. a 1-cell $v : A \to C$;
C3. a 2-cell $\lambda : [1] \to \mathcal{C}(A, X)$, whose source is forced by the naturality condition to be fu, and whose target is gv.

More explicitly, natural transformations $W \Rightarrow \mathcal{C}(A, F\_)$ correspond to squares

$$\begin{array}{ccc} A & \xrightarrow{v} & C \\ {\scriptstyle u}\downarrow & \lambda \nearrow & \downarrow{\scriptstyle g} \\ B & \xrightarrow{f} & X \end{array} \tag{4.44}$$

filled by a 2-cell $\lambda : fu \Rightarrow gv$. The terminal such 2-cell is then $\lim^W F$.

One can now routinely dualise the above construction to define *cocomma objects*: the following example freely uses some notions from Chapter 5.

Example 4.2.9 (Cocomma objects). Let $\mathcal{C}$ be a 2-category, and f, g a span of 1-cells like

$$\begin{array}{ccc} X & \xrightarrow{g} & C \\ {\scriptstyle f}\downarrow & & \\ B & & \end{array} \tag{4.45}$$

This can be regarded as the image of a functor $F : \Lambda^2_0 \to \mathcal{C}$, where Λ^2_0 is the 'generic span' $\{0 \to 2 \leftarrow 1\}$. Let us consider again the weight $W : (\Lambda^2_0)^{\text{op}} \cong \Lambda^2_2 \to \text{Cat}$ whose image is

$$\begin{array}{ccc} & & [0] \\ & & \downarrow{\scriptstyle d_1} \\ [0] & \xrightarrow[d_0]{} & [1] \end{array} \tag{4.46}$$

where $d_i : \{i\} \to \{0 \to 1\}$ chooses the object i. The colimit of F weighted by W is the *cocomma object* of the pair (f, g).

Exercise 5.6 provides you with another proof that the cocomma object in Cat of two functors $f, g : B \leftarrow X \to A$ can be described as the category $\left[\begin{smallmatrix} f \\ g \end{smallmatrix}\right]$ having objects those of $B \sqcup C$, and morphisms $x \to y$ defined as follows:

- $B(b, b')$ if $b = x, b' = y$ are both objects of B;
- $C(c, c')$ is $c = x, c' = y$ are both objects of C;
- if $b = x \in B_o$ and $c = y \in C_o$, the set of morphisms in $\left[\begin{smallmatrix} f \\ g \end{smallmatrix}\right](b, c)$ is the coend
$$\int^x C(gx, c) \times B(b, fx); \tag{4.47}$$
- otherwise, the hom-set is empty.

We shall now present a proof of the universality of $\left[\begin{smallmatrix} f \\ g \end{smallmatrix}\right]$ based solely on the construction of the coend that defines $\left[\begin{smallmatrix} f \\ g \end{smallmatrix}\right](b, c)$.

The intuition that shall guide the reader is that $\left[\begin{smallmatrix} f \\ g \end{smallmatrix}\right](b, c)$ is a set of 'fake arrows', i.e. of triples

$$b \xrightarrow{\varphi} fa \overset{(\xi)}{\dashrightarrow} ga \xrightarrow{\psi} c \tag{4.48}$$

where the arrow $(\xi) : fa \dashrightarrow ga$ is in some suitable sense freely adjoined in the disjoint union $B \sqcup C$.

Inside the coend, we identify two fake arrows $b \xrightarrow{\varphi} fa \overset{(\xi)}{\dashrightarrow} ga \xrightarrow{\psi} c$

and $b \xrightarrow{\varphi} fa' \overset{(\xi')}{\dashrightarrow} ga' \xrightarrow{\psi} c$ precisely when there is a 'hammock' diagram between a, a' of the following form:

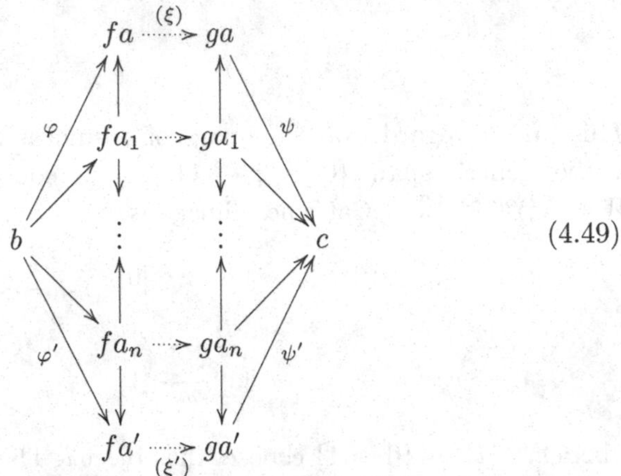

(4.49)

Now we have enough material to discover the universal property of such an object: first of all, there are canonical functors $i_B : B \to \left[\begin{smallmatrix} f \\ g \end{smallmatrix}\right]$ and $i_C : C \to \left[\begin{smallmatrix} f \\ g \end{smallmatrix}\right]$, and a canonical natural transformation $\zeta \colon i_B f \Rightarrow i_C g$ comes from taking the equivalence class of identity arrows $(1_{ga}, 1_{fa}) \in Y(ga, ga) \times X(fa, fa)$ under the composition

$$
\begin{array}{ccc}
\hom(ga, ga) \times \hom(fa, fa) & & \ni (1_{ga}, 1_{fa}) \\
\downarrow & & \Big\downarrow \\
\coprod_{a \in A} \hom(c, ga) \times \hom(fa, b) & & \\
\downarrow & & \\
\left[\begin{smallmatrix} f \\ g \end{smallmatrix}\right](fa, ga) & & \ni [(1_{ga}, 1_{fa})]
\end{array}
\tag{4.50}
$$

as a natural candidate for $\zeta_a \colon fa \to ga$.

Now, suppose we are given a commutative square

$$
\begin{array}{ccc}
X & \xrightarrow{g} & C \\
{\scriptstyle f}\downarrow & \overset{\theta}{\nearrow} & \downarrow{\scriptstyle v} \\
B & \xrightarrow[w]{} & Y
\end{array}
\tag{4.51}
$$

filled by a 2-cell $\theta : wf \Rightarrow vg$. Then define a unique functor $u : \left[\begin{smallmatrix} f \\ g \end{smallmatrix}\right] \to Y$ on objects and true arrows in B or C acting as $v \colon C \to Y, w \colon B \to Y$. The action of u on a fake arrow

$$b \xrightarrow{\varphi} fa \overset{(\xi)}{\dashrightarrow} ga \xrightarrow{\psi} c$$

is induced by the composition

$$wb \xrightarrow{w*\varphi} wfa \xrightarrow{\theta_a} vga \xrightarrow{v*\psi} vc$$

(all arrows exist now!). At this point, all remaining checks are pure routine: u is unique due to the tautological definition of ζ; the cell ζ has a 2-dimensional universal property as well; $\zeta \circ \_$ reflects isomorphisms.

Once this is done, the willing reader can embark on all sorts of instructive computations: exchanging the rôle of B, C in the above construction, one obtains a 2-cell in the opposite direction; in the terminology of the next chapter (see Section 5.2 and (5.4)), $\left[\begin{smallmatrix} f \\ g \end{smallmatrix}\right]$ is the category of elements of the composite profunctor $B \overset{\mathfrak{p}_f}{\rightsquigarrow} X \overset{\mathfrak{p}^g}{\rightsquigarrow} C$ regarded as a presheaf $B^{\text{op}} \times C \to \text{Set}$; what if the functor f or g is the identity?

4.3 Enriched (Co)ends

4.3.1 Preliminaries on Enriched Categories

In the setting of enriched category theory, the property of being complete is stated in terms of an existence result for every weighted limit $\lim^W F$: in short, the reason for this choice is that the categories admitting only (co)limits weighted by terminal presheaves contain too few objects and shall not be considered complete by the internal language of $\underline{\mathcal{V}\text{-Cat}}$.

This can be made precise in the following way. There is a canonical choice of a Yoneda structure (see Section 2.4) on the 2-category of $\mathcal{V}$-enriched categories, where Yoneda maps are given by $\mathcal{V}$-enriched Yoneda embeddings; every 2-category with a Yoneda structure has a notion of a *cocomplete object*, and in that Yoneda structure (co)completeness coincides with having all weighted (co)limits (see [SW78]).

The present section serves the purpose of introducing a sensible definition of *enriched (co)end* in 4.3.9, 4.3.10. In short, using 4.1.13.WC3, given a $\mathcal{V}$-functor $H : \mathcal{C}^{\text{op}} \boxtimes \mathcal{C} \to \mathcal{V}$ we can define the end $\int_C H(C, C)$ to be the limit of H weighted by the enriched hom $\mathcal{V}$-functor $\mathcal{C}^{\text{op}} \boxtimes \mathcal{C} \to \mathcal{V}$ (we will introduce the $\_ \boxtimes =$ product in 4.3.3 below).

In case all (co)ends exist and the codomain category of functors T : $\mathcal{C}^{\text{op}} \times \mathcal{C} \to \mathcal{A}$ has (co)tensors, we can moreover easily characterise the (co)end functor as an adjoint to a '(co)tensor with hom' functor.

Lemma 4.3.1. Let $\mathcal{C}$ be a tensored and cotensored $\mathcal{V}$-category and W a functor to be treated as a weight.

WA1. If $W : \mathcal{J} \to \mathcal{V}$, the $\lim^W \_$ functor has a left adjoint, given by tensoring with the weight W; in other words, there is an adjunction

$$\mathcal{C} \underset{\lim^W \_}{\overset{W\otimes \_}{\rightleftarrows}} \mathrm{Cat}(\mathcal{J},\mathcal{C}) \tag{4.52}$$

where $W \otimes C : \lambda J.WJ \otimes C$.

WA2. Dually, if $W : \mathcal{J}^{\mathrm{op}} \to \mathcal{V}$, the $\mathrm{colim}^W \_$ functor has a right adjoint, given by cotensoring with the weight W; in other words, there is an adjunction

$$\mathrm{Cat}(\mathcal{J},\mathcal{C}) \underset{W\pitchfork \_}{\overset{\mathrm{colim}^W \_}{\rightleftarrows}} \mathcal{C} \tag{4.53}$$

where $W \pitchfork C : \lambda J.WJ \pitchfork C$.

Proof Both arguments consist of an easy computation in coend calculus:

$$\begin{aligned}\mathcal{C}(C, \lim^W F) &\cong \mathcal{C}\Big(C, \int_J WJ \pitchfork FJ\Big)\\ &\cong \int_J \mathcal{C}(WJ \otimes C, FJ)\\ &\cong \mathrm{Cat}(\mathcal{J},\mathcal{C})(W \otimes X, F).\end{aligned}$$

Dually, we have

$$\begin{aligned}\mathcal{C}(\mathrm{colim}^W F, C) &\cong \mathcal{C}\Big(\int^J WJ \otimes FJ, C\Big)\\ &\cong \int_J \mathcal{C}(WJ \otimes FJ, C)\\ &\cong \int_J \mathcal{C}(FJ, WJ \pitchfork C)\\ &\cong \mathrm{Cat}(\mathcal{J},\mathcal{C})(F, W \pitchfork X).\end{aligned}$$

This concludes the proof. □

Remark 4.3.2. This finally sheds a light on our proof of the Fubini theorem in 1.3.1: given that (co)ends are hom weighted (co)limits (see 4.1.13.WC3), the (co)end functor has a right/left adjoint given by tensoring with the weight hom: this is exactly the way in which we proved that $\int^A$: $\mathrm{Cat}(\mathcal{C}^{\mathrm{op}} \times \mathcal{C}, \mathcal{D})$ had a right adjoint, only without explicitly mentioning the technology of weighted (co)limits.

4.3.2 The Theory of Enriched (Co)ends

A fundamental step in laying the foundation of weighted limit theory is the natural isomorphism

$$\mathcal{C}(C, \lim^W F) \cong \mathcal{V}\text{-Cat}(\mathcal{J}, \mathcal{V})(W, \mathcal{C}(C, F\_)) \tag{4.54}$$

valid for functors

$$\mathcal{C} \xleftarrow{F} \mathcal{J} \xrightarrow{W} \text{Set}. \tag{4.55}$$

In order to export this isomorphism to the enriched setting, we shall make (4.54) take place in the base cosmos $\mathcal{V}$; in short, this means that we have to find a way to promote the category $\mathcal{V}$-Cat$(\mathcal{J}, \mathcal{V})$ of $\mathcal{V}$-functors and $\mathcal{V}$-natural transformations as an enriched category $\underline{\mathcal{V}\text{-Cat}}(\mathcal{C}, \mathcal{V})$. This means that every $\mathcal{V}$-Cat$(\mathcal{J}, \mathcal{V})(F, G)$, for functors F, G, must become an *object of enriched natural transformations* in $\mathcal{V}$.

To do this, we will endow $\mathcal{V}$-Cat with a closed symmetric monoidal structure, such that the natural isomorphism

$$\mathcal{V}\text{-Cat}(\mathcal{C} \boxtimes \mathcal{E}, \mathcal{D}) \cong \mathcal{V}\text{-Cat}(\mathcal{E}, \underline{\mathcal{V}\text{-Cat}}(\mathcal{C}, \mathcal{D})) \tag{4.56}$$

holds for $\mathcal{V}$-categories $\mathcal{C}, \mathcal{D}, \mathcal{E}$. The $\mathcal{V}$-category $\underline{\mathcal{V}\text{-Cat}}(\mathcal{C}, \mathcal{D})$ will thus be the internal hom for the closed monoidal structure given by $\boxtimes$.

Definition 4.3.3 (Tensor product of $\mathcal{V}$-categories). Given two $\mathcal{V}$-categories $\mathcal{C}, \mathcal{D}$ we define the $\mathcal{V}$-category $\mathcal{C} \boxtimes \mathcal{D}$ having

- as objects the set $\mathcal{C} \times \mathcal{D}$, and
- as $\mathcal{V}$-object of arrows $(C, D) \to (C', D')$ the object

$$\mathcal{C}(C, C') \otimes \mathcal{D}(D, D') \in \mathcal{V}. \tag{4.57}$$

The free $\mathcal{V}$-category $\mathcal{I}$ associated to the terminal category, having a single object $*$ and where $\mathcal{I}(*, *) = I$, the monoidal unit of $\mathcal{V}$, is the unit object for this monoidal structure.

Various checks are now in order:

- $\mathcal{C} \boxtimes \mathcal{D}$ really is a $\mathcal{V}$-category;
- $\_ \boxtimes = : \mathcal{V}\text{-Cat} \times \mathcal{V}\text{-Cat} \to \mathcal{V}\text{-Cat}$ is a bifunctor, which endows $\mathcal{V}$-Cat with a monoidal structure.

None of these constitutes a conceptual challenge.

Proposition 4.3.4. The monoidal category $(\mathcal{V}\text{-Cat}, \boxtimes)$ can be promoted to a closed monoidal category, with internal hom denoted $\underline{\mathcal{V}\text{-Cat}}(\_, =) : \mathcal{V}\text{-Cat}^{\text{op}} \times \mathcal{V}\text{-Cat} \to \mathcal{V}\text{-Cat}$.

Proof Given $\mathcal{C}, \mathcal{D} \in \mathcal{V}$-Cat we define a $\mathcal{V}$-category whose objects are $\mathcal{V}$-functors $F, G : \mathcal{C} \to \mathcal{D}$ and where (with a little abstraction from 1.4.1 to the enriched setting) the $\mathcal{V}$-object of natural transformations $F \Rightarrow G$ is defined via the end

$$\mathcal{V}\text{-Cat}(\mathcal{C}, \mathcal{D})(F, G) := \int_{C \in \mathcal{C}} \mathcal{D}(FC, GC). \tag{4.58}$$

In the unenriched case, the end was better understood as the equaliser of a pair of arrows:

$$\int_{C \in \mathcal{C}} \mathcal{D}(FC, GC) \cong \text{eq}\Big(\prod_{C \in \mathcal{C}} \mathcal{D}(FC, GC) \rightrightarrows \prod_{C, C'} \prod_{c \to c'} \mathcal{D}(FC, GC') \Big). \tag{4.59}$$

In the enriched case, we can consider the same symbol, and re-interpret the product $\prod_{\mathcal{C}(C,C')}$ as a suitable *power* in $\mathcal{V}$:

$$\int_{C \in \mathcal{C}} \mathcal{D}(FC, GC) \cong \text{eq}\Big(\prod_{C \in \mathcal{C}} \mathcal{D}(FC, GC) \rightrightarrows \prod_{C, C'} \mathcal{C}(C, C') \pitchfork \mathcal{D}(FC, GC') \Big) \tag{4.60}$$

(see [Gra80, §2.3] for more on this definition). It is now a matter of unwinding the definition to show that a $\mathcal{V}$-natural transformation corresponds to a generalised element of $\int_{C \in \mathcal{C}} \mathcal{D}(FC, GC)$; we leave the proof to the reader in Exercise 4.5.

It remains to prove, now, that the isomorphism (4.56) holds. This is rather easy, since in the above notations, any functor $F : \mathcal{C} \boxtimes \mathcal{E} \to \mathcal{D}$ defines a unique functor $\hat{F} : \mathcal{E} \to \underline{\mathcal{V}\text{-Cat}}(\mathcal{C}, \mathcal{D})$. For any two objects $E, E' \in \mathcal{E}$, the collection of arrows

$$\begin{aligned} \mathcal{E}(E, E') \to \mathcal{V}(\mathcal{C}(C, C'), \mathcal{D}(F(C, E), F(C', E'))) \\ = \mathcal{C}(C, C') \pitchfork \mathcal{D}(F(C, E), F(C', E')) \end{aligned}$$

given by the action of F on hom objects is a wedge in the pair (C, C'). Thus, since

$$\int_{CC'} \mathcal{C}(C, C') \pitchfork \mathcal{D}(F(C, E), F(C', E')) \cong \int_C \mathcal{D}(F(C, E), F(C, E')) \tag{4.61}$$

by 4.1.8, we get a correspondence on objects $\hat{F} : E \mapsto (\lambda C.F(C, E))$, and a correspondence on hom objects in the form of maps $\hat{F}_{E,E'} : \mathcal{E}(E, E') \to \int_C \mathcal{D}(\hat{F}E(C), \hat{F}E'(C)) = \underline{\mathcal{V}\text{-Cat}}(\mathcal{C}, \mathcal{D})(\hat{F}E, \hat{F}E')$. The facts that each $\hat{F}(E)$ is a $\mathcal{V}$-functor $\mathcal{C} \to \mathcal{D}$, and that $\hat{F}$, so defined, is a $\mathcal{V}$-functor $\mathcal{E} \to \underline{\mathcal{V}\text{-Cat}}(\mathcal{C}, \mathcal{D})$, are both necessary but tedious checks. As a proof of why we choose the word 'tedious' to describe the process, let us show the reader the argument proving that the triangle

$$\begin{array}{c} I \\ \iota \swarrow \quad \boxed{1} \quad \searrow \iota' \\ \mathcal{E}(E,E) \xrightarrow[\hat{F}_{E,E}]{} \int_C \mathcal{D}(F(C,E),F(C,E)) \end{array} \tag{4.62}$$

commutes, so that $\hat{F}$ preserves the identity arrows $\iota : I \to \mathcal{E}(E;E)$ and $\iota' : I \to \int_C \mathcal{D}(F(C,E),F(C,E))$. First, we have to define ι'. The component at (C,E) is obtained from the universal property of the end at codomain, starting from the wedge induced by F. Second, it is evident that diagram (4.62) commutes if and only if the whiskered diagram

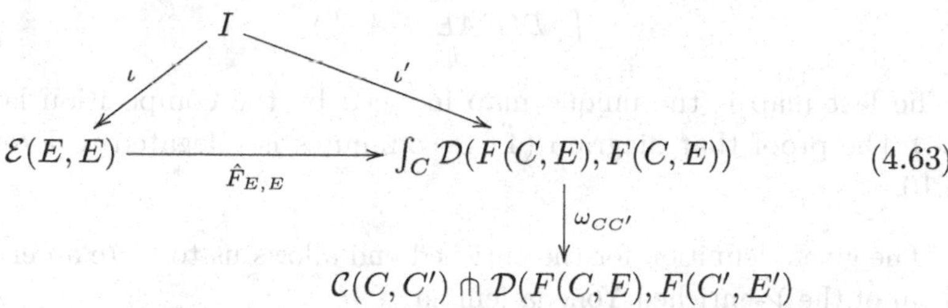

commutes for every choice of $C, C' \in \mathcal{C}$ (we implicitly use (4.61) to describe conveniently the universal wedge). In order to show this last commutativity, consider the diagram

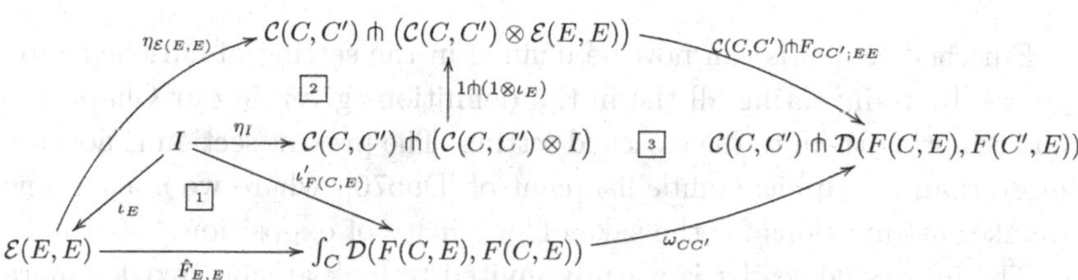

The sub-diagrams $\boxed{2}$ and $\boxed{3}$ commute, respectively because the unit of the adjunction $\mathcal{C}(C,C)\otimes \_ \dashv \mathcal{C}(C,C) \pitchfork \_$ is natural, and because we assumed F was a $\mathcal{V}$-functor. From this, we deduce the desired commutativity.

The fact that $\hat{F}$ preserves composition translates into the commutativity of the square

$$\begin{array}{ccc} \mathcal{E}(E,E')\otimes\mathcal{E}(E',E'') & \xrightarrow{c^{\mathcal{E}}} & \mathcal{E}(E,E'') \\ \downarrow & & \downarrow{\scriptstyle \hat{F}_{E,E''}} \\ \int_A \mathcal{D}(FAE,FAE')\otimes\int_B \mathcal{D}(FBE',FBE'') & \xrightarrow[k]{} & \int_C \mathcal{D}(FCE,FCE'') \end{array} \tag{4.64}$$

where the lower horizontal map k arises from the composition of maps

$$\begin{array}{c} \int_A \mathcal{D}(FAE,FAE')\otimes\int_B \mathcal{D}(FBE',FBE'') \\ \downarrow{\scriptstyle \omega\otimes\omega} \\ \mathcal{D}(FAE,FAE')\otimes\mathcal{D}(FAE',FAE'') \\ \downarrow{\scriptstyle \gamma} \\ \int_A \mathcal{D}(FAE,FAE'') \end{array} \tag{4.65}$$

(The last map is the unique map induced by the composition law of $\mathcal{D}$.) The proof that diagram (4.64) commutes is relegated to Exercise 4.10. □

The given definition for the enriched end allows us to state an elegant form of the $\mathcal{V}$-enriched Yoneda lemma.

Remark 4.3.5 ($\mathcal{V}$-Yoneda lemma). Let $\mathcal{D}$ be a small $\mathcal{V}$-category, $D \in \mathcal{D}$ an object, and $F : \mathcal{D} \to \mathcal{V}$ a $\mathcal{V}$-functor. Then the canonical map

$$FD \to \underline{\mathcal{V}\text{-Cat}}(\mathcal{D},\mathcal{V})(\mathcal{D}(D,\_),F) \tag{4.66}$$

induced by the universal property of the involved end[1] is a $\mathcal{V}$-isomorphism.

Enriched (co)ends can now be defined in the setting of enriched categories, by re-inventing all the initial definitions given in our Chapter 1, and adapting them to the enriched setting. The present section is nothing more than a graphical embellishment of [Dub70], where we make a few blanket assumptions for the sake of simplicity of exposition.

The interested reader is warmly invited to look at said text for more information and more general statements.

[1] Notice that this is an alternative point of view on the proof of the ninja Yoneda lemma 2.2.1: the morphism in (4.66) is induced by a wedge $FD \to \mathcal{V}(\mathcal{D}(D,D'),FD')$ in D', whose members are the mates of the various $\mathcal{D}(D,D') \to \mathcal{V}(FD,FD')$ giving the action of F on arrows.

Notation 4.3.6. Our blanket assumption throughout this section is the following: categories are $\mathcal{V}$-cotensored (see [Dub70]). In the absence of cotensors, the enriched counterpart of a (co)end is not well-behaved enough to be interesting; in such cases, one loses the equivalent description of a (co)end as weighted (co)limit, because our 4.3.1 fails to be true).

Moreover, we sometimes blur the distinction between (di)natural families $D \to T(C,C)$, for a $\mathcal{V}$-functor $T : \mathcal{C}^{\text{op}} \boxtimes \mathcal{C} \to \mathcal{D}$ and $\mathcal{V}$-arrows $I \to \mathcal{D}(D, T(C,C))$. We do this quite liberally especially when drawing commutative diagrams or referring to components of $\mathcal{V}$-natural transformations.

The enriched analogue of extranaturality can be defined as follows.

Definition 4.3.7 (Enriched extranaturality). Let $P : \mathcal{A} \boxtimes \mathcal{B}^{\text{op}} \boxtimes \mathcal{B} \to \mathcal{E}$ and $Q : \mathcal{A} \boxtimes \mathcal{C}^{\text{op}} \boxtimes \mathcal{C} \to \mathcal{E}$ be $\mathcal{V}$-functors; an *extranatural transformation* $\alpha P \Rightarrow Q$ consists of a family of morphisms

$$\alpha_{ABC} : P(A,B,B) \to Q(A,C,C) \tag{4.67}$$

in $\mathcal{E}$, indexed by the objects of $\mathcal{A}, \mathcal{B}, \mathcal{C}$, such that the following three diagrams made by the action on morphisms of P, Q commute:

$$\begin{array}{ccc}
\mathcal{A}(A,A') & \xrightarrow{P(_,B,B)} & \mathcal{E}(P(A,B,B), P(A',B,B)) \\
{\scriptstyle Q(_,C,C)}\downarrow & & \downarrow{\scriptstyle \mathcal{E}(1,\alpha_{A'BC})} \\
\mathcal{E}(Q(A,C,C), Q(A',C,C)) & \xrightarrow[\mathcal{E}(\alpha_{ABC},1)]{} & \mathcal{E}(P(A,B,B), Q(A',C,C))
\end{array}$$

$$\begin{array}{ccc}
\mathcal{B}(B,B') & \xrightarrow{P(A,B',_)} & \mathcal{E}(P(A,B',B), P(A,B',B')) \\
{\scriptstyle P(A,_,B)}\downarrow & & \downarrow{\scriptstyle \mathcal{E}(1,\alpha_{AB'C})} \\
\mathcal{E}(P(A,B',B), P(A,B,B)) & \xrightarrow[\mathcal{E}(1,\alpha_{ABC})]{} & \mathcal{E}(P(A,B',B), Q(A,C,C))
\end{array} \tag{4.68}$$

$$\begin{array}{ccc}
\mathcal{C}(C,C') & \xrightarrow{A(A,C,_)} & \mathcal{E}(Q(A,C,C), Q(A,C,C')) \\
{\scriptstyle Q(A,_,C')}\downarrow & & \downarrow{\scriptstyle \mathcal{E}(\alpha_{ABC},1)} \\
\mathcal{E}(Q(A,C',C'), Q(A,C,C')) & \xrightarrow[\mathcal{E}(\alpha_{ABC'},1)]{} & \mathcal{E}(P(A,B,B), Q(A,C,C'))
\end{array}$$

(A, A', B, B', C, C' are objects of the respective categories, and $\mathcal{E}(u, 1)$ is the image of u under the functor $\mathcal{E}(\_, E) : \mathcal{E}^{\text{op}} \to \mathcal{V}$ as E runs over the objects of $\mathcal{E}$).

We collect the $\mathcal{V}$-extranatural transformations $\alpha : D \Rightarrow T$ into the object

$$\underline{\mathcal{V}\text{-Cat}}_e(\mathcal{C}^{op} \boxtimes \mathcal{C}, \mathcal{D})(\Delta D, T). \tag{4.69}$$

Remark 4.3.8. *Enriched dinaturality* does not seem to appear in the literature. The scope of the present remark is to show why such a notion is almost always useless. In short, for a generic base of enrichment there is no notion of 'constant' $\mathcal{V}$-functor, and thus there is no way to define (co)wedges as dinatural transformations to/from a constant.

First, we define the enriched end of a functor taking value in the base of enrichment.

Definition 4.3.9. [Dub70, I.3.1] Given a $\mathcal{V}$-category $\mathcal{C}$ and a $\mathcal{V}$-functor $T : \mathcal{C}^{op} \boxtimes \mathcal{C} \to \mathcal{V}$, the *end* of T is an object of $\mathcal{V}$, denoted $\int_C T(C, C)$, and a $\mathcal{V}$-natural family of morphisms $\{\int_C T(C, C) \xrightarrow{p_C} T(C, C) \mid C \in \mathcal{C}\}$ such that given any other $\mathcal{V}$-natural family $\{u_C : V \to T(C, C) \mid C \in \mathcal{C}\}$ there exists a unique $V \to \int_C T(C, C)$ such that $p_C \circ \bar{u} = u_C$ in the diagram

$$\begin{array}{ccc} V & \xrightarrow{\bar{u}} & \int_C T(C,C) \\ & {}_{u_C}\searrow \quad \swarrow_{p_C} & \\ & T(C,C) & \end{array} \tag{4.70}$$

The definition of end for a generic codomain is now given representably, following the enriched version of Yoneda–Grothendieck philosophy (see 2.1.1). The enriched Yoneda lemma 4.3.5 draws the connection between the two.

Definition 4.3.10. Let $T : \mathcal{C}^{op} \boxtimes \mathcal{C} \to \mathcal{D}$ be a $\mathcal{V}$-functor; the *end* of T is an object $\int_C T(C, C)$ of $\mathcal{D}$ endowed with a $\mathcal{V}$-natural family of morphisms $\{\int_C T(C, C) \xrightarrow{p_C} T(C, C) \mid C \in \mathcal{C}\}$ such that given any $D \in \mathcal{D}$ the family of arrows

$$\mathcal{D}(D, \textstyle\int_C T(C, C)) \xrightarrow{\mathcal{D}(D, p_C)} \mathcal{D}(D, T(C, C)) \tag{4.71}$$

exhibits the end of $\mathcal{D}(D, T(\_, =)) : \mathcal{C}^{op} \boxtimes \mathcal{C} \to \mathcal{D} \to \mathcal{V}$.

Remark 4.3.11. Equivalent to the universal property above is the fact that there is a natural bijection between the $\mathcal{V}$-wedges $D \xrightarrow{u_C} T(C, C)$ and the underlying set of $\mathcal{D}(D, \int_C T(C, C))$:

$$\underline{\mathcal{V}\text{-Cat}}_e(\mathcal{C}^{op} \boxtimes \mathcal{C}, \mathcal{D})(\Delta D, T) \cong \mathcal{V}\Big(I, \mathcal{D}\Big(D, \int_C T(C, C)\Big)\Big). \tag{4.72}$$

Definition 4.3.12. Let $T : \mathcal{C}^{\text{op}} \boxtimes \mathcal{C} \to \mathcal{D}$ be a $\mathcal{V}$-functor; the universal property of $\int_C T(C,C)$ yields a unique morphism $G\big(\int_C T(C,C)\big) \to \int_C GT(C,C)$ for every $\mathcal{V}$-functor $G : \mathcal{D} \to \mathcal{E}$. This is the unique morphism closing the diagram

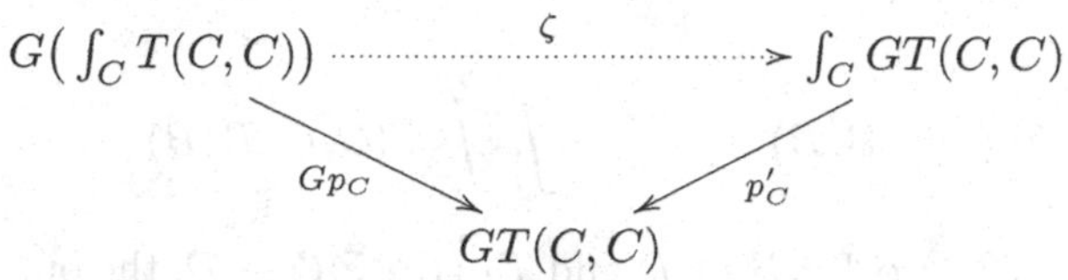

$$\begin{array}{ccc} G\big(\int_C T(C,C)\big) & \overset{\zeta}{\dashrightarrow} & \int_C GT(C,C) \\ & {\scriptstyle Gp_C}\searrow \quad \swarrow{\scriptstyle p'_C} & \\ & GT(C,C) & \end{array} \tag{4.73}$$

to a commutative one, where p'_C is the terminal wedge of GT. We say that G *preserves* the end of T if whenever $\int_C T$ exists, so does $\int_C GT$, and the above comparison morphism ζ_G is invertible.

Remark 4.3.13 (Parametric ends of $\mathcal{V}$-functors)**.** Let $T : \mathcal{C}^{\text{op}} \boxtimes \mathcal{C} \boxtimes \mathcal{E} \to \mathcal{D}$ be a $\mathcal{V}$-functor; the monoidal closed structure of $\mathcal{V}$-Cat gives T a mate $\hat{T} : \mathcal{C}^{\text{op}} \boxtimes \mathcal{C} \to [\mathcal{E}, \mathcal{D}]$. The *parametric end* of T, provided it exists, consists of the end of $\hat{T}$, promoted to a $\mathcal{V}$-functor $\mathcal{E} \to \mathcal{D}$.

In more detail, this means that the parametric end of T exists if for every $E \in \mathcal{E}$ the end of $T(\_, =; E) : \mathcal{C}^{\text{op}} \boxtimes \mathcal{C} \to \mathcal{D}$ exists and there is a unique morphism

$$\mathcal{E}(E, E') \to \mathcal{D}\Big(\int_C T(C,C;E), \int_C T(C,C;E')\Big) \tag{4.74}$$

giving to $\int_C T(C,C;\_)$ the structure of a $\mathcal{V}$-functor $\mathcal{E} \to \mathcal{D}$, in such a way that $p_{(C),E} : \int_C T(C,C;E) \to T(C,C;E)$ is a $\mathcal{V}$-wedge for $T(\_,\_;E)$, and it is $\mathcal{V}$-natural in E.

The above remark is based on the fact that if $\mathcal{D}$ is a complete $\mathcal{V}$-category, namely a $\mathcal{V}$-category such that every limit of $F : \mathcal{J} \to \mathcal{E}$ weighted by $W : \mathcal{J} \to \mathcal{V}$ exists, then $[\mathcal{E}, \mathcal{D}]$ is complete as well, because limits can be computed pointwise. Spelled out precisely, this means that given a $\mathcal{V}$-functor $F : \mathcal{J} \to [\mathcal{E}, \mathcal{D}]$, the correspondence

$$E \mapsto \int_J WJ \pitchfork F(J,E) \tag{4.75}$$

defines a $\mathcal{V}$-functor, and such a $\mathcal{V}$-functor has the universal property of the limit $\lim^W F$.

This statement follows at once from a simple computation with the involved end: it just remains to see that the above definition gives a well defined $\mathcal{V}$-functor.

Let now $\mathcal{E} = \mathcal{B}^{\mathrm{op}} \boxtimes \mathcal{B}$; then an enriched Fubini rule holds for functors

$$T : \mathcal{C}^{\mathrm{op}} \boxtimes \mathcal{C} \boxtimes \mathcal{B}^{\mathrm{op}} \boxtimes \mathcal{B} \to \mathcal{E}.$$

Theorem 4.3.14 (Enriched Fubini rule). Whenever both inner parametric ends

$$\int_C \int_B T(C, C; B, B), \qquad \int_B \int_C T(C, C; B, B) \tag{4.76}$$

exist as functors $T_{\mathcal{B}} : \mathcal{B}^{\mathrm{op}} \boxtimes \mathcal{B} \to \mathcal{D}$ and $T_{\mathcal{C}} : \mathcal{C}^{\mathrm{op}} \boxtimes \mathcal{C} \to \mathcal{D}$, the outer ends exist if and only if either one of them exists, and they are canonically isomorphic objects, in turn isomorphic to the end of the rearranged functor $(\mathcal{B} \boxtimes \mathcal{C})^{\mathrm{op}} \boxtimes (\mathcal{B} \boxtimes \mathcal{C}) \to \mathcal{D}$.

Exercises

4.1 Prove all the statements in 4.1.13.

4.2 This is a corollary to 4.3.1, where we take the tensored and cotensored $\mathcal{V}$-category $\mathcal{A} = \mathcal{V}$. Prove that the adjunction reduces to the adjunction $\mathrm{Lan}_y W \dashv N_W$ of 3.2.2. Do this in two ways: first, through coend calculus, and then exhibiting unit and counit of an adjunction $W \otimes \_ \dashv \mathcal{V}(W\_, =) : [\mathcal{C}, \mathcal{V}] \leftrightarrows \mathcal{V}$.

4.3 [Kel89] Mimic the argument in 4.2.7 to give a characterisation based on coend calculus of the following weighted (co)limits.

(a) The *equifier* of a pair of functors $f, g : B \rightrightarrows C$ and two natural transformations $\alpha, \beta : f \Rightarrow g$; it is defined as the weighted limit $\lim^W F$ where $W : \mathcal{J} \to \mathrm{Cat}$ sends the 2-category

$$\mathcal{J} = 0 \underset{\longrightarrow}{\overset{\longrightarrow}{\Downarrow \quad \Downarrow}} 1$$

to the diagram of 2-cells $[0] \underset{t}{\overset{s}{\underset{\longrightarrow}{\overset{\longrightarrow}{\Downarrow\delta_0\ \Downarrow\delta_1}}}} [1]$ and $F : \mathcal{J} \to \mathrm{Cat}$

sends $\mathcal{J}$ to the diagram of 2-cells $B \underset{g}{\overset{f}{\underset{\longrightarrow}{\overset{\longrightarrow}{\Downarrow\alpha\ \Downarrow\beta}}}} C$ and enjoys

the following universal property (see [Kel89, 4.5]). There is a

universal diagram

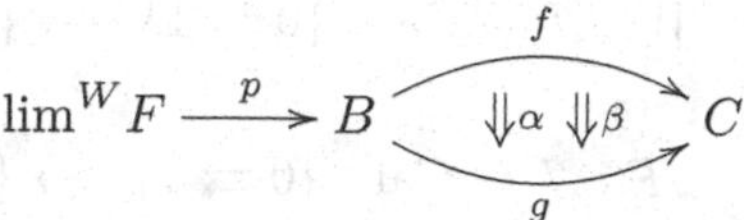

such that $\alpha * p = \beta * p$ and given any other $q : A \to B$ such that $\alpha * q = \beta * q$ there is a unique $\bar{q} : A \to \lim^W F$ such that $q = p \circ \bar{q}$ in the diagram

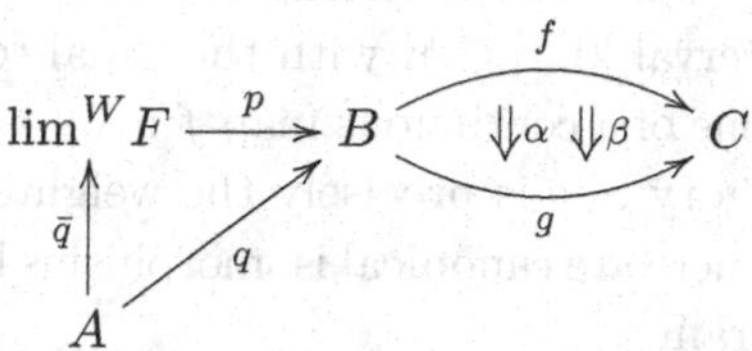

Moreover, given $h, k : A \to \lim^W F$ and a 2-cell $\mu : ph \Rightarrow pk$, there is a unique 2-cell $\bar{\mu}$ such that $p * \bar{\mu} = \mu$.

(b) The *co*equifier of a pair of functors $f, g : A \rightrightarrows B$. This is defined as the weighted colimit $\operatorname{colim}^W F$ for the same W, F above, and enjoys the dual universal property (write it down in detail).

(c) The *lax limit* of a functor $f : A \to B$. This is defined as the weighted limit $\lim^W F$ where $W : \{0 < 1\} \to \text{Cat}$ chooses the functor $\{0\} \xrightarrow{0} \{0 \to 1\}$ and $F : \mathcal{J} \to \text{Cat}$ chooses the functor $f : B \to C$. The object $\lim^W F$ has the following universal property. There exists a pair (u, v) of 1-cells and a 2-cell $\lambda : fu \Rightarrow v$ in a diagram

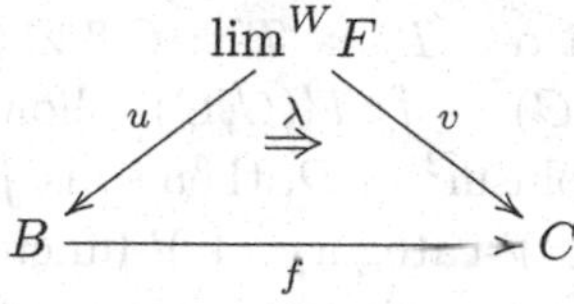

terminal with respect to this property (write down the universal property in detail).

(d) The *pseudolimit* of a functor $f : A \to B$, where $\mathcal{J}, F$ are the same, and W is instead the embedding of the domain in the generic isomorphism, i.e. the functor $\{0\} \xrightarrow{0} \{0 \cong 1\}$.

4.4 Let $\mathcal{J} = \{0 \rightrightarrows 1\}$, let $[n]$ denote the category $\{0 < 1 < \cdots < n\}$

and

$$W : \mathcal{J} \to \mathrm{Cat} : \{0 \rightrightarrows 1\} \longmapsto \{[1] \underset{d_2}{\overset{d_0}{\rightrightarrows}} [2]\}$$

$$F : \mathcal{J} \to \mathrm{Cat} : \{0 \rightrightarrows 1\} \longmapsto \{B \underset{g}{\overset{f}{\rightrightarrows}} C\}$$

where $d_i : [n] \to [n+1]$ avoids the ith element. What is the universal property of $\lim^W F$?

4.5 Complete the proof of 4.3.4.

4.6 Let $W : S^0 \hookrightarrow D^1$ be the canonical inclusion of the endpoints $\{0, 1\}$ into the interval $[0, 1] \subset \mathbb{R}$ with the usual topology; prove that the mapping cone of a continuous map $f : X \to Y$ regarded as a functor to the category Spc is precisely the weighted colimit $\mathrm{colim}^W f$.

4.7 Show that there are canonical isomorphisms $\lim^W FJ \cong \lim^{\mathrm{Lan}_J W} F$, in the diagram

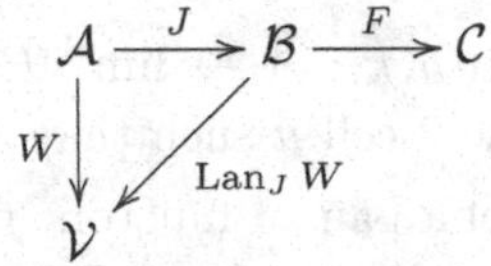

and dually $\mathrm{colim}^{WJ} F \cong \mathrm{colim}^W \mathrm{Lan}_J F$, in the diagram

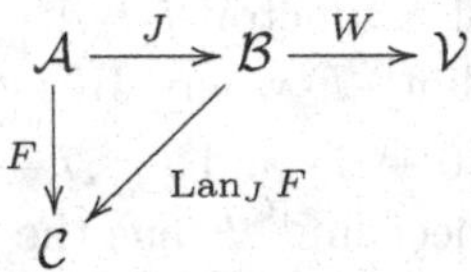

4.8 Is there a Fubini rule for weighted (co)limits?

4.9 Use the universal property of 4.3.10 to show that every $\mathcal{V}$-natural transformation $\alpha : T \Rightarrow T' : \mathcal{C}^{\mathrm{op}} \boxtimes \mathcal{C} \to \mathcal{D}$ induces an arrow $\int_C \alpha : \int_C T(C, C) \to \int_C T'(C, C)$. Show that if α is a componentwise monomorphism[2] in $\mathcal{D}$, then so is $\int_C \alpha$.

4.10 Prove that the $\mathcal{V}$-category of $\mathcal{V}$-functors $\mathcal{C} \to \mathcal{D}$ is indeed a $\mathcal{V}$-category, in the sense that the axioms of A.7.1 hold. More precisely, the claim here is that given $\mathcal{V}$-functors $F, G : \mathcal{C} \to \mathcal{D}$, the $\mathcal{V}$-category with hom objects $\int_C \mathcal{D}(FC, GC)$ is well defined and indeed a $\mathcal{V}$-category. Verifying the axioms once and for all is instructive, but things get a little bit hairy.

[2] We say that $u : D \to D'$ is a $\mathcal{V}$-*monomorphism* if the image of u under the enriched Yoneda embedding $よ(u) : よD \Rightarrow よD'$ is a monomorphism in $\mathcal{V}\text{-Cat}(\mathcal{D}^{\mathrm{op}}, \mathcal{V})$.

For example, this is the commutative diagram witnessing that composition has a right identity induced by $\iota^{GX} : I \to \mathcal{D}(GX, GX)$:

$$\begin{array}{ccc}
\left(\int_C \mathcal{D}(FC,GC)\right) \otimes I & \xrightarrow{1\otimes \iota^G} & \int_C \mathcal{D}(FC,GC) \otimes \int_B \mathcal{D}(GB,GB) \\
\downarrow {\scriptstyle \omega_C^{FG}\otimes I} & & \downarrow {\scriptstyle \omega_C^{FG}\otimes\omega_C^{GG}} \\
\mathcal{D}(FC,GC) \otimes I & \xrightarrow[1\otimes\iota^{GC}]{} & \mathcal{D}(FC,GC) \otimes \mathcal{D}(GC,GC) \\
& \searrow^{\cong} & \downarrow {\scriptstyle \mathrm{comp}} \\
& & \mathcal{D}(FC,GC)
\end{array}$$

where ι^G is the unique map such that $\omega_C^{GG} \circ \iota^G = \iota^{GC}$. The associativity property is a bit worse, but the appropriate diagram can be reshaped and broken according to the following scheme:

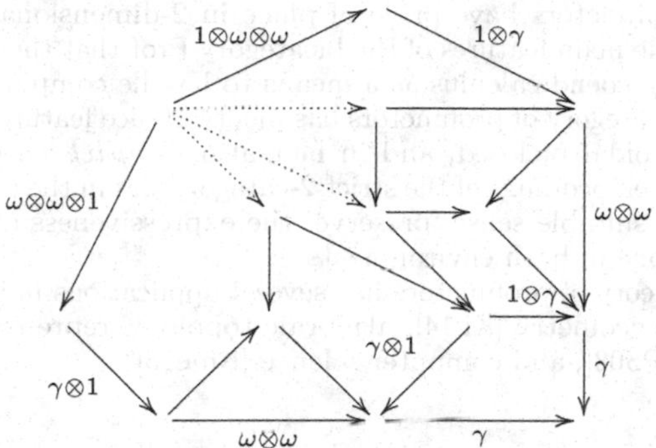

where γ are suitable composition maps, ω are suitable cowedge maps, and the dotted arrows are induced as canonical maps, say,

$$\left(\int_C \mathcal{D}(FC,GC)\right) \otimes X \to \int_C \big(\mathcal{D}(FC,GC) \otimes X\big). \tag{4.77}$$

Starting from this, prove that diagram (4.64) commutes.

4.11 Prove 4.3.14 with the aid of [Dub70, I.3.4].

4.12 Prove the following: if $T : \mathcal{C}^{\mathrm{op}} \boxtimes \mathcal{C} \boxtimes \mathcal{E} \to \mathcal{D}$ is a $\mathcal{V}$-functor, we consider $T(C, C; \_) : \mathcal{E} \to \mathcal{D}$ as a $\mathcal{V}$-functor, and we let $W : \mathcal{C} \to \mathcal{V}$ be a weight. Then

$$\int_C \lim\nolimits^W T(C, C; E) \cong \lim\nolimits^W \int_C T(C, C; E).$$

Dualise for coends.

5
Profunctors

5.1 The 2-Category of Profunctors

SUMMARY. The present chapter introduces the theory of *profunctors*. Regarded as a generalisation of presheaves and modules over rings, profunctors have pride of place in 2-dimensional algebra. We explore the main features of the bicategory Prof that they form, heavily employing coend calculus as a means to handle computations.

The bicategory of profunctors has plenty of nice features: for example, it is monoidal biclosed, and in fact also *compact* closed. There are canonical embeddings of the strict 2-category Cat in the bicategory Prof, that in a suitable sense 'preserve' the expressiveness of the category theory done in both environments.

The theory of profunctors has several applications in algebra [GJ17], algebraic geometry [KL14], algebraic topology, representation theory [Tam06, PS08], and computer science [Kme18].

> On peut regarder une pièce d'un puzzle pendant trois jours et croire tout savoir de sa configuration et de sa couleur sans avoir le moins du monde avancé : seule compte la possibilité de relier cette pièce à d'autres pièces
>
> G. Perec, *La vie, mode d'emploi*

The lucid presentation in the notes [BS00] and in [Bor94b], [CP89, §4] are standard references for following this chapter. (Co)end calculus provides a closer look and a unified description for the material appearing in [BS00]; here some arguments become neater, some others are made more explicit or computationally evident.

First, recall from 3.2.12 and 3.2.13 that we can define the tensor product of modules as a coend.

Example (The tensor product of modules as a coend). Any ring R can be regarded as an Ab-enriched category having a single object: under this identification, the category of left modules over R is just the category of functors $R \to \mathrm{Ab}$, and dually, the category of right R-modules is the category of contravariant functors, $R^{\mathrm{op}} \to \mathrm{Ab}$

$$\begin{aligned}\mathrm{Mod}_R &\cong \mathrm{Cat}(R^{\mathrm{op}}, \mathrm{Ab}) \\ {}_R\mathrm{Mod} &\cong \mathrm{Cat}(R, \mathrm{Ab}).\end{aligned} \tag{5.1}$$

Moreover, given functors $A : R^{\mathrm{op}} \to \mathrm{Ab}, B : R \to \mathrm{Ab}$, there is a canonical isomorphism between the functor tensor product $A \otimes B$ defined as the coend of $A \otimes_{\mathbb{Z}} B$, and $A \otimes_R B$. In short, there is a coequaliser diagram

$$\bigoplus_{r \in R} A \otimes_{\mathbb{Z}} B \underset{1 \otimes r}{\overset{r \otimes 1}{\rightrightarrows}} A \otimes_{\mathbb{Z}} B \longrightarrow A \otimes_R B. \tag{5.2}$$

Remark 5.1.1. We can define a bicategory (see A.7.11) Mod having

- 0-cells the rings $R, S, \ldots$;
- 1-cells $R \to S$ the modules ${}_R M_S$, regarded as functors $M : R \times S^{\mathrm{op}} \to \mathrm{Ab}$;
- 2-cells $f : {}_R M_S \Rightarrow {}_R N_S$ the module homomorphisms $f : M \to N$.

The notion of profunctor arises from a massive, but fairly natural, generalisation of this construction.

The parallel here is motivated by the fact that categories are certain monoid objects, and the features of such monoids are captured by their *categories of action*: in this perspective a (left) action of a category on a set is merely a functor $\mathcal{C} \to \mathrm{Set}$. The analogy with a group action is once again evident. In fact, a (left) group action of G on a set X is merely a presheaf $G \to \mathrm{Set}$, when G is regarded as a category.

This allows us to state the following definition.

Definition 5.1.2 (The bicategory of profunctors). There exists a bicategory Prof having

- 0-cells the (small) categories $\mathcal{A}, \mathcal{B}, \ldots$;
- 1-cells $\mathfrak{p}, \mathfrak{q} \ldots$, denoted as arrows $\mathfrak{p} : \mathcal{A} \rightsquigarrow \mathcal{B}$, the functors
$$\mathcal{A}^{\mathrm{op}} \times \mathcal{B} \to \mathrm{Set}; \tag{5.3}$$
- 2-cells $\alpha : \mathfrak{p} \Rightarrow \mathfrak{q}$ natural transformations between functors.

Given two contiguous 1-cells $\mathcal{A} \overset{\mathfrak{p}}{\rightsquigarrow} \mathcal{B} \overset{\mathfrak{q}}{\rightsquigarrow} \mathcal{C}$ we define their composition $\mathfrak{q} \diamond \mathfrak{p}$ as the coend

$$\mathfrak{q} \diamond \mathfrak{p}(A, C) := \int^{B \in \mathcal{B}} \mathfrak{p}(A, B) \times \mathfrak{q}(B, C). \tag{5.4}$$

Definition 5.1.3. This definition works well also with Set replaced by an arbitrary *Bénabou cosmos* $\mathcal{V}$, i.e. in any symmetric monoidal closed, complete and cocomplete category. In this case we speak of $\mathcal{V}$-profunctors in the bicategory $\mathsf{Prof}(\mathcal{V})$.

Remark 5.1.4 (Naming a category)**.** Since their first introduction, profunctors have been called many other names, depending on the leading perspective that guided their definition.

PN1. The 1-cells of Prof are called *profunctors*, because they generalise functors. We will see that some profunctors are *representable*; they are the ones of the form $\mathcal{B}(B, FA)$ or $\mathcal{B}(B, FA)$ for some functor $F : \mathcal{A} \to \mathcal{B}$ between categories. A *pro*functor thus works 'on behalf' of a functor (this is one of the meanings of the prefix *pro*).

PN2. In the same vein, *relations* are generalised functions too: this is why some people (among which, A. Joyal) prefer to call the 1-cells of Prof *relators*. Just as a func*tion* is a special kind of rela*tion*, a func*tor* is a special kind of rela*tor*. This analogy is deeper than it seems: we will see in 5.1.5 that there is a bicategory of relations, and this bicategory is a $\mathsf{Prof}(\mathcal{V})$ for a certain choice of $\mathcal{V}$.

PN3. Following the idea that *distributions* are generalised functions in functional analysis, the 1-cells of Prof are also called *distributors*, when we follow the intuition that funct*ions* are to funct*ors* as distribut*ions* are to distribut*ors*. As the nLab [nLa21] says,

> Jean Bénabou, who invented the term and originally used "profunctor", now prefers "distributor", which is supposed to carry the intuition that a distribut*or* generalises a funct*or* in a similar way to how a distribut*ion* generalises a funct*ion*.

Again, this intuition is deeper than it seems: in [Law07] Lawvere defined a notion of 'distribution between toposes', in such a way that the points of a topos $p : \mathrm{Set} \to \mathcal{E}$ behave like Dirac delta functions, and such that distributions between presheaf toposes are exactly profunctors. We discuss the analogy between Dirac deltas and representable functors in 2.2.6.

PN4. The 1-cells of Prof are sometimes called *correspondences*: consider the case when $\mathcal{V} = \{0, 1\}$, i.e. where $\mathcal{A}, \mathcal{B}$ are sets regarded as discrete categories, and see 5.1.5 below.

PN5. Drawing from 3.2.13, the 1-cells of Prof are sometimes called *bimodules*: indeed, the bicategory Mod is precisely the subcategory of $\mathsf{Prof}(\mathrm{Ab})$ made by one-object Ab-categories. See [KL14] for applications in algebraic geometry, and [Gen15] for applications of bimodules in homological algebra.

Category theorists know well that an elephant can have different names according to the angle it is observed from; we accept this situation, and we tacitly stick to calling the 1-cells of Prof 'profunctors' without making further mention of this variety of names. However, we invite the reader to maintain clear the intuition conveyed by all these names, in order to appreciate the variety of contexts in which the notion of profunctor naturally arises.

Example 5.1.5 (Profunctors as generalised relations). We consider categories enriched over the monoidal category $\mathcal{V} = \{0 < 1\}$.

By definition, a profunctor between $\{0,1\}$-categories is a function between sets $A^{\text{op}} \times B \to \{0,1\}$, or more precisely a function $A \times B \to \{0,1\}$ (a $\{0,1\}$-enriched category is merely a set, and dualisation on a discrete category has no effect), that is to say a *relation* regarded as a subset $R \subseteq A \times B$.

The standpoint regarding profunctors as generalised relations is what Lawvere [Law73, §4,5] calls *generalised logic*, and it regards the coend in 5.1.2, as well as the product $\mathfrak{p}(A,X) \times \mathfrak{q}(X,B)$ therein, as a *generalised existential quantification* and a *generalised conjunction* respectively, giving a composition rule for generalised relations. The coend stands on the same ground as the composition rule for relations (even more, the two constructions have the same universal property), in such a way that the composition of relations $R \subseteq Y \times Z, S \subseteq X \times Y$ is given by the rule

$$(x,z) \in R \circ S \iff \exists y \in Y : \big((x,y) \in S\big) \wedge \big((y,z) \in R\big). \tag{5.5}$$

$(x,z) \in S \circ R$	$\iff$	$\exists y \in Y$	$\big((x,y) \in S\big)$	$\wedge$	$\big((y,z) \in R\big)$
$(\mathfrak{q} \diamond \mathfrak{p})(X,Z)$	$=$	$\int^{Y \in \mathcal{Y}}$	$\mathfrak{p}(Y,Z)$	$\times$	$\mathfrak{q}(X,Y)$

The analogy between the composition of profunctors between categories and the composition of relations between sets gives rise to what Lawvere calls *generalised logic*.

Example 5.1.6. Let again A, B be two sets, considered as discrete categories. A profunctor $\mathfrak{p} : A \rightsquigarrow B$ is then simply a collection of sets P_{ab}, one for each $a \in A, b \in B$. Profunctor composition then results in a 'categorified' matrix multiplication, since the coend in (5.4) boils down to a mere coproduct of sets: given $\mathfrak{p} : A \to B$, $\mathfrak{q} : B \to C$ we have

$$(\mathfrak{p} \diamond \mathfrak{q})_{ac} = \coprod_{b \in B} P_{ab} \times Q_{bc} \tag{5.6}$$

if $P_{ab} = \mathfrak{p}(A, B)$ and $Q_{bc} = \mathfrak{q}(b, c)$.

Remark 5.1.7. There is an alternative but equivalent definition for the composite profunctor $\mathfrak{q} \diamond \mathfrak{p}$ which exploits the universal property of $\mathrm{Cat}(\mathcal{C}^{\mathrm{op}}, \mathrm{Set})$ as a free cocompletion. By definition, the category of profunctors $\mathfrak{p} : \mathcal{A} \rightsquigarrow \mathcal{B}$ and natural transformations fits into an isomorphism of categories

$$\mathrm{Cat}(\mathcal{A}^{\mathrm{op}} \times \mathcal{B}, \mathrm{Set}) \cong \mathrm{Cat}(\mathcal{B}, \mathrm{Cat}(\mathcal{A}^{\mathrm{op}}, \mathrm{Set})); \tag{5.7}$$

under this isomorphism, $\mathfrak{p}$ corresponds to a functor $\widehat{\mathfrak{p}} : \mathcal{B} \to \mathrm{Cat}(\mathcal{A}^{\mathrm{op}}, \mathrm{Set})$ obtained as $B \mapsto \mathfrak{p}(\_, B)$.

We can thus define the composition $\mathcal{A} \overset{\mathfrak{p}}{\rightsquigarrow} \mathcal{B} \overset{\mathfrak{q}}{\rightsquigarrow} \mathcal{C}$ to be $\mathrm{Lan}_{\text{よ}}\, \widehat{\mathfrak{p}} \circ \widehat{\mathfrak{q}}$ ($\circ$ is the usual composition of functors):

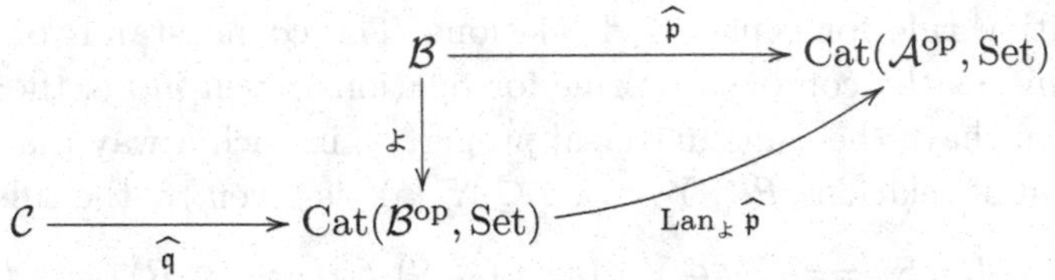

$$\begin{array}{ccccc} & & \mathcal{B} & \xrightarrow{\ \widehat{\mathfrak{p}}\ } & \mathrm{Cat}(\mathcal{A}^{\mathrm{op}}, \mathrm{Set}) \\ & & \downarrow{\scriptstyle \text{よ}} & \nearrow{\scriptstyle \mathrm{Lan}_{\text{よ}} \widehat{\mathfrak{p}}} & \\ \mathcal{C} & \xrightarrow[\ \widehat{\mathfrak{q}}\ ]{} & \mathrm{Cat}(\mathcal{B}^{\mathrm{op}}, \mathrm{Set}) & & \end{array} \tag{5.8}$$

Note that this is in line with the fact that, given a category $\mathcal{C}$, presheaves on $\mathcal{C}$ correspond to profunctors $\mathcal{C} \rightsquigarrow 1$; covariant functors $\mathcal{C} \to \mathrm{Set}$ instead correspond to profunctors $1 \rightsquigarrow \mathcal{C}$, where 1 is the terminal category.

Equation (5.8) above is equivalent to the previous definition of profunctor composition, in view of the characterisation of a left Kan extension as a coend in $\mathrm{Cat}(\mathcal{A}^{\mathrm{op}}, \mathrm{Set})$ that we have given in 2.27:

$$\mathrm{Lan}_{\text{よ}}\, \widehat{\mathfrak{p}} \cong \int^{B} \mathrm{Cat}(\mathcal{B}^{\mathrm{op}}, \mathrm{Set})(\text{よ}(B), \_) \otimes \widehat{\mathfrak{p}}(B). \tag{5.9}$$

We have

$$\mathrm{Lan}_{\text{よ}}\, \widehat{\mathfrak{p}}(\widehat{\mathfrak{q}}(C)) \cong \int^{B} \mathrm{Cat}(\mathcal{B}^{\mathrm{op}}, \mathrm{Set})(\text{よ}(B), \widehat{\mathfrak{q}}(C)) \otimes \widehat{\mathfrak{p}}(B)$$

$$\cong \int^{B} \widehat{\mathfrak{q}}(C)(B) \otimes \widehat{\mathfrak{p}}(B)$$
$$\cong \int^{B} \mathfrak{p}(\_, B) \times \mathfrak{q}(B, C).$$

Remark 5.1.8 (Prof is a bicategory)**.** The properties of (strong) associativity and unitality for the composition of profunctors follow directly from the associativity of cartesian product, its co-continuity as a functor of a fixed variable, and from the ninja Yoneda lemma 2.2.1, as shown by the following computations.

- Composition of profunctors is associative up to isomorphism. Given three profunctors $\mathcal{B} \overset{\mathfrak{r}}{\rightsquigarrow} \mathcal{X} \overset{\mathfrak{q}}{\rightsquigarrow} \mathcal{Y} \overset{\mathfrak{p}}{\rightsquigarrow} \mathcal{A}$, giving the *associator* of a bicategory structure:

$$\begin{aligned}
\mathfrak{p} \diamond (\mathfrak{q} \diamond \mathfrak{r}) &= \int^{Y} \mathfrak{p}(Y, A) \times (\mathfrak{q} \diamond \mathfrak{r})(B, Y) \\
&= \int^{Y} \mathfrak{p}(Y, A) \times \Big(\int^{X} \mathfrak{q}(X, Y) \times \mathfrak{r}(B, X) \Big) \\
&\cong \int^{XY} \mathfrak{p}(Y, A) \times \Big(\mathfrak{q}(X, Y) \times \mathfrak{r}(B, X) \Big) \\
(\mathfrak{p} \diamond \mathfrak{q}) \diamond \mathfrak{r} &= \int^{X} (\mathfrak{p} \diamond \mathfrak{q})(X, A) \times \mathfrak{r}(B, X) \\
&\cong \int^{XY} \Big(\mathfrak{p}(Y, A) \times \mathfrak{q}(X, Y) \Big) \times \mathfrak{r}(B, X)
\end{aligned}$$

  (we freely employ most of the rules of (co)end calculus we have learned so far, in particular the Fubini rule 1.3.1) these results are clearly isomorphic, and naturally so.
- Every object $\mathcal{A}$ has an identity arrow, given by the hom profunctor $\mathcal{A}(\_, =) : \mathcal{A}^{\mathrm{op}} \times \mathcal{A} \to \mathrm{Set}$: the fact that $\mathfrak{p} \diamond \hom \cong \mathfrak{p}$, $\hom \diamond \mathfrak{q} \cong \mathfrak{q}$ simply rewrites the ninja Yoneda lemma 2.2.1.

Remark 5.1.9. The isomorphism above is part of the data turning Prof into a bicategory; the *associator* realises the identification between different parenthesisations of 1-cells, and the *unitor* realises the identification between $\mathfrak{p} \diamond \hom \cong \mathfrak{p}$.

In order to get a bicategory, some coherence conditions have to be imposed, precisely those of A.7.11.

It is immediate to observe that the validity of the pentagon identity in the case of the cartesian monoidal structure of Set, and the naturality thereof, ensure that the associator (whose components are) $(\mathfrak{p} \diamond \mathfrak{q}) \diamond \mathfrak{r} \Rightarrow$

$\mathfrak{p} \diamond (\mathfrak{q} \diamond \mathfrak{r})$ satisfies the pentagon identity. A similar argument shows that the unitor satisfies similar (left and right) triangular identities, as a consequence of the naturality of the ninja Yoneda lemma 2.2.1.

The reader might find instructive the exercise of showing by means of the bare universal property of the product that the isomorphisms $A \times (B \times C) \cong (A \times B) \times C$ in a cartesian category satisfy the pentagon equation.

Notation 5.1.10 (Einstein notation). The reader might have noticed, at this point, that computations with coends can become rather heavy-handed, and that sometimes multiple indices have to be considered at the same time. We shall introduce here a useful notation to shorten computations a bit, which is particularly evocative when dealing with profunctors; we choose to call it the *Einstein convention* for evident reasons.[1]

Let $\mathfrak{p} : \mathcal{A} \rightsquigarrow \mathcal{B}$, $\mathfrak{q} : \mathcal{B} \rightsquigarrow \mathcal{C}$ be two composable profunctors. If we adopt the notation $\mathfrak{p}^A_B, \mathfrak{q}^B_C$ to denote the images $\mathfrak{p}(A, B), \mathfrak{q}(B, C) \in \text{Set}$ (keeping track that superscripts are contravariant and subscripts are covariant components), then composition of profunctors acquires the form of a product of tensor components:

$$\mathfrak{p} \diamond \mathfrak{q}(A, C) = \int^B \mathfrak{p}^A_B \times \mathfrak{q}^B_C = \int^B \mathfrak{p}^A_B \mathfrak{q}^B_C. \tag{5.10}$$

The convention is then defined as follows. We denote as superscripts and subscripts the objects a bifunctor depends on; whenever an object $B \in \mathcal{B}$ appears once covariantly, say in $\mathfrak{p}^X_B$, and once contravariantly, say in $\mathfrak{q}^B_Y$, the result can be integrated as a coend into an object $\int^B \mathfrak{p}^X_B \mathfrak{q}^B_Y$ or as an end $\int_B \mathfrak{p}^X_B \mathfrak{q}^B_Y$.

From now on, we freely employ the Einstein summation convention when typesetting long computations.

5.2 Embeddings and Adjoints

In the present section we define two canonical identity-on-objects embeddings $\mathsf{Cat} \to \mathsf{Prof}$: the *covariant* one reverses the direction of 1-cells, but maintains the direction of 2-cells; the *contravariant* one does the

[1] To the knowledge of the author, this notation was first adopted also in [RV14], a valuable reading in itself for more than this single reason, not least because it provides a description of 'Reedy calculus' in homotopical algebra that heavily, albeit without an explicit focus, employs coends.

opposite.[2] The two correspondences are pictorially represented by the assignments

$$\mathcal{C} \quad \xrightarrow{\mathsf{Cat}^{\mathrm{op}}\to\mathsf{Prof}} \quad \mathcal{C} \qquad\qquad \mathcal{C} \quad \xrightarrow{\mathsf{Cat}^{\mathrm{co}}\to\mathsf{Prof}} \quad \mathcal{C} \tag{5.11}$$

$$F \overset{\alpha}{\Rightarrow} G \quad \mapsto \quad \mathfrak{p}_F \overset{\mathfrak{p}_\alpha}{\Rightarrow} \mathfrak{p}_G \qquad\qquad F \overset{\alpha}{\Rightarrow} G \quad \mapsto \quad \mathfrak{p}^F \overset{\mathfrak{p}^\alpha}{\Leftarrow} \mathfrak{p}^G$$

$$\mathcal{D} \qquad\qquad \mathcal{D} \qquad\qquad\qquad \mathcal{D} \qquad\qquad \mathcal{D}$$

and they will usually be nameless 2-functors throughout the discussion, just called 'the embedding' of Cat in Prof.

Given a functor $F : \mathcal{A} \to \mathcal{B}$, we define

E1. $\mathfrak{p}_F$ to be the profunctor $\mathcal{B}(1, F) : \mathcal{B} \rightsquigarrow \mathcal{A}$ defined as sending $(B, A) \mapsto \mathcal{B}(B, FA)$,

E2. $\mathfrak{p}^F$ to be the profunctor $\mathcal{B}(F, 1) : \mathcal{A} \rightsquigarrow \mathcal{B}$ defined as sending $(A, B) \mapsto \mathcal{B}(FA, B)$.

The 1-cells $\mathfrak{p}_F, \mathfrak{p}^F$ will usually be called *representable* profunctors induced by $F : \mathcal{A} \to \mathcal{B}$. If the context forces us to distinguish between $\mathfrak{p}^F$ and $\mathfrak{p}_F$ we call the first the *co*representable one (mnemonic: sending F to its corepresentable profunctor is a 2-functor $\mathsf{Cat}^{\mathrm{co}} \to \mathsf{Prof}$). This is however uncommon, and which of the two 'representable profunctor induced by F' is usually clear from the context.

Both choices define (pseudo) functors of the appropriate variance, since the isomorphisms

- $\mathfrak{p}_{FG} \cong \mathfrak{p}_F \diamond \mathfrak{p}_G$, and $\mathfrak{p}^{FG} \cong \mathfrak{p}^G \diamond \mathfrak{p}^F$ and
- $\mathfrak{p}_{\mathrm{id}_\mathcal{A}} = \mathfrak{p}^{\mathrm{id}_\mathcal{A}} = \mathcal{A}(-, =)$

can be easily established using elementary coend calculus, and they hold for every composable pair of functors F, G and every category $\mathcal{A}$. For example,

$$\begin{aligned}(\mathfrak{p}_F \diamond \mathfrak{p}_G)(C, A) &\cong \int^B \mathcal{B}(B, FA) \times \mathcal{C}(C, GB)\\ &\cong \mathcal{C}(C, GFA) = \mathfrak{p}_{FG}(C, A)\end{aligned}$$

by the ninja Yoneda lemma 2.2.1.

Natural transformations $\alpha : F \Rightarrow G$ are obviously sent to 2-cells in

[2] Recall from chapter 2 that given a 2-category $\mathcal{K}$ there is a 2-category $\mathcal{K}^{\mathrm{op}}$, where 1-cells are reversed and 2-cells are untouched, and a 2-category $\mathcal{K}^{\mathrm{co}}$, where 2-cells are reversed and 1-cells are untouched; of course, $\mathcal{K}^{\mathrm{coop}} = (\mathcal{K}^{\mathrm{op}})^{\mathrm{co}} = (\mathcal{K}^{\mathrm{co}})^{\mathrm{op}}$.

Prof, and the covariance of this assignment is evident from the definition of $\mathfrak{p}^F$ and $\mathfrak{p}_F$.

Finally, the Yoneda lemma entails that both functors $F \mapsto \mathfrak{p}^F$ and $F \mapsto \mathfrak{p}_F$ are locally fully faithful, meaning that there are bijections

$$\mathsf{Cat}(\mathcal{A}, \mathcal{B})(F, G) \cong \mathsf{Prof}(\mathcal{B}, \mathcal{A})(\mathfrak{p}_F, \mathfrak{p}_G) \tag{5.12}$$

for every $F, G : \mathcal{A} \to \mathcal{B}$. If $\mathfrak{p}^F \cong \mathfrak{p}^G$. This means we have a natural bijection $\mathcal{B}(FA, B) \cong \mathcal{B}(GA, B)$, which entails the existence of a natural isomorphism $\alpha : F \Rightarrow G$.

The most important fact about representable and corepresentable profunctors associated to the same functor is that they are *adjoint* 1-cells in Prof.

Remark 5.2.1. There is a tight link between the 1-cells $\mathfrak{p}_F$, $\mathfrak{p}^F$: they are *adjoint* 1-cells in the bicategory Prof. Indeed, for every $F \in \mathrm{Cat}(\mathcal{A}, \mathcal{B})$ we can define 2-cells

$$\begin{aligned} \epsilon_F &: \mathfrak{p}_F \diamond \mathfrak{p}^F \Longrightarrow \mathcal{B}(-, =) \\ \eta_F &: \mathcal{A}(-, =) \Longrightarrow \mathfrak{p}^F \diamond \mathfrak{p}_F \end{aligned} \tag{5.13}$$

(*counit* and *unit* of the adjunction), satisfying the zig-zag identities. We choose a very explicit, set-theoretic proof for this statement, leaving it to the reader to generalise the argument to a generic base of enrichment.

- The components of the counit ϵ are determined by inspecting the coend $\mathfrak{p}_F \diamond \mathfrak{p}^F$ as the quotient set

$$\int^X \mathcal{B}(A, FX) \times \mathcal{B}(FX, B) = \Big(\coprod_{x \in \mathcal{A}} \mathcal{B}(A, FX) \times \mathcal{B}(FX, B) \Big) / \simeq \tag{5.14}$$

where $\simeq$ is the equivalence relation generated by $(A \xrightarrow{u} FX, FX \xrightarrow{v} B) \simeq (A \xrightarrow{u'} FY, FY \xrightarrow{v'} B)$ if there is $t : X \to Y$ such that $v' = Ft \circ v$ and $Ft \circ u = u'$. This can be visualised as the commutativity of the square

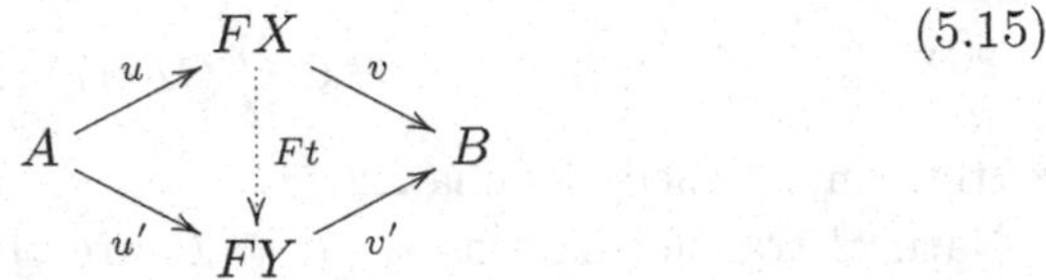

Now it is easily seen that sending $(A \xrightarrow{u} FX, FX \xrightarrow{v} B)$ in the composition $v \circ u$ descend to the quotient with respect to $\simeq$, hence

there is a well defined map $\epsilon : \mathfrak{p}_F \diamond \mathfrak{p}^F \to \mathcal{B}(\_, =)$. All boils down to the fact that the composition

$$c : \mathcal{B}(A, FX) \times \mathcal{B}(FX, B) \to \mathcal{B}(A, B) \tag{5.16}$$

defines a cowedge in the variable X.

- The components of the unit η are determined as functions

$$\eta : \mathcal{A}(A, B) \Longrightarrow \mathfrak{p}^F \diamond \mathfrak{p}_F(A, B). \tag{5.17}$$

Since $\mathfrak{p}^F \diamond \mathfrak{p}_F(A, B) = \int^X \mathcal{B}(FA, X) \times \mathcal{B}(X, FB) \cong \mathcal{B}(FA, FB)$ (as a consequence of the ninja Yoneda lemma), these are simply determined by the action of F on arrows, $\mathcal{A}(A, B) \to \mathcal{B}(FA, FB)$.

We now have to verify that the triangle identities (see [Bor94a, 3.1.5.(2)]) hold:

$$(\mathfrak{p}^F \diamond \epsilon) \circ (\eta \diamond \mathfrak{p}^F) = \mathrm{id}_{\mathfrak{p}^F}$$
$$(\epsilon \diamond \mathfrak{p}_F) \circ (\mathfrak{p}_F \diamond \eta) = \mathrm{id}_{\mathfrak{p}_F}.$$

As for the first, we must verify that the diagram

$$\begin{array}{ccccc} \mathfrak{p}^F & \xrightarrow{\sim} & \mathcal{B}(\_, =) \diamond \mathfrak{p}^F & \xrightarrow{\eta \diamond \mathfrak{p}^F} & (\mathfrak{p}^F \diamond \mathfrak{p}_F) \diamond \mathfrak{p}^F \\ \| & & & & \downarrow \cong \\ \mathfrak{p}^F & \xleftarrow[\sim]{} & \mathfrak{p}^F \diamond \mathcal{B}(\_, =) & \xleftarrow[\mathfrak{p}^F \diamond \epsilon]{} & \mathfrak{p}^F \diamond (\mathfrak{p}_F \diamond \mathfrak{p}^F) \end{array} \tag{5.18}$$

commutes. One has to send $h \in \mathfrak{p}^F(u, v) = \mathcal{B}(Fu, v)$ in the equivalence class $[(\mathrm{id}_u, h)] \in \int^X \mathcal{B}(U, X) \times \mathcal{B}(FX, V)$, which must go under $\eta \diamond \mathfrak{p}^F$ in the equivalence class $[(F(\mathrm{id}_u), h)] \in \int^{XY} \mathcal{B}(FA, X) \times \mathcal{B}(X, FY) \times \mathcal{B}(FY, B)$, canonically identified with the coend $\int^Y \mathcal{B}(FA, FY) \times \mathcal{B}(FY, B)$. Now $\mathfrak{p}^F \diamond \epsilon$ acts composing the two arrows, and one obtains $F(\mathrm{id}_A) \circ h = h$ back.

Similarly, to prove the second identity, the diagram

$$\begin{array}{ccccc} \mathfrak{p}_F & \xrightarrow{\sim} & \mathfrak{p}_F \diamond \mathcal{B}(\_, =) & \xrightarrow{\mathfrak{p}_F \diamond \eta} & \mathfrak{p}_F \diamond (\mathfrak{p}^F \diamond \mathfrak{p}_F) \\ \| & & & & \downarrow \cong \\ \mathfrak{p}_F & \xleftarrow[\sim]{} & \mathcal{B}(\_, =) \diamond \mathfrak{p}_F & \xleftarrow[\epsilon \diamond \mathfrak{p}_F]{} & (\mathfrak{p}_F \diamond \mathfrak{p}^F) \diamond \mathfrak{p}_F \end{array} \tag{5.19}$$

must commute (all the unlabelled isomorphisms are the canonical ones).

This translates into

$$\left(a \xrightarrow{u} FB\right) \longmapsto (u, \mathrm{id}_B)_\sim \longmapsto (u, F(\mathrm{id}_B)) \longmapsto u \circ F(\mathrm{id}_B) = u, \tag{5.20}$$

which is what we want; hence $\mathfrak{p}_F \dashv \mathfrak{p}^F$. □

Remark 5.2.2. Two functors $F : \mathcal{A} \leftrightarrows \mathcal{B} : G$ are adjoints if and only if $\mathfrak{p}_F \cong \mathfrak{p}^G$ (and therewith $G \dashv F$) or $\mathfrak{p}_G \cong \mathfrak{p}^F$ (and therewith $F \dashv G$).

The covariant embedding of Cat in Prof gives every 1-cell $F : \mathcal{A} \to \mathcal{B}$ a left adjoint. It turns out that many properties of functors can be translated into properties of the associated representable profunctor. As an elementary example, we take the fact that a functor is full and faithful.

Remark 5.2.3. It is a well-known fact (see [Bor94a, dual of Prop. 3.4.1]) that if $F \dashv G$, then F is fully faithful if and only if the unit of the adjunction $\eta : 1 \to GF$ is an isomorphism.

This criterion can be extended also to functors that do not admit a 'real' right adjoint, once noticed that F is fully faithful if and only if $\mathcal{A}(A, B) \cong \mathcal{B}(FA, FB)$ for any two $A, B \in \mathcal{A}$, i.e. if and only if the unit $\eta : \hom_{\mathcal{A}} \Rightarrow \mathfrak{p}^F \diamond \mathfrak{p}_F$ is an isomorphism in the bicategory Prof.

It is possible to define an *action* of profunctors on functors; more precisely, we can give the following definition.

Definition 5.2.4. Given a profunctor $\mathfrak{p} : \mathcal{A} \rightsquigarrow \mathcal{B}$ and a functor $F : \mathcal{B} \to \mathcal{D}$ we can define $\mathfrak{p} \otimes F$ to be the functor $\mathcal{A} \to \mathcal{D}$ given by $\mathrm{Lan}_{よ} F \circ \widehat{\mathfrak{p}}$, where $\widehat{\mathfrak{p}} : \mathcal{B} \to \mathrm{Cat}(\mathcal{A}^{\mathrm{op}}, \mathrm{Set})$ is the mate of $\mathfrak{p}$, as in (5.8).

More explicitly,

$$\mathfrak{p} \otimes F(A) = \int^{B} [よ_B, \mathfrak{p}(\_, A)] \otimes FB \cong \int^{B} \mathfrak{p}_A^B \otimes F_B. \tag{5.21}$$

Note that a conceptual definition for this operation is the following: $\mathfrak{p} \otimes F$, evaluated on $A \in \mathcal{A}$, is the weighted colimit $\mathrm{colim}^{\mathfrak{p}_A} F$. Exploiting this definition, it is possible to prove using coend calculus that this operation can be rightly called a left action $\mathrm{Prof}(\mathcal{A}, \mathcal{B}) \times \mathrm{Cat}(\mathcal{B}, \mathcal{D}) \to \mathrm{Cat}(\mathcal{A}, \mathcal{D})$:

- $\hom_{\mathcal{B}} \otimes F \cong F$ as a consequence of the ninja Yoneda lemma;
- if $\mathcal{C} \overset{\mathfrak{q}}{\rightsquigarrow} \mathcal{A} \overset{\mathfrak{p}}{\rightsquigarrow} \mathcal{B} \xrightarrow{F} \mathcal{X}$, then $(\mathfrak{p} \diamond \mathfrak{q}) \otimes F \cong \mathfrak{q} \otimes (\mathfrak{p} \otimes F)$. Indeed

$$\begin{aligned}[(\mathfrak{p} \diamond \mathfrak{q}) \otimes F] &= \int^{B} (\mathfrak{p} \diamond \mathfrak{q})^B \times F_B \\ &\cong \int^{BX} \mathfrak{p}_X^B \times \mathfrak{q}^X \times F_B\end{aligned}$$

$$\cong \int^X \mathfrak{q}^X \times \left(\int^B \mathfrak{p}_X^B \times F_B\right)$$
$$\cong \int^X \mathfrak{q}^X \times (F \otimes \mathfrak{p})_X = [\mathfrak{q} \otimes (\mathfrak{p} \otimes F)].$$

The composition of profunctors is a 'closed structure' in the following sense: both functors $\mathfrak{p} \diamond \_$ and $\_ \diamond \mathfrak{q}$ have right adjoints, respectively given by a right lifting and a right extension operation (see 2.1.3).

Example 5.2.5 (Kan extensions in Prof). Every profunctor $\mathfrak{p}$ has a right Kan extension $\mathrm{Ran}_{\mathfrak{p}}$ in the sense that the notion has in any bicategory, where composition of functors or natural transformations is replaced by composition of 1- or 2-cells.

One has the following chain of isomorphisms in Prof:

$$\begin{aligned}
\mathsf{Prof}(\mathfrak{g} \diamond \mathfrak{p}, \mathfrak{r}) &\cong \int_{AB} \mathrm{Set}\big((\mathfrak{g} \diamond \mathfrak{p})(A,B), \mathfrak{r}(A,B)\big) \\
&\cong \int_{AB} \mathrm{Set}\Big(\int^X \mathfrak{g}(A,X) \times \mathfrak{p}(X,B), \mathfrak{r}(A,B)\Big) \\
&\cong \int_{ABX} \mathrm{Set}\big(\mathfrak{g}(A,X), \mathrm{Set}(\mathfrak{p}(X,B), \mathfrak{r}(A,B))\big) \\
&\cong \int_{AX} \mathrm{Set}\Big(\mathfrak{g}(A,X), \int_B \mathrm{Set}(\mathfrak{p}(X,B), \mathfrak{r}(A,B))\Big) \\
&\cong \int_{AX} \mathrm{Set}\Big(\mathfrak{g}(A,X), \langle \mathfrak{p}/\mathfrak{r}\rangle(A,X)\Big) \\
&\cong \mathsf{Prof}(\mathfrak{g}, \langle \mathfrak{p}/\mathfrak{r}\rangle)
\end{aligned}$$

when we define $\langle \mathfrak{p}/\mathfrak{r}\rangle(A,X)$ to be the set of natural transformations $\mathfrak{p}(X,\_) \Rightarrow \mathfrak{r}(A,\_)$. This yields that $\langle \mathfrak{p}/\mathfrak{r}\rangle$ has the universal property of the right extension of $\mathfrak{r}$ along $\mathfrak{p}$.

Similarly, we can prove that a suitable end $\langle \mathfrak{g}\backslash\mathfrak{r}\rangle$ defines the right lifting of $\mathfrak{r}$ along $\mathfrak{g}$. This is the content of Exercise 5.2.

5.3 The Structure of Prof

The bicategory of profunctors has an extremely rich structure; we briefly give an account of its main properties, drawing from [CW96].

As 5.1.5 shows, a relation $R \subseteq A \times B$, regarded as a function $A \times B \to \{0,1\}$, is a 1-cell of $\mathsf{Prof}(\{0,1\})$; but what are the domain and codomain of such an R? It is evident that the symmetry of the product

$\sigma_{AB} : A \times B \cong B \times A$ induces an identity-on-objects identification

$$\mathsf{Prof}(\{0,1\}) \cong \mathsf{Prof}(\{0,1\})^{\mathrm{op}} \tag{5.22}$$

(see 5.1.3 for the notation $\mathsf{Prof}(\mathcal{V})$) sending a relation $R \subseteq A \times B$, seen as a 1-cell $R : A \rightsquigarrow B$, to the relation $\sigma_{AB} \circ R \subseteq B \times A$, regarded as a 1-cell $R : B \rightsquigarrow A$.

This is a general phenomenon: every $\mathsf{Prof}(\mathcal{V})$ admits a 'tautological' bi-equivalence with $\mathsf{Prof}(\mathcal{V})^{\mathrm{op}}$, determined as follows (see [CW96]).

Proposition 5.3.1 (The canonical dualiser of Prof)**.** The identification $\mathrm{Cat}(\mathcal{A}^{\mathrm{op}} \times \mathcal{B}, \mathrm{Set}) \cong \mathrm{Cat}(\mathcal{B} \times \mathcal{A}^{\mathrm{op}}, \mathrm{Set})$ determines an equivalence $(\_)^{\circ} : \mathsf{Prof}^{\mathrm{op}} \to \mathsf{Prof}$ determined by the following correspondences.

- On objects, a 0-cell $\mathcal{A}$ of Prof goes to $\mathcal{A}^{\circ} = \mathcal{A}^{\mathrm{op}}$.
- The 1-cell $\mathfrak{p} : \mathcal{A} \rightsquigarrow \mathcal{B}$ goes to 'itself', $\mathfrak{p}^{\circ} : \mathcal{B}^{\mathrm{op}} \rightsquigarrow \mathcal{A}^{\mathrm{op}}$, under the identification (isomorphism of categories)

$$\mathsf{Prof}(\mathcal{A},\mathcal{B}) = \mathrm{Cat}(\mathcal{A}^{\mathrm{op}} \times \mathcal{B}, \mathrm{Set}) \cong \mathrm{Cat}(\mathcal{B} \times \mathcal{A}^{\mathrm{op}}, \mathrm{Set}) = \mathsf{Prof}(\mathcal{B}^{\mathrm{op}}, \mathcal{A}^{\mathrm{op}}). \tag{5.23}$$

- The 2-cell $\alpha : \mathfrak{p} \Rightarrow \mathfrak{q}$ goes to 'itself', again under the identification

$$\mathsf{Prof}(\mathcal{A},\mathcal{B})(\mathfrak{p},\mathfrak{q}) \cong \mathsf{Prof}(\mathcal{B}^{\mathrm{op}}, \mathcal{A}^{\mathrm{op}})(\mathfrak{p}^{\mathrm{op}}, \mathfrak{q}^{\mathrm{op}}). \tag{5.24}$$

We pair the existence of a *dualiser* on Prof with the following result.

Proposition 5.3.2. [CW96, 4.3.7] The bifunctor

$$\_ \times = : \mathsf{Prof} \times \mathsf{Prof} \to \mathsf{Prof} \tag{5.25}$$

defined as the product functor in each dimension, i.e. the correspondence sending

- two 0-cells $\mathcal{A}, \mathcal{B} \in \mathsf{Prof}$ into the product category $\mathcal{A} \times \mathcal{B}$,
- two 1-cells $\mathfrak{p} : \mathcal{A} \rightsquigarrow \mathcal{B}, \mathfrak{q} : \mathcal{C} \rightsquigarrow \mathcal{D}$ to

$$\mathfrak{p} \times \mathfrak{q} : \mathcal{A} \times \mathcal{C} \rightsquigarrow \mathcal{B} \times \mathcal{D} : (A, C; B, D) \mapsto \mathfrak{p}(A, B) \times \mathfrak{q}(C, D), \tag{5.26}$$

- two 2-cells $\alpha : \mathfrak{p} \Rightarrow \mathfrak{q}$ and $\beta : \mathfrak{l} \Rightarrow \mathfrak{r}$ to their product components

$$(\alpha \times \beta)_{A,C;B,D} : \mathfrak{p} \times \mathfrak{l}(A, C; B, D) \Rightarrow \mathfrak{q} \times \mathfrak{r}(A, C; B, D) \tag{5.27}$$

equips Prof with a monoidal structure.

Remark 5.3.3. [CW96, 4.3.3] A somewhat odd feature of Prof is that (pseudo)products and (pseudo)coproducts both exist, they coincide on objects, and exchange place on 1-cells.

- The span
$$\mathcal{A} \overset{p_{\mathcal{A}}}{\leftsquigarrow} \mathcal{A} \coprod \mathcal{B} \overset{p_{\mathcal{B}}}{\rightsquigarrow} \mathcal{B} \tag{5.28}$$
exhibits the universal property of the pseudo-product $\mathcal{A} \,\&\, \mathcal{B}$ of $\mathcal{A}$ and $\mathcal{B}$ in Prof, if the 'projection' maps are defined as $\left[\begin{smallmatrix}\hom_{\mathcal{A}}\\ \varnothing\end{smallmatrix}\right] : (\mathcal{A}^{\text{op}} \times \mathcal{A}) \amalg (\mathcal{B}^{\text{op}} \times \mathcal{A}) \to \text{Set}$ and $\left[\begin{smallmatrix}\varnothing\\ \hom_{\mathcal{B}}\end{smallmatrix}\right] : (\mathcal{A}^{\text{op}} \times \mathcal{B}) \amalg (\mathcal{B}^{\text{op}} \times \mathcal{B}) \to \text{Set}$.
- The cospan
$$\mathcal{A} \overset{i_{\mathcal{A}}}{\rightsquigarrow} \mathcal{A} \coprod \mathcal{B} \overset{i_{\mathcal{B}}}{\leftsquigarrow} \mathcal{B} \tag{5.29}$$
exhibits the universal property of the pseudo-coproduct $\mathcal{A} \oplus \mathcal{B}$ of $\mathcal{A}$ and $\mathcal{B}$ in Prof, if the 'injection' maps are defined as $\left[\begin{smallmatrix}\hom_{\mathcal{A}}\\ \varnothing\end{smallmatrix}\right] : (\mathcal{A}^{\text{op}} \times \mathcal{A}) \amalg (\mathcal{A}^{\text{op}} \times \mathcal{B}) \to \text{Set}$ and $\left[\begin{smallmatrix}\varnothing\\ \hom_{\mathcal{B}}\end{smallmatrix}\right] : (\mathcal{B}^{\text{op}} \times \mathcal{A}) \amalg (\mathcal{B}^{\text{op}} \times \mathcal{B}) \to \text{Set}$.

Note in particular that the monoidal structure in 5.3.2 is not the cartesian one. Since the empty diagram carries no information on 1- and 2-cells, it is thus possible to use this construction to show that Prof has a pseudo-zero object, namely the initial category $\varnothing$. This appears as [CW96, 4.3.5].

Remark 5.3.4. Taken together, these results allow us to prove that the bicategory Prof is *compact closed*. Given that the monoidal unit for the structure in 5.3.2 is the terminal category, we shall find suitable profunctors
$$\eta : 1 \rightsquigarrow \mathcal{A}^{\text{op}} \times \mathcal{A} \qquad \epsilon : \mathcal{A} \times \mathcal{A}^{\text{op}} \rightsquigarrow 1 \tag{5.30}$$
such that the following two *yanking equations* are satisfied:
$$A \cong A \times I \xrightarrow{\mathcal{A}\times\eta} \mathcal{A} \times (\mathcal{A}^\circ \times \mathcal{A}) \cong (\mathcal{A} \times \mathcal{A}^\circ) \times \mathcal{A} \xrightarrow{\epsilon\times\mathcal{A}} I \times \mathcal{A} \cong \mathcal{A} \tag{5.31}$$
$$\mathcal{A}^\circ \cong I \times \mathcal{A}^\circ \xrightarrow{\eta\times\mathcal{A}^\circ} (\mathcal{A}^\circ \times \mathcal{A}) \times \mathcal{A}^\circ \cong \mathcal{A}^\circ \times (\mathcal{A} \times \mathcal{A}^\circ) \xrightarrow{\mathcal{A}^\circ\times\epsilon} \mathcal{A}^\circ \times I \cong \mathcal{A}^\circ \tag{5.32}$$
It is evident that suitable hom functors (in fact, the same hom functor of $\mathcal{A}$ and its dualised copy $\hom^\circ$) do the job.

Remark 5.3.5. The coend operation $\int^{\mathcal{C}}$: $\text{Cat}(\mathcal{C}^{\text{op}} \times \mathcal{C}, \text{Set})$ endows Prof with a *traced structure*: this means that the category Prof is monoidal with respect to the product defined in 5.3.2, and that there exists a family of functions
$$\text{tr}^{\mathcal{C}}_{\mathcal{A},\mathcal{B}} : \mathsf{Prof}(\mathcal{A} \times \mathcal{C}, \mathcal{B} \times \mathcal{C}) \to \mathsf{Prof}(\mathcal{A}, \mathcal{B}) \tag{5.33}$$
satisfying suitable coherence conditions, expressed in [JSV96]. We record these conditions in Exercise 5.12 and leave the proof to the reader.

The map $\mathrm{tr}^{\mathcal{C}}_{\mathcal{A},\mathcal{B}}$ is defined as the composition

$$\begin{aligned}\mathrm{Cat}((\mathcal{A}\times\mathcal{C})^{\mathrm{op}}\times\mathcal{B}\times\mathcal{C},\mathrm{Set}) &\cong \mathrm{Cat}(\mathcal{A}^{\mathrm{op}}\times\mathcal{B},\mathrm{Cat}(\mathcal{C}^{\mathrm{op}}\times\mathcal{C},\mathrm{Set}))\\ &\xrightarrow{\int^{\mathcal{C}}} \mathrm{Cat}(\mathcal{A}^{\mathrm{op}}\times\mathcal{B},\mathrm{Set}) \qquad (5.34)\end{aligned}$$

(a few completely straightforward identifications have been omitted).

5.4 A More Abstract Look at Prof

Remark 5.4.1. The bicategory of profunctors can be promoted to a *multi-bicategory* in the sense of [CKS03, 1.4]; this means that we exploit the composition operation to specify a class of multi-morphisms $\eta : \mathfrak{p}_1,\dots,\mathfrak{p}_n \rightsquigarrow \mathfrak{q}$, depicted as diagrams

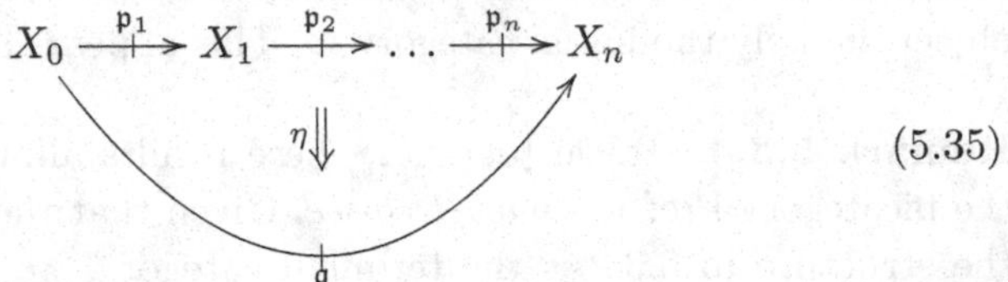

(5.35)

whose composition, associativity, and unitality follow at once from pasting laws for 2-cells in 2-categories [Kel05a] (try to outline them explicitly as a straightforward exercise). More generally, the bicategory of profunctors is a toy example of an *fc-multicategory* [Lei99] or, in more modern terms, a *virtual double category* (vdc) [CS10].

A VDC $\mathbf{C}$ as defined in [CS10] is a category-like structure whose 'cells' α have the form

$$\begin{array}{ccccccc} X_0 & \xrightarrow{p_1} & X_1 & \xrightarrow{p_2} & \cdots & \xrightarrow{p_n} & X_n \\ \big\downarrow & & & \Downarrow\alpha & & & \big\downarrow \\ Y_0 & & & \xrightarrow[q]{} & & & Y_1 \end{array} \qquad (5.36)$$

where the vertical arrows are the morphisms of a category $\mathbf{C}_v$ called the *vertical category* of the vdc, and the cell α has an n-tuple (for $n \geq 0$, the non-negative integer n is called the *arity* of the cell) of *horizontal* morphisms $(p_1, p_2, \dots, p_n)$ as horizontal domain, and a single q as horizontal codomain.

These cells are subject to certain straightforward coherence conditions of associativity. First of all, vertical morphisms compose like they do in a category; second, the cells can be *grafted* as follows. Every

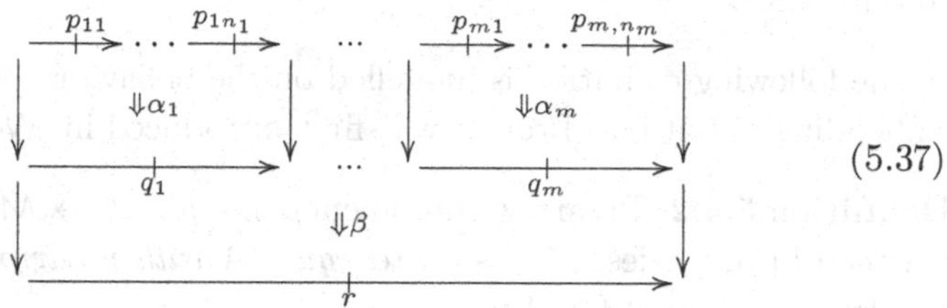

(5.37)

can be composed into a single cell

$$\begin{array}{ccc} & \xrightarrow{p_{11}} \cdots \xrightarrow{p_{1n_1}} \quad \cdots \quad \xrightarrow{p_{m1}} \cdots \xrightarrow{p_{m,n_m}} & \\ \downarrow & \Downarrow \beta \odot (\alpha_1, \dots, \alpha_m) & \downarrow \\ & \xrightarrow[r]{} & \end{array} \tag{5.38}$$

so that this operation is compatible with the rest of the data. The *0-ary* cells are determined by diagrams like

$$\begin{array}{ccc} & X & \\ \swarrow & & \searrow \\ Y_0 & \longrightarrow & Y_1. \end{array} \tag{5.39}$$

The standard choice to make $\mathsf{Prof}(\mathcal{V})$ a VDC is to take

- as objects the small categories,
- as vertical category the category Cat,
- as n-ary cells $\alpha : (p_1, \dots, p_n) \Rightarrow q$ having vertical domain $F : X_0 \to Y_0$ and vertical codomain $G : X_n \to Y_1$ the natural transformations

$$\alpha : \mathfrak{p}_G \diamond \mathfrak{p}_n \diamond \cdots \diamond \mathfrak{p}_1 \Rightarrow \mathfrak{q} \diamond \mathfrak{p}_F. \tag{5.40}$$

The reader will routinely check that all coherences stated in [CS10] hold.

Of course, whenever we consider composable profunctors, every n-ary cell can be reduced to a 1-ary cell by composing the n-tuple of horizontal arrows that form its horizontal domain; but in situations where the composition does not exist, for example for categories enriched in a non-cocomplete base, it is still possible to define the VDC of these generalised profunctors.

The notion of VDC can thus be considered as the 'correct' generalisation of categories $\mathsf{Prof}(\mathcal{V})$ enriched over a generic base, and exhibits many of its nice features even without horizontal compositions. Examples of vdcs abound inside and outside category theory: the reader is invited to consult [CS10].

The following definition is modelled on the behaviour of the canonical embedding of Cat into Prof; it was first introduced in [Woo82].

Definition 5.4.2 (Proarrow equipment)**.** Let $p^* : \mathcal{A} \to \mathcal{M}$ be a 2-functor between bicategories; p^* is said to *equip* $\mathcal{A}$ *with proarrows*, or to be a *proarrow equipment for* $\mathcal{A}$ if

PE1. p^* is locally fully faithful,

PE2. for every arrow $f \in \mathcal{A}$, $p^* f$ has a right adjoint in $\mathcal{M}$.

It is clear how the embedding $\mathsf{Cat}^{\mathrm{op}} \to \mathsf{Prof}$ equips Cat with proarrows (see (5.11); our choice for the direction of 1-cells in Prof forces us to treat the embedding as contravariant).[3]

Remark 5.4.3 (Exact squares and profunctors)**.** ☻☻ Let us consider Cat as a 2-category; a square

$$\begin{array}{ccc} \mathcal{A} & \xrightarrow{T} & \mathcal{Y} \\ {\scriptstyle S}\downarrow & \nearrow\alpha & \downarrow{\scriptstyle V} \\ \mathcal{X} & \xrightarrow[U]{} & \mathcal{B} \end{array} \tag{5.41}$$

filled by a 2-cell α will be called a *carré* (see [Gui80]). Any carré induces a 2-cell

$$\begin{array}{ccc} \mathcal{A} & \xleftarrow{\mathfrak{p}^T} & \mathcal{Y} \\ {\scriptstyle \mathfrak{p}_S}\downarrow & \searrow\alpha^\flat & \downarrow{\scriptstyle \mathfrak{p}_V} \\ \mathcal{X} & \xleftarrow[\mathfrak{p}^U]{} & \mathcal{B} \end{array} \tag{5.42}$$

defined by the universal property of profunctor composition via the cowedge

$$\alpha^\flat_{(A),XY} : \mathcal{X}(X, SA) \times \mathcal{Y}(TA, Y) \longrightarrow \mathcal{B}(UX, VY) \tag{5.43}$$

sending the pair $(f, g) = \left(\left[\begin{smallmatrix} X \\ \downarrow \\ SA \end{smallmatrix}\right], \left[\begin{smallmatrix} TA \\ \downarrow \\ Y \end{smallmatrix}\right]\right)$ into $Vg \circ \alpha_A \circ Uf$.

[3] A rather unexpected tight connection between vdcs and proarrow equipments is the following: equipments arise precisely as those vdcs where horizontal arrows can all be composed, and have local horizontal identities for every object. This is [CS10, §7] and in particular its Definition 7.6.

We say that a carré is *exact* if $\alpha^\flat$ is invertible, i.e. if $\int^A \mathcal{X}(X, SA) \times \mathcal{Y}(TA, Y) \cong \mathcal{B}(UX, VY)$.

[Gui80] observes that there is a criterion for a carré to be exact that does not involve profunctors. Let us consider a square as above, and the induced diagram

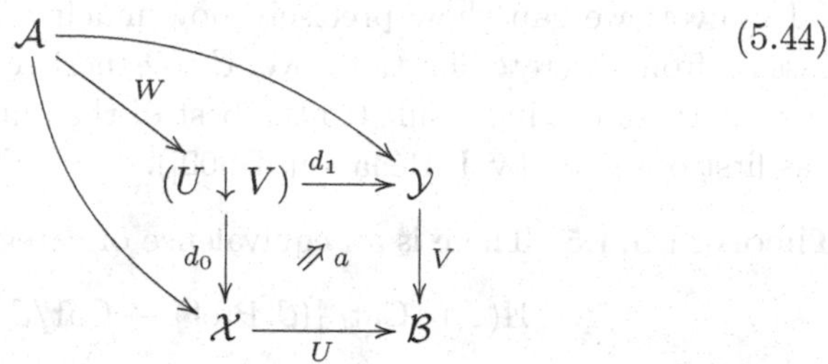

(5.44)

where $(U \downarrow V)$ is the *comma category* of the cospan $\mathcal{X} \xrightarrow{U} \mathcal{B} \xleftarrow{V} \mathcal{Y}$ and W is the unique induced functor to $(U \downarrow V)$. For each pair of functors $\mathcal{X} \xrightarrow{P} \mathcal{Z} \xleftarrow{Q} \mathcal{Y}$ we have the following identifications of sets:

$$\begin{array}{ccc}
\mathrm{Cat}(U \downarrow V, \mathcal{Z})(Pd_0, Qd_1) & \longrightarrow & \mathrm{Cat}(\mathcal{A}, \mathcal{Z})(PS, QT) \\
\| & & \| \\
\mathsf{Prof}(\mathcal{Y}, \mathcal{X})(\mathfrak{p}_{d_0} \diamond \mathfrak{p}^{d_1}, \mathfrak{p}^P \diamond \mathfrak{p}_Q) & & \\
\| & & \| \\
\mathsf{Prof}(\mathcal{Y}, \mathcal{X})(\mathfrak{p}^U \diamond \mathfrak{p}_V, \mathfrak{p}^P \diamond \mathfrak{p}_Q) & \longrightarrow & \mathsf{Prof}(\mathcal{Y}, \mathcal{X})(\mathfrak{p}_A \diamond \mathfrak{p}^T, \mathfrak{p}^P \diamond \mathfrak{p}_Q)
\end{array} \tag{5.45}$$

where the horizontal arrows are induced by whiskering. The carré is exact if and only if for each P, Q there is a bijection

$$\mathrm{Cat}(\mathcal{A}, \mathcal{Z})(PS, QT) \cong \mathrm{Cat}(U \downarrow V, \mathcal{Z})(Pd_0, Qd_1). \tag{5.46}$$

Remark 5.4.4 (Displayed category). The bicategory of profunctors appears in a generalised form of Grothendieck construction, as described in our A.5.14. There, we draw an equivalence of categories

$$p_- : \mathrm{DFib}(\mathcal{C}) \leftrightarrows \mathrm{Cat}(\mathcal{C}^{\mathrm{op}}, \mathrm{Set}) : \mathcal{C} \textstyle\int - \tag{5.47}$$

between discrete opfibrations over $\mathcal{C}$ and presheaves over $\mathcal{C}$ (see A.5.12). In recent years it has become popular to call the pair of adjuncts that realises this equivalence respectively 'straightening' and 'unstraightening'.

A good question to ask, now, is whether there is some kind of (un)straightening for generic functors $q : \mathcal{E} \to \mathcal{C}$ over a base that do not satisfy

A.5.12. Since the fibration condition on $p : \mathcal{E} \to \mathcal{C}$ is meant exactly to ensure that each morphism $C \to C'$ in the base induces a function/-functor between the fibres $p^{\leftarrow}C' \to p^{\leftarrow}C$; for a generic functor $q : \mathcal{E} \to \mathcal{C}$ the straightened $\Gamma_q : \mathcal{C}^{\mathrm{op}} \to \mathcal{K}$ exists, but it will not be strictly functorial, nor will ensure functorial correspondences among the fibres.

However, we can show precisely how much regularity is lost when passing from discrete fibrations over $\mathcal{C}$ to generic categories lying over $\mathcal{C}$, proving the following result (to the best of the author's knowledge, this was first observed by J. Bénabou [BS00]).

Theorem 5.4.5. There is an equivalence of categories

$$\Pi(\_) : \mathrm{Cat}_{l,1}(\mathcal{C}, \mathsf{Prof}) \to \mathrm{Cat}/\mathcal{C} : \Gamma \tag{5.48}$$

where $\mathrm{Cat}_{l,1}(\mathcal{X}, \mathcal{Y})$ denotes the category of *normal lax functors* (see A.7.6; recall that a lax functor is *normal* if it preserves identities) between bicategories $\mathcal{X}, \mathcal{Y}$. The correspondence is defined by a 'generalised Grothendieck construction', in that the functor Π sends $F \in \mathrm{Cat}_{l,1}(\mathcal{C}^{\mathrm{op}}, \mathsf{Prof})$ to the (strict) pullback

$$\begin{array}{ccc} \Pi(F) & \longrightarrow & \mathsf{Prof}_* \\ \big\downarrow & \lrcorner & \big\downarrow U \\ \mathcal{C} & \xrightarrow[F]{} & \mathsf{Prof} \end{array} \tag{5.49}$$

where $U : \mathsf{Prof}_* \to \mathsf{Prof}$ is the forgetful functor from *pointed profunctors* that forgets the basepoint.[4]

Notation 5.4.6. The category $\Pi(F)$ corresponding to a normal lax functor $F : \mathcal{C} \to \mathsf{Prof}$ is no longer *fibred*; we call it a *displayed* category.

Proof As for the Π correspondence, there is nothing to check apart from its functoriality. This is easy, as $\Pi(F)$ is a category, as $\mathcal{C}$ is a category. Functoriality goes as follows: given a 2-cell $\alpha : F \Rightarrow G : \mathcal{C} \to \mathsf{Prof}$ between normal lax functors we can induce a morphism $\Pi(F) \to \Pi(G)$ of categories over $\mathcal{C}$ using the defining universal property of Π.

We shall now define the functor Γ and prove that it lands on the declared domain. First, we define the following correspondences.

- A correspondence on objects, $p : \mathcal{E} \to \mathcal{C}$, that sends such p into normal lax functors $\Gamma_p : \mathcal{C} \to \mathsf{Prof}$;

[4] A pointed profunctor $\mathfrak{p} : (\mathcal{C}, C) \rightsquigarrow (\mathcal{D}, D)$ of pointed categories is a functor $\mathfrak{p} : (\mathcal{C}^{\mathrm{op}} \times \mathcal{D}, (C, D)) \to \mathrm{Set}$ with a specified element of $\mathfrak{p}(C, D)$. The functor U forgets this specified element and keeps only the functor.

GP1. under Γ_p, an object C goes to the category

$$p^{\leftarrow}(C) = \begin{bmatrix} X \in \mathcal{E} : pX = C, \\ f : C \to C' : pf = \mathrm{id}_C \end{bmatrix}, \tag{5.50}$$

i.e. to the *fibre* of p over $C \in \mathcal{C}$;

GP2. a morphism $f : C \to C'$ goes to a profunctor $\Gamma_p(f) : p^{\leftarrow}(C)^{\mathrm{op}} \times p^{\leftarrow}(C') \to \mathrm{Set}$, defined by sending the pair $(X, Y) \in p^{\leftarrow}(C)^{\mathrm{op}} \times p^{\leftarrow}(C')$ to the set of all $u : X \to Y$ such that $pu = f$.

- A correspondence on morphisms over $\mathcal{C}$ that sends a functor $h : \mathcal{E} \to \mathcal{E}'$ of categories over $\mathcal{C}$, i.e. such that $p' \circ h = p$ for projections p, p', into a morphism of normal lax functors

$$\Gamma_h : \mathcal{C} \underset{\Gamma_{p'}}{\overset{\Gamma_p}{\Downarrow}} \mathsf{Prof}. \tag{5.51}$$

  It is evident how a functor h as given above induces a morphism of this kind simply because it respects the fibres of p and p' over the same object.

Now, we shall show that $\Gamma_p : \mathcal{C} \to \mathsf{Prof}$ is indeed a normal lax functor: for the moment we have nothing but the bare definition.

It is however quite easy to prove that the correspondence $\Gamma_p(f)$ is functorial and contravariant in the first component: if $\alpha : X \to X'$, there is an obvious function $\Gamma_p(X', Y) \to \Gamma_p(X, Y)$. Similarly, $\Gamma_p(f)$ is covariant in the second component.

We shall now show that $\Gamma_p(\_)$ is a normal lax functor. In order to do so we shall show the following.

- The functor is indeed normal: by definition $\Gamma_p(\mathrm{id}_C)$ is a functor

$$p^{\leftarrow}(C)^{\mathrm{op}} \times p^{\leftarrow}(C) \to \mathrm{Set} \tag{5.52}$$

  that sends a pair (X, Y) into the set of all $u : X \to Y$ such that $pu = \mathrm{id}_C$; but this is no less than the set of *all* morphisms $X \to Y$ in the fibre $p^{\leftarrow}(C)$, so that $\Gamma_p(\mathrm{id}_C)$ is the hom functor of the fibre of p over C, i.e. the identity profunctor $p^{\leftarrow}(C) \rightsquigarrow p^{\leftarrow}(C)$.
- Given a pair of composable morphisms $C \xrightarrow{g} C' \xrightarrow{f} C''$, we shall find a 2-cell filling the diagram

$$\begin{array}{ccccc} & & p^{\leftarrow}(C') & & \\ & {}^{\Gamma_p(g)}\nearrow & \Downarrow & \searrow^{\Gamma_p(f)} & \\ p^{\leftarrow}(C) & & \xrightarrow[\Gamma_p(fg)]{} & & p^{\leftarrow}(C'') \end{array} \tag{5.53}$$

The correspondences $\Gamma_p(f)$ and $\Gamma_p(g)$ are respectively defined by

$$p^{\leftarrow}(C')^{\mathrm{op}} \times p^{\leftarrow}(C'') \xrightarrow{\Gamma_p(f)} \mathrm{Set}$$
$$(X,Y) \overset{\Gamma_p(f)}{\longmapsto} \{u : X \to Y \mid pu = f\};$$
$$p^{\leftarrow}(C)^{\mathrm{op}} \times p^{\leftarrow}(C') \xrightarrow{\Gamma_p(g)} \mathrm{Set}$$
$$(Z,X) \overset{\Gamma_p(g)}{\longmapsto} \{v : Z \to X \mid pu = g\}.$$

Composition in the categories $p^{\leftarrow}(C), p^{\leftarrow}(C'), p^{\leftarrow}(C'')$ now forms a cowedge that induces a unique morphism

$$\begin{array}{c} \int^{X \in p^{\leftarrow}(C')} \Gamma_p(g)(Z,X) \times \Gamma_p(f)(X,Y) \\ \Big\downarrow c \\ \Gamma_p(fg)(Z,Y) = \{w : Z \to Y \mid pw = fg\} \end{array} \tag{5.54}$$

by the universal property of the coend involved in the definition of $\Gamma_p(f) \diamond \Gamma_p(g)$. We leave to the reader the routine verification that this is indeed part of the laxity constraint of a lax functor $\Gamma_p : \mathcal{C} \to \mathsf{Prof}$.

This concludes the proof, up to some fine details that we leave to the reader. (Additional exercise: try to find sufficient conditions fror the laxity cell $\Gamma_p(f) \diamond \Gamma_p(g) \Rightarrow \Gamma_p(fg)$ to be invertible.) □

5.5 Addendum: Fourier Theory

According to our Section 5.3, the bicategory Prof is monoidal with respect to the pseudocartesian structure (similarly, every $\mathsf{Prof}(\mathcal{V})$ inherits a symmetric monoidal structure from a symmetric monoidal structure on $\mathcal{V}$-Cat).

This means that we can consider internal monoids in Prof: objects endowed with maps

$$\mathcal{M} \times \mathcal{M} \overset{\mathfrak{m}}{\longrightarrow\!\!\!\!\!\!\!+} \mathcal{M} \qquad 1 \overset{\mathfrak{j}}{\longrightarrow\!\!\!\!\!\!\!+} \mathcal{M} \tag{5.55}$$

in Prof that witness the fact that $\mathcal{M}$ is an internal (pseudo)monoid.

Such internal monoids are called *promonoidal categories.*

Informally speaking, a promonoidal category is what we obtain if we replace every occurrence of the word *functor* with the word *profunctor* in the definition of monoidal category (of course, taking care of the coherence conditions imposed by the weak 2-category structure of Prof).

More precisely, we can give the following definition.

Definition 5.5.1 (Promonoidal structure). Let $\mathcal{C}$ be a category. A *promonoidal structure* consists of a tuple

$$\mathfrak{P} = (\mathcal{C}, P, J, \alpha, \lambda, \rho) \tag{5.56}$$

where

PM1. $\mathcal{C}$ is a category endowed with
PM2. a bi-profunctor $P : \mathcal{C} \times \mathcal{C} \rightsquigarrow \mathcal{C}$ (the monoidal *multiplication*),
PM3. a profunctor $J : 1 \rightsquigarrow \mathcal{C}$ (the monoidal unit), such that the following two diagrams

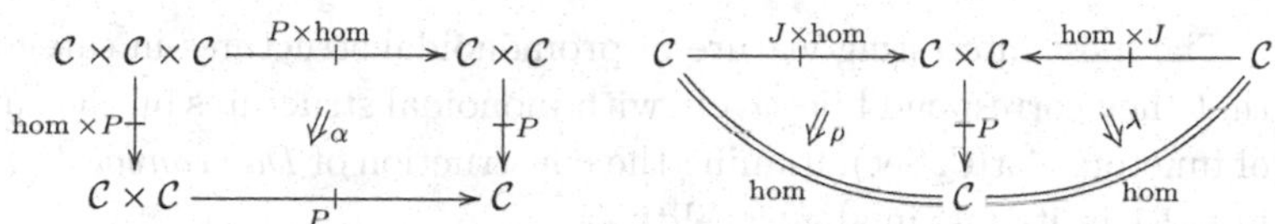

are filled by the indicated 2-cells,
PM4. 2-cells, respectively called the *associator*

$$\alpha : P \diamond (P \times \hom) \cong P \diamond (\hom \times P) \tag{5.57}$$

and the *left* and *right unitors*

$$\lambda : P \diamond (\hom \times J) \cong \hom \qquad \rho : P \diamond (J \times \hom) \cong \hom\,. \tag{5.58}$$

Remark 5.5.2. Coend calculus allows us to turn (5.57) and (5.58) into diagrammatic relations.

- The associator amounts to an isomorphism linking the two sets below (note that each component α_D^{ABC} has four arguments, three contravariant and one covariant, whereas P has components $P_C^{AB} = P(A, B; C)$ as a functor $\mathcal{C}^{\mathrm{op}} \times \mathcal{C}^{\mathrm{op}} \times \mathcal{C} \to \mathrm{Set}$):

$$\begin{aligned}(P \diamond (\hom \times P))_{ABC;D} &= \int^{XY} P_D^{XY} H_A^X P_Y^{BC}\\ &\cong \int^{Y} z\Big(\int^{X} P_D^{XY} H_A^X\Big) P_Y^{BC}\\ &\cong \int^{Z} P_D^{AY} P_Y^{BC}\end{aligned}$$

$$(P \diamond (P \times \hom))_{ABC;D} \cong \int^{XY} P^{XY}_D H^C_Y P^{AB}_X$$
$$\cong \int^Z P^{AB}_X P^{XC}_D .$$

- The left unit axiom is equivalent to the isomorphism between the functor

$$(A,B) \mapsto \int^{YZ} J_Z H^A_Y P^{YZ}_B$$
$$\cong \int^Z J_Z\Big(\int^Y H^A_Y P^{YZ}_B\Big)$$
$$\cong \int^Z J_Z P^{AZ}_B$$

and the hom functor $(A,B) \mapsto \mathcal{C}(A,B)$.

The most interesting feature of promonoidal structures in categories is that they correspond bijectively with monoidal structures on the category of functors $\mathrm{Cat}(\mathcal{C}, \mathrm{Set})$, framing the construction of *Day convolution* given in 6.2.1 in its maximal generality.

Proposition 5.5.3. Let $\mathfrak{P} = (P, J, \alpha, \rho, \lambda)$ be a promonoidal structure on the category $\mathcal{C}$; then we can define a $\mathfrak{P}$-*convolution* monoidal structure on the category $\mathrm{Cat}(\mathcal{C}, \mathrm{Set})$, via

$$[F *_{\mathfrak{P}} G]C = \int^{AB} P(A,B;C) \times FA \times GB \tag{5.59}$$
$$J_{\mathfrak{P}} = J \tag{5.60}$$

and this turns out to be a monoidal structure on $\mathrm{Cat}(\mathcal{C}, \mathrm{Set})$. We denote the monoidal structure $(\mathrm{Cat}(\mathcal{C}, \mathrm{Set}), *_{\mathfrak{P}}, J_{\mathfrak{P}})$ in brief as $[\mathcal{C}, \mathrm{Set}]_{\mathfrak{P}}$.

Remark 5.5.4. The same definition, changing the cartesian structure with the monoidal structure of $\mathcal{V}$, yields a notion of $\mathfrak{P}$-convolution on $\mathcal{V}$-$\mathrm{Cat}(\mathcal{C}, \mathcal{V})$ for a $\mathcal{V}$-category $\mathcal{C}$.

Definition 5.5.5. A functor $\Phi : [\mathcal{A}, \mathrm{Set}]_{\mathfrak{P}} \to [\mathcal{B}, \mathrm{Set}]_{\mathfrak{Q}}$ is said to *preserve the convolution product* if the obvious isomorphisms hold in $[\mathcal{B}, \mathrm{Set}]_{\mathfrak{Q}}$:

- $\Phi(F *_{\mathfrak{P}} G) \cong \Phi(F) *_{\mathfrak{Q}} \Phi(G)$;
- $\Phi(J_{\mathfrak{P}}) = J_{\mathfrak{Q}}$;

in other words Φ is a strong monoidal functor with respect to convolution product. When Φ is a colimit preserving functor, this condition is equivalent to the requirement that Φ defines a *multiplicative kernel* between $\mathcal{A}, \mathcal{B}$ regarded as objects of Prof.

Remark 5.5.6. It is observed in [IK86] that for a monoidal $\mathcal{A}$ the category of presheaves $[\mathcal{A}^{\text{op}}, \mathcal{V}]$ endowed with the convolution monoidal structure is the *free monoidal cocompletion* of $\mathcal{A}$, having in Mon (monoidal categories, monoidal functors and monoidal natural transformations) the same universal property that $[\mathcal{A}^{\text{op}}, \mathcal{V}]$ has in Cat.

There is a bijection between the promonoidal structures on $\mathcal{C}$, and monoidal structure on $\text{Cat}(\mathcal{C}, \text{Set})$; this is the content of Exercise 5.14.

5.5.1 Fourier Transforms via Coends

For the rest of the section, $\mathcal{V}$ is assumed to be a complete and cocomplete *-autonomous category.

Definition 5.5.7. Let $\mathcal{A}, \mathcal{C}$ be two promonoidal categories (thus implicitly regarded as objects of Prof) with promonoidal structures $\mathfrak{P}$ and $\mathfrak{Q}$ respectively; a *multiplicative kernel* from $\mathcal{A}$ to $\mathcal{C}$ consists of a profunctor $K : \mathcal{A} \rightsquigarrow \mathcal{C}$ endowed with two natural isomorphisms

$$\int^{YZ} K_Y^A K_Z^B P_X^{YZ} \cong \int^C K_X^C P_C^{AB} \tag{5.61}$$

$$\int^C K_X^C J_C \cong J_X. \tag{5.62}$$

These isomorphisms say that K mimics the behaviour of the hom functor (in fact, the hom functor $\hom_{\mathcal{A}}$ is the *identity* multiplicative kernel $\mathcal{A} \rightsquigarrow \mathcal{A}$: the isomorphisms above follow from 2.2.1).

We define a *multiplicative* natural transformation $\alpha : K \to H$ between two kernels as a 2-cell in Prof commuting with the structural isomorphisms given in 5.5.7. This, together with the fact that multiplicative kernels compose, yields a category of kernels $\ker(\mathcal{A}, \mathcal{C})$.

Definition 5.5.8. Let $K : \mathcal{A} \rightsquigarrow \mathcal{C}$ be a multiplicative kernel between promonoidal categories. We define the *K-Fourier transform* $f \mapsto \hat{K}(f) : \mathcal{C} \to \text{Set}$, obtained as the image of $f : \mathcal{A} \to \text{Set}$ under the left Kan extension $\text{Lan}_{よ} K : [\mathcal{A}, \text{Set}] \to [\mathcal{C}, \text{Set}]$. Explicitly, this is the coend

$$\mathfrak{F}_K(f) : X \mapsto \int^A K(A, X) \otimes fA. \tag{5.63}$$

We can also define the dual Fourier transform:

$$\mathfrak{F}^\vee(g) : Y \mapsto \int_A [K(A,X), gA] \tag{5.64}$$

and find the relation $\mathfrak{F}^\vee_K(g) \cong \mathfrak{F}_K(g^*)^*$.

The following results are easily proved using standard (co)end calculus:

Proposition 5.5.9. Let $K : \mathcal{A} \rightsquigarrow \mathcal{X}$ be a multiplicative kernel, and let $\mathcal{A}$ be a promonoidal category. Then

MK1. $\mathfrak{F}_K$ preserves the upper $\mathfrak{P}_\mathcal{A}$-convolution of presheaves f, g, defined as

$$f \mathbin{\bar{*}} g = \int^{AA'} fA \otimes gA' \otimes P(A, A', \_); \tag{5.65}$$

MK2. dually, $\mathfrak{F}^\vee_K$ preserves the lower $\mathfrak{P}_\mathcal{A}$-convolution of presheaves f, g, defined as

$$f \mathbin{\underline{*}} g = \int_{AA'} \Big(fA^* \otimes (gA')^* \otimes P(A, A', \_)\Big)^*. \tag{5.66}$$

Observe that (as stated in [Day11]), both the upper and lower convolution products yield associative and unital monoidal structures on the functor category $\mathcal{V}\text{-Cat}(\mathcal{A}, \mathcal{V})$; the upper product preserves $\mathcal{V}$-colimits in each variable, while the lower product preserves $\mathcal{V}$-limits in each variable.

The lower and upper convolutions transform into each other under the equivalence of $\mathcal{V}$-categories

$$\mathcal{V}\text{-Cat}(\mathcal{A}, \mathcal{V})^{\text{op}} \cong \mathcal{V}\text{-Cat}(\mathcal{A}^{\text{op}}, \mathcal{V}^{\text{op}}). \tag{5.67}$$

This means that under the above equivalence $(f \mathbin{\bar{*}} g)^* \cong f^* \mathbin{\underline{*}} g^*$.

Theorem 5.5.10. Let $\mathcal{V}$ be a *-autonomous monoidal base; then we can define the pairing $\mathcal{V}\text{-Cat}(\mathcal{A}, \mathcal{V}) \times \mathcal{V}\text{-Cat}(\mathcal{A}, \mathcal{V}) \to \mathcal{V}$ as the twisted form of functor tensor product (as defined in 3.2.14)

$$\langle f, g \rangle = \int^A fA^* \otimes gA. \tag{5.68}$$

If K is a kernel such that the Fourier transform $\mathrm{Lan}_{\text{よ}} K$ is fully faithful, we have an analogue of the *Parseval formula*:

$$\langle f, g \rangle \cong \langle \mathfrak{F}_K(f), \mathfrak{F}_K(g) \rangle. \tag{5.69}$$

Fourier theory is linked to the theory of Joyal's *combinatorial species* (see 2.3.11). Let $E : \mathcal{A} \rightsquigarrow \mathcal{X}$ be a multiplicative kernel; if we write $E(A, X) := X^A$ (without any reference to a tensor operation between

A and X), the E-Fourier transform can be expressed as an $\mathcal{A}$-indexed formal power series as follows:

$$\begin{aligned}\mathfrak{F}_E(f) &= \int^A fA \otimes X^A \\ &\cong \sum_{A\in\mathcal{A}} fA \otimes_{\mathcal{A}} X^A \end{aligned} \tag{5.70}$$

(it is understood that $f : \mathcal{A} \to \mathcal{V}$ is a fixed combinatorial species). This can be made precise as follows. Let $\mathcal{A}$ be the (free $\mathcal{V}$-category on the) permutation category of 6.1.1; then $E(n, X) := X^{\otimes n} = X \otimes \cdots \otimes X$ is a multiplicative kernel and an E-analytic functor results as the left Kan extension

$$\begin{aligned}FX &= \int^n f(n) \otimes X^{\otimes n} \\ &\cong \sum_{n\in\mathcal{P}} f(n) \otimes_{\mathrm{Sym}(n)} X^{\otimes n}. \end{aligned} \tag{5.71}$$

Given two combinatorial species $f, g : \mathcal{A} \to \mathcal{V}$ and the associated analytic functors F, G, the convolution $F * G$ is again an analytic functor, and its generating combinatorial species is the upper convolution product of f, g (with respect to the implicit promonoidal structure of $\mathcal{A}$):

$$\begin{aligned}F * G(X) &\cong \int^{AB} FA \otimes GB \otimes p(A, B; X) \\ &\cong \int^{ABUV} fU \otimes E(U, A) \otimes gV \otimes E(V, B) \otimes p(A, B; X) \\ &\cong \int^{UVC} fU \otimes gV \otimes p(U, V; C) \otimes E(C, X) \qquad \text{(k)} \\ &\cong \int^{C} (f \,\bar{*}\, g)(C) \otimes E(C, X). \end{aligned} \tag{5.72}$$

(Note that in (k) we used the fact that E is a multiplicative kernel.)

5.6 Addendum: Tambara Theory

Definition 5.6.1. Let $\mathcal{C}$ be a monoidal category with monoidal unit I. A (*left*) *Tambara module* on $\mathcal{C}$ consists of the following.

- A profunctor $P : \mathcal{C}^{\mathrm{op}} \times \mathcal{C} \to \mathrm{Set}$.

- A family of functions $\tau_A(X,Y) : P(X,Y) \longrightarrow P(A \otimes X, A \otimes Y)$ natural in X, Y and a wedge in A, satisfying the two equations:

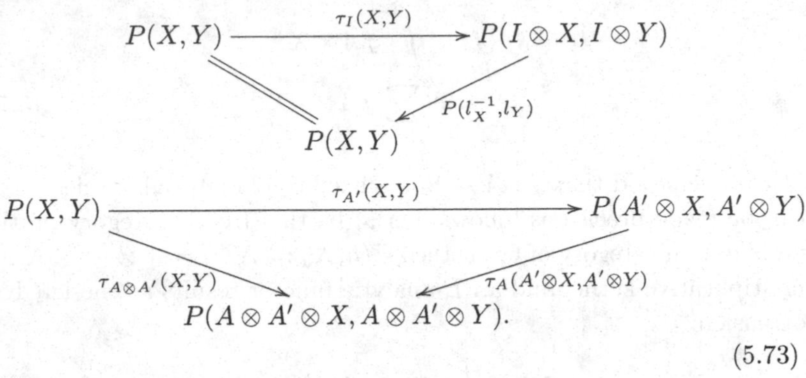

(5.73)

The notion of *right* Tambara module is given in a similar fashion, using maps $\nu_A(X,Y) : P(X,Y) \longrightarrow P(X \otimes A, Y \otimes A)$ satisfying the relations

LT1. $P(r_X, r_X^{-1}) \circ \nu_A(X,Y) = \mathrm{id}_{P(X,Y)}$,
LT2. $\nu_A(X \otimes A', Y \otimes A') \circ \nu_{A'}(X,Y) = \nu_{A' \otimes A}(X,Y)$.

The definition can be given for profunctors enriched over any other monoidal base different from Set; for example, in the original work by Tambara, [Tam06], categories are assumed to be enriched over vector spaces. In the paper [PS08] the enrichment base is completely arbitrary (i.e. it is just a symmetric monoidal closed category).

Definition 5.6.2. We define the category $\mathrm{Tamb}(\mathcal{C})$ as follows.

- The objects are Tambara modules (P, τ) consisting of a profunctor $P : \mathcal{C}^{\mathrm{op}} \times \mathcal{C} \to \mathrm{Set}$ and Tambara structures $\tau_A(X,Y)$.
- The morphisms $(P,\tau) \to (Q,\sigma)$ are natural transformations $\gamma : P \Rightarrow Q$ such that for all A, X, Y the following diagram commutes:

$$\begin{array}{ccc} P(X,Y) & \xrightarrow{\tau_A(X,Y)} & P(A \otimes X, A \otimes Y) \\ {\scriptstyle \gamma(X,Y)}\downarrow & & \downarrow{\scriptstyle \gamma(A \otimes X, A \otimes Y)} \\ Q(X,Y) & \xrightarrow[\sigma_A(X,Y)]{} & Q(A \otimes X, A \otimes Y) \end{array} \tag{5.74}$$

There is a functor to the category of endoprofunctors on $\mathcal{C}$,

$$\iota : \mathrm{Tamb}(\mathcal{C}) \longrightarrow \mathsf{Prof}(\mathcal{C}, \mathcal{C}) \tag{5.75}$$

which forgets the Tambara structure.

The codomain of ι is monoidal under composition of 1-cells, as is every category of endomorphisms in a 2-category. It turns out that Tambara modules can be composed, and that ι is strong monoidal with respect to this monoidal structure.

Remark 5.6.3. The category $\mathrm{Tamb}(\mathcal{C})$ has a monoidal structure defined as follows.

- The *unit* is the hom functor $\hom_{\mathcal{C}} : \mathcal{C}^{\mathrm{op}} \times \mathcal{C} \to \mathrm{Set}$ with its canonically associated Tambara structure:

$$\mathcal{C}(X,Y) \longrightarrow \mathcal{C}(A \otimes X, A \otimes Y). \tag{5.76}$$

- The profunctor composition of (P,τ) and (Q,σ) given by the coend

$$(P \diamond Q)(X,Y) = \int^{Z} P(X,Z) \times Q(Z,Y) \tag{5.77}$$

  has a Tambara structure $(P \diamond Q)(X,Y) \to (P \diamond Q)(A \otimes X, A \otimes Y)$ induced by the maps

$$\tau_A \times \sigma_A : P(X,Z) \times Q(Z,Y) \to P(A \otimes X, A \otimes Z) \times Q(A \otimes Z, A \otimes Y) \tag{5.78}$$

  using the universal property of the coend.

This makes the functor $\iota : \mathrm{Tamb}(\mathcal{C}) \to \mathsf{Prof}(\mathcal{C},\mathcal{C})$ strong monoidal.

Proposition 5.6.4. The forgetful functor $\iota : \mathrm{Tamb}(\mathcal{C}) \to \mathsf{Prof}(\mathcal{C},\mathcal{C})$ forms part of an adjoint triple:

$$\mathrm{Tamb}(\mathcal{C}) \underset{\varphi}{\overset{\theta}{\leftrightarrows}} \mathsf{Prof}(\mathcal{C},\mathcal{C}). \quad (\text{with } \iota \text{ in the middle}) \tag{5.79}$$

- The left adjoint $\varphi : \mathsf{Prof}(\mathcal{C},\mathcal{C}) \to \mathrm{Tamb}(\mathcal{C})$ constructs the *free Tambara module* from a profunctor. This is given by the formula

$$\varphi_P(X,Y) = \int^{C,U,V} \mathcal{C}(X, C \otimes U) \times \mathcal{C}(C \otimes V, Y) \times P(U,V) \tag{5.80}$$

  with Tambara module structure given by

$$\begin{array}{c} \mathcal{C}(X, C \otimes U) \times \mathcal{C}(C \otimes V, Y) \times P(U,V) \\ \downarrow \\ \mathcal{C}(A \otimes X, A \otimes C \otimes U) \times \mathcal{C}(A \otimes C \otimes V, A \otimes Y) \times P(U,V) \end{array} \tag{5.81}$$

  together with the coprojection $q_{A \otimes C}$, using the universal property of the coend.

- The right adjoint $\theta : \mathsf{Prof}(\mathcal{C},\mathcal{C}) \to \mathrm{Tamb}(\mathcal{C})$ constructs the *cofree Tambara module* from a profunctor. This is given by the formula

$$\theta_P(X,Y) = \int_C P(C \otimes X, C \otimes Y) \tag{5.82}$$

with Tambara module structure given by $\theta_P(X,Y) \to \theta_P(A\otimes X, A\otimes Y)$ is induced by the projection functions,

$$p_{C\otimes A} : \int_C P(C \otimes X, C \otimes Y) \to P(C \otimes A \otimes X, C \otimes A \otimes Y)$$
$$p_C : \int_C P(C \otimes A \otimes X, C \otimes A \otimes Y) \to P(C \otimes A \otimes X, C \otimes A \otimes Y) \tag{5.83}$$

using the universal property of the end.

The proof of the following proposition (a 'recognition principle' for Tambara modules) goes by inspection using the definition of coalgebra. Such a map is determined by the following:

- an object P of $\mathsf{Prof}(\mathcal{C},\mathcal{C})$ given by $P : \mathcal{C}^{\mathrm{op}} \times \mathcal{C} \to \mathrm{Set}$;
- a structure map given by a natural transformation $\tau : P \Rightarrow \theta_P$ in $\mathsf{Prof}(\mathcal{C},\mathcal{C})(P,\theta_P)$ whose components $\tau(X,Y) : P(X,Y) \to \theta_P(X,Y)$ are given by:

$$\tau(X,Y) : P(X,Y) \longrightarrow \int_A P(A \otimes X, A \otimes Y); \tag{5.84}$$

- given the universal property of the end at codomain, the structure map is determined by a wedge in A,

$$\tau_A(X,Y) : P(X,Y) \longrightarrow P(A \otimes X, A \otimes Y) \tag{5.85}$$

that is moreover natural in X, Y.

Proposition 5.6.5. The adjunction $\iota \dashv \theta$ yields a comonad

$$\Theta : \mathsf{Prof}(\mathcal{C},\mathcal{C}) \to \mathsf{Prof}(\mathcal{C},\mathcal{C}) \tag{5.86}$$

whose category of coalgebras is isomorphic to $\mathrm{Tamb}(\mathcal{C})$. Dually, the adjunction $\varphi \dashv \iota$ yields a monad $\Phi : \mathsf{Prof}(\mathcal{C},\mathcal{C}) \to \mathsf{Prof}(\mathcal{C},\mathcal{C})$ whose category of algebras is isomorphic to $\mathrm{Tamb}(\mathcal{C})$. Moreover, there is an adjunction

$$\mathsf{Prof}(\mathcal{C},\mathcal{C}) \underset{\Theta}{\overset{\Phi}{\underset{\longleftarrow}{\overset{\longrightarrow}{\bot}}}} \mathsf{Prof}(\mathcal{C},\mathcal{C}) \tag{5.87}$$

between the resulting monad $\Phi = \iota \circ \varphi$ and comonad $\Theta = \iota \circ \theta$ on $[\mathcal{C}^{\mathrm{op}} \times \mathcal{C}, \mathrm{Set}]$.

Profunctors $\mathcal{A} \rightsquigarrow \mathcal{B}$ can equivalently be described as left adjoints $\mathrm{Cat}(\mathcal{B}^{\mathrm{op}}, \mathrm{Set}) \to \mathrm{Cat}(\mathcal{A}^{\mathrm{op}}, \mathrm{Set})$; thus we obtain the following corollary.

Corollary 5.6.6. The left adjoint

$$\Phi : \mathrm{Cat}(\mathcal{C}^{\mathrm{op}} \times \mathcal{C}, \mathrm{Set}) \to \mathrm{Cat}(\mathcal{C}^{\mathrm{op}} \times \mathcal{C}, \mathrm{Set}) \tag{5.88}$$

is equivalent to the *endoprofunctor* $\check{\Phi} : \mathcal{C}^{\mathrm{op}} \times \mathcal{C} \rightsquigarrow \mathcal{C}^{\mathrm{op}} \times \mathcal{C}$ whose action on objects is given by the coend

$$\check{\Phi}(X, Y, U, V) = \int^{C} \mathcal{C}(X, C \otimes U) \times \mathcal{C}(C \otimes V, Y). \tag{5.89}$$

This endoprofunctor $\check{\Phi} : \mathcal{C}^{\mathrm{op}} \times \mathcal{C} \rightsquigarrow \mathcal{C}^{\mathrm{op}} \times \mathcal{C}$ actually underlies a *promonad* (see Exercise 5.10) in the bicategory Prof. From the formal theory of monads [Str72] it is known that the bicategory Prof admits the Kleisli construction for promonads, so we can ask what is the Kleisli category of $\check{\Phi}$. Such a category is called the (*left*) *double* of the monoidal category $\mathcal{C}$ and it is denoted $\mathrm{Db}(\mathcal{C})$; it has the same objects as $\mathcal{C}^{\mathrm{op}} \times \mathcal{C}$, and hom-sets defined by the coend

$$\mathrm{Db}(\mathcal{C})\big((X, Y), (U, V)\big) = \int^{C} \mathcal{C}(X, C \otimes U) \times \mathcal{C}(C \otimes V, Y). \tag{5.90}$$

This formula provides the foundation for all of *profunctor optics* (see [CEG+, Mil, PGW17]). [PS08] proves that there is an equivalence of categories:

$$\mathrm{Tamb}(\mathcal{C}) \simeq \mathrm{Cat}(\mathrm{Db}(\mathcal{C}), \mathrm{Set}). \tag{5.91}$$

Exercises

5.1 Describe the bicategory of profunctors between monoids, regarded as one-object categories; describe the bicategory of profunctors between posets regarded as thin categories; similarly, the bicategory Mod of modules.

5.2 Dualise 5.2.5: given $\mathfrak{r} : \mathcal{D} \rightsquigarrow \mathcal{A}$ and $\mathfrak{l} : \mathcal{E} \rightsquigarrow \mathcal{A}$ we can define

$$\mathfrak{l} \rtimes \mathfrak{r} : \mathcal{E} \rightsquigarrow \mathcal{D} : (D, E) \mapsto \int_{A} [\mathfrak{r}(D, A), \mathfrak{l}(E, A)].$$

Show that this second operation is a right *Kan lifting* (we spell out explicitly the definition that can be evinced from 2.1.3):. Given 1-cells $p : B \to C$, $f : A \to C$ in a 2-category K, a *right Kan lift* of f through p, denoted $\mathrm{Rift}_p f$, is a 1-cell $\mathrm{Rift}_p(f) : A \to B$ equipped with a 2-cell

$$\varepsilon : p \circ \mathrm{Rift}_p(f) \Rightarrow f$$

satisfying the following universal property. Given any pair $(g : A \to B, \alpha : p \circ g \Rightarrow f)$, there exists a unique 2-cell

$$\zeta : g \Rightarrow \mathrm{Rift}_p(f)$$

such that the following diagram of 2-cells commutes for a unique $\zeta : g \Rightarrow \mathrm{Rift}_p(f)$

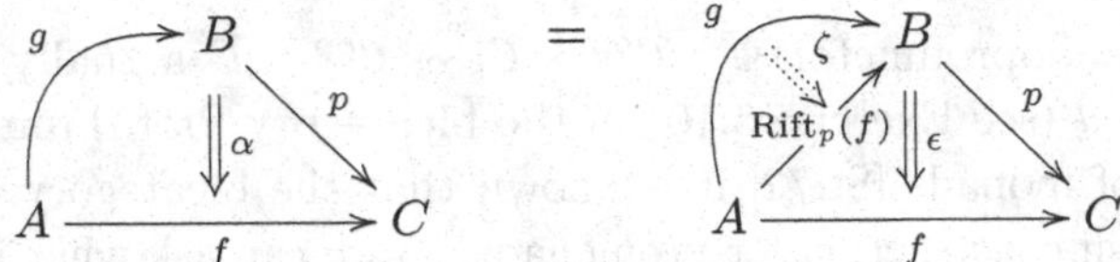

i.e. there is a unique factorisation $\varepsilon \circ (p * \zeta) = \alpha$.

5.3 Show that the structure on Prof given by $\diamond$ is *biclosed* (i.e. $\diamond$ is a bifunctor $\mathsf{Prof}(\mathcal{A}, \mathcal{B}) \times \mathsf{Prof}(\mathcal{B}, \mathcal{C}) \to \mathsf{Prof}(\mathcal{A}, \mathcal{C})$ and each $\mathfrak{p} \diamond \_$, as well as each $\_ \diamond \mathfrak{q}$, has right adjoints).

5.4 The *collage* of two categories $\mathcal{A}, \mathcal{B}$ along a profunctor $\mathfrak{p} : \mathcal{A} \rightsquigarrow \mathcal{B}$ is defined as the category $\mathcal{A} \uplus_{\mathfrak{p}} \mathcal{B}$ with the same objects as $\mathcal{A} \amalg \mathcal{B}$ and morphisms given by the rule

$$\mathcal{A} \uplus_{\mathfrak{p}} \mathcal{B}(X, Y) = \begin{cases} \mathcal{A}(X, Y) & \text{if } X, Y \in \mathcal{A} \\ \mathcal{B}(X, Y) & \text{if } X, Y \in \mathcal{B} \\ \mathfrak{p}(X, Y) & \text{if } X \in \mathcal{A}, Y \in \mathcal{B} \end{cases}$$

and empty in every other case. Show that $\mathcal{A} \uplus_{\mathfrak{p}} \mathcal{B}$ has the universal property of the category of elements of $\mathfrak{p}$, regarded as a presheaf. Find an isomorphism between the category $\mathcal{A} \uplus_{\mathfrak{p}} \mathcal{B}$ so defined and the coend

$$\int^{(A,B)} (\mathcal{A}^{\mathrm{op}} \times \mathcal{B})/(A, B) \otimes \mathfrak{p}(A, B)$$

of 4.2.2 ($\otimes$ is the Set-tensor of Cat).

5.5 Show that the composition laws $\mathfrak{p}(A, B) \times \mathcal{B}(B, B') \to \mathfrak{p}(A, B')$, $\mathcal{A}(A, A') \times \mathfrak{p}(A', B) \to \mathfrak{p}(A, B)$ of arrows in the collage $\mathcal{A} \uplus_{\mathfrak{p}} \mathcal{B}$ are universal cowedges of a coend.

5.6 The *cocomma object* $[{F \atop G}]$ of two functors $\mathcal{X} \xleftarrow{F} \mathcal{A} \xrightarrow{G} \mathcal{Y}$ is defined to be the pushout of

$$\begin{array}{ccc} \mathcal{A} \amalg \mathcal{A} & \longrightarrow & \mathcal{A} \times [1] \\ \downarrow & & \\ \mathcal{X} \amalg \mathcal{Y} & & \end{array}$$

in Cat, where the horizontal arrow is the 'cylinder' embedding when $\mathcal{A} \amalg \mathcal{A}$ is identified with $\mathcal{A}^{\{0,1\}} = \mathrm{Cat}(\{0,1\}, \mathcal{A})$. Show that $[{F \atop G}]$ is the collage of $\mathcal{X}$ and $\mathcal{Y}$ along the profunctor $\mathfrak{p}^G \diamond \mathfrak{p}_F : \mathcal{X} \rightsquigarrow \mathcal{Y}$.

5.7 Given profunctors $\mathcal{A} \overset{\mathfrak{p}}{\rightsquigarrow} \mathcal{B} \overset{\mathfrak{q}}{\rightsquigarrow} \mathcal{C}$ consider the categories $\mathcal{A} \uplus_{\mathfrak{p}} \mathcal{B}$ and $\mathcal{B} \uplus_{\mathfrak{q}} \mathcal{C}$. Describe the pushout

$$\begin{array}{ccc} \mathcal{B} & \longrightarrow & \mathcal{A} \uplus_{\mathfrak{p}} \mathcal{B} \\ \downarrow & & \downarrow \\ \mathcal{B} \uplus_{\mathfrak{q}} \mathcal{C} & \longrightarrow & \mathcal{H} \end{array}$$

in Cat. Is there a relation between $\mathcal{H}$ and the collage $\mathcal{A} \uplus \mathcal{C}$ along $\mathfrak{q} \diamond \mathfrak{p}$?

5.8 Isbell duality (3.2.18) can be regarded as an adjunction between the categories

$$\mathsf{Prof}(1, \mathcal{C})^{\mathsf{op}} \leftrightarrows \mathsf{Prof}(\mathcal{C}, 1)$$

where 1 is the terminal category. Is it possible to extend this result to an adjunction $\mathsf{Prof}(\mathcal{D}, \mathcal{C})^{\mathsf{op}} \leftrightarrows \mathsf{Prof}(\mathcal{C}, \mathcal{D})$? (Hint: use right Kan extensions in Prof.)

5.9 Show that there is a canonical isomorphism

$$\mathrm{Cat}(\mathcal{X}, \mathcal{Y})(FG, HK) \cong \mathsf{Prof}(\mathcal{W}, \mathcal{Z})(\mathfrak{p}_G \diamond \mathfrak{p}^K, \mathfrak{p}^F \diamond \mathfrak{p}_H)$$

for each square

$$\begin{array}{ccc} \mathcal{X} & \xrightarrow{G} & \mathcal{Z} \\ {\scriptstyle K}\downarrow & \swArrow\alpha & \downarrow{\scriptstyle F} \\ \mathcal{W} & \xrightarrow[H]{} & \mathcal{Y} \end{array}$$

filled by a 2-cell α. This map sends α to a 2-cell $\alpha^\sharp$ in Prof. Does this equivalence restrict to strongly commutative squares, giving strongly commutative squares in Prof?

5.10 A *promonad* is a monad $T : \mathcal{A} \rightsquigarrow \mathcal{A}$ over an object of Prof; this means that there are maps $T \diamond T \Rightarrow T$ and hom $\Rightarrow T$ fitting in diagrams similar to A.6.3.

PM1. Regard a set A as a discrete category; show that every promonad $T : A \rightsquigarrow A$ determines and is determined by a category structure for A, i.e. that a promonad on a discrete category amounts exactly as a category having A as its set of objects.

PM2. ◉◉ Show a similar result for non-discrete categories. More precisely, observe that a promonad on $\mathcal{A}$ in Prof corresponds to a monad on $\mathrm{Cat}(\mathcal{A}^{\mathrm{op}}, \mathrm{Set})$ whose underlying endofunctor preserves colimits. Show that every category $\mathcal{B}$ that admits an identity-on-objects functor $F : \mathcal{A} \to \mathcal{B}$ induces such a monad; show that every colimit preserving monad between presheaf categories induces an identity-on-object functor $\mathcal{A} \to \mathcal{B}$.

5.11 Show that the equivalence

$$\mathrm{LAdj}(\mathrm{Cat}(\mathcal{X}^{\mathrm{op}}, \mathrm{Set}), \mathrm{Cat}(\mathcal{X}^{\mathrm{op}}, \mathrm{Set})) \cong \mathsf{Prof}(\mathcal{X}, \mathcal{X})$$

of 3.1.1 is monoidal, i.e. colimit preserving monads T on $\mathrm{Cat}(\mathcal{X}^{\mathrm{op}}, \mathrm{Set})$ go to promonads $\tau_T : \mathcal{X} \rightsquigarrow \mathcal{X}$.

5.12 Gather from Chapter 1 the results you need to prove that the trace operator tr^U_{XY} in 5.3.5 really defines a traced monoidal structure.

TM1. *Naturality in X and Y*: for every $f : X \otimes U \to Y \otimes U$ and $g : X' \to X$,

$$\mathrm{tr}^U_{X',Y}(f \circ (g \otimes \mathrm{id}_U)) = \mathrm{tr}^U_{X,Y}(f) \circ g,$$

and for every $f : X \otimes U \to Y \otimes U$ and $g : Y \to Y'$,

$$\mathrm{tr}^U_{X,Y'}((g \otimes \mathrm{id}_U) \circ f) = g \circ \mathrm{tr}^U_{X,Y}(f).$$

TM2. *Dinaturality in U*: for every $f : X \otimes U \to Y \otimes U'$ and $g : U' \to U$

$$\mathrm{tr}^U_{X,Y}((\mathrm{id}_Y \otimes g) \circ f) = \mathrm{tr}^{U'}_{X,Y}(f \circ (\mathrm{id}_X \otimes g)).$$

TM3. *Two vanishing conditions*: for every $f : X \otimes I \to Y \otimes I$, (with $\rho_X : X \otimes I \cong X$ being the right unitor),

$$\mathrm{tr}^I_{X,Y}(f) = \rho_Y \circ f \circ \rho_X^{-1},$$

and for every $f : X \otimes U \otimes V \to Y \otimes U \otimes V$

$$\mathrm{tr}^U_{X,Y}(\mathrm{tr}^V_{X \otimes U, Y \otimes U}(f)) = \mathrm{tr}^{U \otimes V}_{X,Y}(f).$$

TM4. *Superposing*: for every $f : X \otimes U \to Y \otimes U$ and $g : W \to Z$,

$$g \otimes \mathrm{tr}^U_{X,Y}(f) = \mathrm{tr}^U_{W\otimes X, Z\otimes Y}(g \otimes f).$$

TM5. *Yanking*: $\mathrm{tr}^X_{X,X}(\gamma_{X,X}) = \mathrm{id}_X$ (where γ is the symmetry of the monoidal category).

5.13 Prove Proposition 5.5.3 using associativity and unitality for $\mathfrak{P}$.

5.14 Let $* : \mathrm{Cat}(\mathcal{C}, \mathrm{Set}) \times \mathrm{Cat}(\mathcal{C}, \mathrm{Set}) \to \mathrm{Cat}(\mathcal{C}, \mathrm{Set})$ be a monoidal structure with monoidal unit $u : \mathcal{C} \to \mathrm{Set}$. Show that the assignment

$$P(A, B; C) := (よ^\vee(A) * よ^\vee(B))(C) \qquad JA := uA \tag{5.92}$$

is a promonoidal structure on $\mathcal{C}$, regarded as an object of Prof. An elegant result of Day shows that this sets up a bijection between the ways in which $\mathrm{Cat}(\mathcal{C}, \mathrm{Set})$ is a (pseudo)monoid in Cat, and the ways in which $\mathcal{C}$ is a (pseudo)monoid in Prof: prove this.

5.15 Outline the promonoidal structure $\mathfrak{P}$ giving the Day convolution described in 6.2.1. If $\mathcal{C}$ is any small category, we define $P(A, B; C) = \mathcal{C}(A, C) \times \mathcal{C}(B, C)$ and define J to be the terminal functor $\mathcal{C} \to \mathrm{Set}$. Outline the convolution product on $\mathrm{Cat}(\mathcal{C}, \mathrm{Set})$, called the *Cauchy convolution*, obtained from this promonoidal structure.

5.16 Is the composition of two kernels (see 5.5.7) again a kernel? Define the category of multiplicative kernels $\mathrm{ker}(\mathcal{A}, \mathcal{C}) \subset \mathrm{Prof}(\mathcal{A}, \mathcal{C})$.

5.17 Show that a profunctor $K : \mathcal{A} \rightsquigarrow \mathcal{C}$ is a multiplicative kernel if and only if the cocontinuous functor $\mathrm{Lan}_よ K = \hat{K} : [\mathcal{A}, \mathrm{Set}] \to [\mathcal{C}, \mathrm{Set}]$ corresponding to $\bar{K} : \mathcal{A} \to \mathcal{V}\text{-}\mathrm{Cat}(\mathcal{C}, \mathcal{V})$ under the construction in 5.1.7 is monoidal with respect to the convolution monoidal structure on both $[\mathcal{A}, \mathrm{Set}]_{\mathfrak{P}}$ and $[\mathcal{C}, \mathrm{Set}]_{\mathfrak{Q}}$.

Describe the isomorphisms k_1, k_2 when $\mathfrak{P}$ is the Day convolution.

5.18 Show that a functor $F : (\mathcal{A}, \otimes_{\mathcal{A}}, I) \to (\mathcal{C}, \otimes_{\mathcal{C}}, J)$ between monoidal categories is strong monoidal if and only if $\mathfrak{p}^F = \hom(F, 1)$ is a multiplicative kernel.

Dually, show that for $\mathcal{A}, \mathcal{C}$ promonoidal, $F : \mathcal{C} \to \mathcal{A}$ preserves convolution on $[\mathcal{A}, \mathrm{Set}]_{\mathfrak{P}}, [\mathcal{C}, \mathrm{Set}]_{\mathfrak{Q}}$ precisely if $\mathfrak{p}_F = \hom(1, F)$ is a multiplicative kernel.

5.19 Show the following properties of the K-Fourier transform.

- There is the canonical isomorphism

$$\hat{K}(f) \cong \int^A K(A, \_) \times f(A).$$

- $\hat{K}$ preserves the convolution monoidal structure (this is the *Parseval identity* for the Fourier transform).

- $\hat{K}$ has a right adjoint defined by

$$\check{K}(g) \cong \int_x [K(\_, X), g(X)].$$

5.20 ◉◉ Prove the following statements.

- There is a monad $\tilde{S}$ on the category of profunctors, such that the following square is commutative:

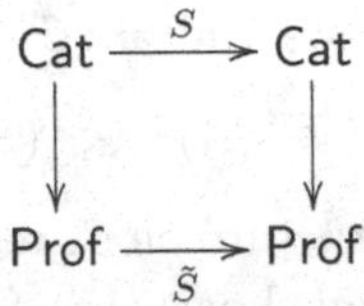

where S sends a category $\mathcal{A}$ to the *free monoidal category* on $\mathcal{A}$ and Cat → Prof is the canonical embedding of Section 5.2 (on which cells it is contravariant?).
- Prove that an $\tilde{S}$-algebra is the same thing as a monad in the bicategory of $\tilde{S}$-profunctors
- Show that the following conditions are equivalent, for a profunctor $\mathfrak{p} : \mathcal{A} \rightsquigarrow \mathcal{B}$ between promonoidal categories $(\mathcal{A}, \mathfrak{P}, J_\mathcal{A}), (\mathcal{B}, \mathfrak{Q}, J_\mathcal{B})$.

  PA1. $\mathfrak{p}$ is a pseudo-$\tilde{S}$-algebra morphism.

  PA2. The cocontinuous left adjoint associated to $\mathfrak{p}$, $\hat{\mathfrak{p}} : [\mathcal{B}^{\mathrm{op}}, \mathrm{Set}] \to [\mathcal{A}^{\mathrm{op}}, \mathrm{Set}]$ is strong monoidal with respect to the convolution monoidal product on presheaf categories.

  PA3. $\mathfrak{p}$ is endowed with arrows

$$\mathfrak{p}^{A_1}_{B_1} \times \mathfrak{p}^{A_2}_{B_2} \times \mathfrak{Q}^{B_1 B_2}_{B} \xrightarrow{\quad\gamma\quad} \int^A \mathfrak{P}^{A_1 A_2}_{A} \times \mathfrak{p}^{A}_{B}$$

$$(\mathfrak{p}^{\mathrm{op}})^{B_1}_{A_1} \times (\mathfrak{p}^{\mathrm{op}})^{B_2}_{A_2} \times \mathfrak{P}^{A_1 A_2}_{A} \xrightarrow{\quad\sigma\quad} \int^B (\mathfrak{p}^{\mathrm{op}})^{B}_{A} \times \mathfrak{Q}^{B_1 B_2}_{B}$$

$$\mathfrak{p}^{A}_{B} \times J_{\mathcal{A}A} \xrightarrow{\quad\delta\quad} J_{\mathcal{B}B}$$

  or more precisely, $\gamma_{B_1 B_2 [A_1 A_2]; A}, \sigma_{B_1 B_2 [A_1 A_2]; A}, \delta_{[A] B}$, exhibiting universal cowedges in the bracketed variables (i.e. the maps induced on coends are isomorphisms), and natural in the others.

- Assume the promonoidal structures $\mathfrak{P}, \mathfrak{Q}$ on $\mathcal{A}, \mathcal{B}$ are representable; then, the conditions above are in turn equivalent to the following.
  Both arrows $\mathfrak{p}^{\triangleleft} : \mathcal{A} \to [\mathcal{B}^{\mathrm{op}}, \mathrm{Set}]$ and $\mathfrak{p}^{\triangleright} : \mathcal{B} \to [\mathcal{A}, \mathrm{Set}]^{\mathrm{op}}$ obtained from $\mathfrak{p}$ under adjunction are strong monoidal with respect to convolution on their codomains.

6

Operads

SUMMARY. We introduce the theory of *operads* employing (co)end calculus; the material is entirely classical and draws from [Kel05b] and equally classical sources. A (symmetric) operad is a collection $O(n)$ of objects of a monoidal category whose objects are natural numbers, and endowed with maps

$$O(n_1) \otimes \cdots \otimes O(n_k) \otimes O(k) \to O\big(\sum n_i\big)$$

satisfying suitable axioms of associativity and compatibility with a natural action of the symmetric group on each component. Each $O(n)$ models a set of generalised n-ary operations serving to describe neatly and intrinsically the equipment of a 'set' with 'structure'. The notion of operad lies at the very core of modern approaches to universal and categorical algebra.

Operads are monoid objects in the presheaf category of representations of the groupoid of natural numbers. Drawing from [Cur12] we provide a standard characterisation of operads as the coKleisli category of a comonad on the bicategory Prof. More precisely, an operad is an object in the coKleisli category of a comonad generated by the presheaf construction $\boldsymbol{P}$ and by the 'free symmetric monoidal category' functor S. This allows for plenty of generalisations to other kinds of operads, by changing the role of S in the same formal argument.

> The sixth [is] the method of returning the letters to their prime-material state and giving them form in accordance with the power of wisdom that confers form. [...]
> Regarding this method, it is stated in the *Sefer Yetzirah*:
> 'Twenty-two cardinal letters; He engraved them and hewed them and weighed them and permuted and combined them and formed by their means the souls of all formed beings.'
>
> A. Abulafia, *Oẓar Eden Ganuz*, quoted in [Ide89]

6.1 Introduction

Operads are mathematical structures of manifold nature: they appear in algebra, topology, algebraic geometry, logic, and in each of these settings they model the notion of 'set endowed with operations' providing an extremely powerful conceptual tool to categorise the old discipline known as universal algebra.

Operads were introduced by P. May in his [May72] to solve a purely algebraic-topological problem. Topologists are often interested in classifying spaces Y which are homotopy equivalent to a loop space ΩX; every such space carries the structure of a group up to homotopy. But they are much more interested in spaces $Y \simeq \Omega^n X$ for higher n, as such spaces carry the structure of an *n-fold commutative* group. Now, there are spaces Y that arise as infinitely many looped X's, i.e. spaces Y such that $Y \simeq \Omega X$ for an X which is $\Omega X'$, for an X' which is $\Omega X'' \dots$: these are called *infinite loop spaces*, and in the world of homotopy coherent operations they behave like abelian groups.

The notion of operad, introduced in [May72], offers a way to recognise infinite loop spaces among all spaces, as they are *algebras* for a suitable operad. The reader is invited to consult [Ada78] for more information; Adams' book is one of the nicest introductions to the topic.

In short, an operad is a family of spaces $O(n)$, one for each natural number n, subject to suitable axioms, one of which is that for every $k \in \mathbb{N}$, and every k-tuple of numbers $n_1, \dots, n_k$, there is a map

$$\gamma_{k,\vec{n}} : O(k) \times O(n_1) \times \cdots \times O(n_k) \to O(n_1 + \cdots + n_k). \tag{6.1}$$

Since their very introduction it has been clear that operads are *monoid-like objects* in some category of functors, and that the maps above behave like multiplications of some sort. This is the reason why they can naturally act on other objects, and why the *algebras* for an operad are so important.[1] Making this analogy a precise statement, using the power of (co)end calculus, is the content of a seminal paper by Max Kelly [Kel05b], which the present chapter follows extremely closely.

(Co)end calculus is a perfect bookkeeping tool in otherwise extremely involved combinatorial arguments involving quotients of sets of n-tuples by the action of a symmetric group.

Unfortunately, a thorough introduction to the theory of operads ex-

[1] This principle echoes Mt 7:20: *Wherefore by their fruits ye shall know them.* The subject of the sentence is, of course, monoids; and the fruits are the monoid actions other objects can carry.

ceeds the aims of the present chapter: beginners (as authors, we are undoubtedly among them) may feel rather disoriented when approaching any book on the subject: algebraists might feel baffled by a geometric approach motivated by topology, and geometers might feel the same way reading about 'algebraic' operads. So, it is extremely difficult to recommend a single, comprehensive reference. Among classical textbooks, we cannot help but mention [May72], and more recent monographs like [LV12, MSS02] written respectively from the algebraist's and topologist's/-geometer's point of view. Among less classical and yet extremely valid points of view, the author profited a lot from a lucid, and unfortunately still unfinished, online draft [Tri] written by T. Trimble.

We shall provide a glance at the use of operads in algebra in Section 6.4 below; the exposition is brief, but keeps to a minimum the cognitive overload and tries to employ ideas and notation from the previous section.

6.1.1 Local Conventions

- We will denote by the letter $\mathcal{P}$ the *groupoid of natural numbers*, i.e. the category having objects the non-empty sets $\{1,\dots,n\}$ (denoted as n for short, assuming that $0 = \varnothing$) where $\mathcal{P}(m,n) = \varnothing$ if $n \neq m$ and S_n (the group of bijections of n-element sets) if $n = m$. It is evident that the groupoid $\mathcal{P}$ is the disjoint union of symmetric groups $\coprod_{n\geq 0} \mathrm{Aut}(n) = \coprod_{n\geq 0} S_n$.[2]
- As elsewhere, $\mathcal{V}$ is a fixed (symmetric) monoidal closed category, having all (weighted) (co)limits needed to cast the relevant (co)ends.
- We make a moderate use of λ-notation. A function $x \mapsto Fx$ will be denoted as $\lambda x.Fx$ as if it were a λ-term; the usual rules of α-conversion and β-reduction straightforwardly apply.
- When needed, we freely employ the 'Einstein notation' defined in 5.1.10; this will allow us to maintain in a single line a few involved computations.

Remark 6.1.1. Following the Einstein convention, we will always denote variables over which we integrate as subscript-superscript pairs. Moreover, monoidal structure symbols $\otimes$ are suppressed when this does not create

[2] A subtlety that will never be mentioned in the discussion is that we blur the distinction between $\mathcal{P}$ and the *free* $\mathcal{V}$*-category* $\overline{\mathcal{P}}^{\mathcal{V}}$ on $\mathcal{P}$, defined as having the same objects, and where the hom object between n and m is the $\mathcal{P}(n,m)$-fold tensor of the monoidal unit $I \in \mathcal{V}$; to motivate the name 'free', think of the case when $\mathcal{V} = \mathrm{Ab}$ is the category of abelian groups. Exercise 6.1 gives the proper definition.

ambiguity; thus, for example, the convolution (6.3) of two presheaves in Einstein notation has the following form:

$$F * G = \lambda X. \int^{CC'} \mathcal{C}_X^{C\otimes C'} F_C G_{C'}.$$

Remark. We record that $\mathcal{P}$ has a natural choice of symmetric monoidal structure, with tensor the sum of natural numbers

$$(n, m) \mapsto n + m = n \sqcup m; \tag{6.2}$$

the action on arrows is given by $(\sigma, \tau) \mapsto \sigma + \tau$ defined as acting as σ on the set $\{1, \dots, m\}$ and as τ on the set $\{m + 1, \dots, m + n\}$ (these permutations are called *shuffles*).

6.2 The Convolution Product

The convolution product can be thought as a categorification of the convolution of regular functions: let G be a topological group, and $C(G)$ the set of 'regular', i.e. continuous, complex-valued, functions $f : G \to \mathbb{C}$. Then, the set $C(G)$ can be endowed with a *convolution* product, given by the integral $(f, g)(x) = \int_G f(xy^{-1})g(y)dy$ (not a coend!), once a suitable left-invariant measure has been chosen on G.

In a similar fashion, if $\mathcal{C}$ is a *monoidal* category, we can endow the category of functors $F : \mathcal{C} \to \mathcal{V}$ with a monoidal structure, which is in general different from the pointwise one. This is called the *convolution product* of functors.

Proposition 6.2.1 (Day convolution). Let $\mathcal{C}$ be a symmetric monoidal $\mathcal{V}$-category with monoidal product $\oplus$; the functor category $\mathcal{V}\text{-Cat}(\mathcal{C}, \mathcal{V})$ is itself a symmetric monoidal category, (and in fact a cosmos if $\mathcal{V}$ is such) with respect to the monoidal structure given by *Day convolution*. Given $F, G \in \mathcal{V}\text{-Cat}(\mathcal{C}, \mathcal{V})$ we define

$$F * G := \int^{CC'} \mathcal{C}(C \oplus C', \_) \cdot FC \otimes GC' \tag{6.3}$$

where we recall (see 2.2.3) that $X \cdot V$ for $X \in \text{Set}, V \in \mathcal{V}$ is the *copower* (or *tensor*) $X \cdot V$ such that

$$\mathcal{V}(X \cdot V, W) \cong \text{Set}(X, \mathcal{V}(V, W)). \tag{6.4}$$

Proof We have to show that this really defines a monoidal structure.

- Associativity follows from the associativity of the tensor product on $\mathcal{C}$ and the ninja Yoneda lemma 2.2.1:[3]

$$\begin{aligned}F * (G * H) &= \lambda X. \int^{AB} \mathcal{C}_X^{A\oplus B} F_A (G * H)_B \\ &\cong \lambda X. \int^{AB} \mathcal{C}_X^{A\oplus B} \int^{CD} \mathcal{C}_B^{C\oplus D} F_A G_C H_D \\ &\cong \lambda X. \int^{ABCD} \mathcal{C}_X^{A\oplus B} \mathcal{C}_B^{C\oplus D} F_A G_C H_D \\ &\cong \lambda X. \int^{ACD} \mathcal{C}_X^{A\oplus(C\oplus D)} F_A G_C H_D.\end{aligned}$$

A similar computation shows that

$$(F * G) * H \cong \lambda X. \int^{ACD} \mathcal{C}_X^{(A\oplus C)\oplus D} F_A G_C H_D. \tag{6.5}$$

The associativity isomorphism of $(\mathcal{C}, \oplus)$ now entails that $F * (G * H) \cong (F * G) * H$.

- (Right) unitality: we show that $J = よ_0 = \mathcal{C}(0, \_)$ plays the role of monoidal unit for the convolution $*$, if 0 is the monoidal unit for $\oplus$. Again, the ninja Yoneda lemma yields

$$\begin{aligned}F * J &\cong \lambda X. \int^{CD} \mathcal{C}_X^{C\oplus D} F_C J_D \\ &\cong \lambda X. \int^{CD} \mathcal{C}_X^{C\oplus D} \mathcal{C}_D^0 F_C \\ &\cong \lambda X. \int^{C} \mathcal{C}_X^{C\oplus 0} F_C \cong F.\end{aligned}$$

Similarly, we obtain left unitality. □

Example 6.2.2 (Subdivision and joins as convolutions)**.** Compare Example 3.2.17 and the definition of join of augmented[4] simplicial sets given

[3] See 6.1.1 above for how we employ Einstein notation; here and elsewhere, it is also harmless to suppress the distinction between monoidal products in $\mathcal{V}$ and $\mathcal{V}$-tensors, since once the infix symbol has been removed to become mere juxtaposition, the two operations behave similarly; in all cases it can be easily devised which operation is which with a 'dimensionality check'.

[4] The category $\mathbf{\Delta}$ would have ordinal sum as monoidal structure, but it lacks an initial object $[-1] = \varnothing$ as monoidal unit; if we add such an object, we get a category $\mathbf{\Delta}_+$, and an *augmented* simplicial set is a presheaf on $\mathbf{\Delta}_+$; the category of augmented simplicial sets is denoted sSet_+. There is a triple of adjoints induced by the inclusion $i : \mathbf{\Delta} \subset \mathbf{\Delta}_+$ and linking the categories of simplicial and augmented simplicial sets.

in [Joy08b]. Given $X, Y \in \mathrm{sSet}_+$ we define

$$X \star Y = \int^{p,q} X_p \times Y_q \times \mathbf{\Delta}(\_, p \oplus q) \tag{6.6}$$

where $\oplus$ is the *ordinal sum* operation.

Proposition 6.2.3. The convolution product of $F, G : \mathcal{C} \to \mathcal{V}$ has the universal property of the following left Kan extension:

$$\begin{array}{ccccc} \mathcal{C} \times \mathcal{C} & \xrightarrow{F \times G} & \mathcal{V} \times \mathcal{V} & \xrightarrow{\otimes} & \mathcal{V} \\ \oplus \downarrow & & & \nearrow & \\ \mathcal{C} & & \cdots\cdots F * G \cdots\cdots & & \end{array} \tag{6.7}$$

Proof Just recognise that the coend expression given in (6.3) coincides with the coend formula for $\mathrm{Lan}_\oplus(F \times G)$ given in (2.27). $\square$

Finding an explicit expression for the unit η of this left extension is the scope of Exercise 6.2.

Remark 6.2.4. The category $\mathcal{V}\text{-Cat}(\mathcal{C}, \mathcal{V})$ is left and right closed: the internal hom $[\![G, H]\!]$ (or rather the functor $[\![G, \_]\!]$ which is right adjoint to $\_ * G$) is given by

$$[\![G, H]\!] := \lambda X. \int_C [GC, H(C \oplus X)] \tag{6.8}$$

where $[\_, =]$ is the internal hom in $\mathcal{V}$.

(When $\mathcal{C}$ is not symmetric monoidal, we shall distinguish between a *right* internal hom and a *left* internal hom.)

Proof We can compute directly:

$$\begin{aligned} \mathcal{V}\text{-Cat}(\mathcal{C}, \mathcal{V})(F * G, H) &\cong \int_C \mathcal{V}((F * G)C, HC) \\ &\cong \int_C \mathcal{V}\left(\int^{AB} \mathcal{C}_C^{A\oplus B} F_A G_B,\ H_C\right) \\ &\cong \int_{ABC} \mathcal{V}(\mathcal{C}_C^{A\oplus B} F_A G_B,\ H_C) \\ &\cong \int_{ABC} \mathcal{V}(F_A, [\mathcal{C}_C^{A\oplus B} G_B, H_C]) \\ &\cong \int_{ABC} \mathcal{V}(F_A, [G_B, [\mathcal{C}_C^{A\oplus B}, H_C]]) \end{aligned}$$

$$
\begin{aligned}
&\cong \int_{AB} \mathcal{V}\Big(F_A, [G_B, \int_C [\mathcal{C}_C^{A\oplus B}, H_C]]\Big) \\
&\cong \int_{AB} \mathcal{V}(F_A, [G_B, H_{A\oplus B}]) \\
&\cong \int_A \mathcal{V}\Big(F_A, \int_B [G_B, H_{A\oplus B}]\Big) \\
&\cong \int_A \mathcal{V}(F_A, [\![G, H]\!]_A) \\
&\cong \mathcal{V}\text{-Cat}(\mathcal{C}, \mathcal{V})\big(F, [\![G, H]\!]\big).
\end{aligned}
$$

This concludes the proof. □

Remark 6.2.5. We shall now specialise the above result to the particular case $\mathcal{C} = \mathcal{P}$ with the sum monoidal structure of 6.1.1; this means that $\mathcal{V}$-Cat$(\mathcal{P}, \mathcal{V})$ is monoidal closed if we define

$$
\begin{aligned}
(F * G)_k &:= \int^{mn} \mathcal{P}(m+n, k) \cdot F_m \otimes G_n \\
[\![F, G]\!]_k &:= \int_n [F_n, G_{n+k}].
\end{aligned}
$$

In particular, we have a formula for the iterated convolution of $F_1, \dots, F_n \in [\mathcal{C}, \mathcal{V}]$, that we will make frequent use of:

$$
F_1 * \cdots * F_n = \int^{k_1,\dots,k_n} \mathcal{P}(\textstyle\sum k_i, \_) \cdot F_1 k_1 \otimes \cdots \otimes F_n k_n. \tag{6.9}
$$

Representing the tuple $k_1, \dots, k_n$ as a vector $\vec{k}$, we can also write (6.9) in a more compact fashion as

$$
F_1 * \cdots * F_n = \int^{\vec{k}} \mathcal{P}(\textstyle\sum \vec{k}, \_) \cdot F\vec{k}.
$$

6.3 Substitution Product and Operads

The gist of the definition of a $\mathcal{V}$-operad relies on the possibility of endowing $\mathcal{V}$-Cat$(\mathcal{P}, \mathcal{V})$ with an additional monoidal structure, defined by means of the Day convolution: this is called the *substitution* product.

Definition 6.3.1 (Substitution product on $\mathcal{V}$-Cat$(\mathcal{P}, \mathcal{V})$). Let $F, G \in \mathcal{V}$-Cat$(\mathcal{P}, \mathcal{V})$. Define

$$
F \odot G := \int^m Fm \otimes G^{*m}, \tag{6.10}
$$

where $G^{*m} := G * \cdots * G$ is the iterated convolution.

Associativity exploits the following result.

Lemma 6.3.2. There exists a natural equivalence $(F \odot G)^{*m} \cong F^{*m} \odot G$.

Proof This is a formal manipulation based on (6.9) and the Yoneda lemma: the vector $\vec{n}$ denotes the m-tuple of integers $(n_1, \ldots, n_m)$ so that (6.9) becomes $\ast_{i=1}^{n} F_i \cong \int^{\vec{k}} \mathcal{P}\big(\sum k_i, \_\big) \cdot \bigotimes_{i=1}^{n} F_i k_i$.

$$\begin{aligned}
(F \odot G)^{*m} &= \int^{\vec{n}} \mathcal{P}\Big(\sum n_i, _\Big) \cdot (F \odot G)n_1 \otimes \cdots \otimes (F \odot G)n_m \\
&\cong \int^{\vec{n},\vec{k}} \mathcal{P}\Big(\sum n_i, _\Big) \cdot Fk_1 \otimes G^{*k_1} n_1 \otimes \cdots \otimes Fk_m \otimes G^{*k_m} n_m \\
&\cong \int^{\vec{n},\vec{k}} Fk_1 \otimes \cdots \otimes Fk_m \otimes \mathcal{P}\Big(\sum n_i, _\Big) \cdot G^{*k_1} n_1 \otimes \cdots \otimes G^{*k_m} n_m \\
&\cong \int^{\vec{k}} Fk_1 \otimes \cdots \otimes Fk_m \otimes (G^{*k_1} * \cdots * G^{*k_m}) \\
&\cong \int^{\vec{k}} Fk_1 \otimes \cdots \otimes Fk_m \otimes G^{*\sum k_i} \\
2.2.1 &\cong \int^{\vec{k},r} \mathcal{P}\Big(\sum k_i, r\Big) \otimes Fk_1 \otimes \cdots \otimes Fk_m \otimes G^{*r} \\
&\cong \int^{r} F^{*m} r * G^{*r} = F^{*m} \odot G.
\end{aligned}$$

Note that the Yoneda lemma has been used in the form

$$G^{*n} \cong \int^{r} \mathcal{P}(n, t) \cdot Gt = \mathcal{P}(n, \_) \odot G, \tag{6.11}$$

because $(n, G) \mapsto G^{*n}$ is easily seen to be a bifunctor $\mathcal{P} \times [\mathcal{P}, \mathcal{V}] \to [\mathcal{P}, \mathcal{V}]$. □

Associativity of the substitution product now follows at once: we have

$$\begin{aligned}
F \odot (G \odot H) &= \lambda k. \int^{m} Fm \otimes (G \odot H)^{*m} k \\
&\cong \lambda k. \int^{m} Fm \otimes (G^{*m} \odot H)k \\
&\cong \lambda k. \int^{m,l} Fm \otimes G^{*m} l \otimes H^{*l} k \\
&\cong \lambda k. \int^{l} (F \odot G)l \otimes H^{*l} k \\
&= (F \odot G) \odot H.
\end{aligned}$$

A unit object for the $\odot$-product is $J = \mathcal{P}(1, \_) \cdot I$; indeed $J(1) = I$, $J(n) = \varnothing_{\mathcal{V}}$ for any $n \neq 1$ and the ninja Yoneda lemma applies on both sides to show unitality rules.

- On the left one has
$$J \odot F = \int^m Jm \otimes F^{*m} = \int^m \mathcal{P}(1, m) \cdot F^{*m} \cong F^{*1} = F. \quad (6.12)$$
- On the right, $G \odot J \cong G$ once noticed that $J^{*m} \cong \mathcal{P}(m, \_) \cdot I$ since
$$\begin{aligned} J^{*m} &= \int^{\vec{n}} \mathcal{P}\Big(\sum n_i, \_\Big) \cdot \mathcal{P}(1, n_1) \cdot \dots \cdot \mathcal{P}(1, n_m) \cdot I \\ 2.2.1 &\cong \mathcal{P}(1 + \dots + 1, \_) \cdot I \\ &= \mathcal{P}(m, \_) \cdot I \end{aligned}$$
because of the Fubini rule 1.3.1 and the Yoneda lemma, which says
$$\begin{aligned} &\int^{\vec{n}} \mathcal{P}\left(n_1 + \dots + n_m, \_\right) \cdot \mathcal{P}(1, n_i) \\ &\qquad \cong \mathcal{P}(n_1 + \dots + n_{i-1} + 1 + n_{i+1} + \dots + n_m, \_), \quad (6.13) \end{aligned}$$
for any $1 \leq i \leq m$. One has then
$$G \odot J = \int^m Gm \otimes J^{*m} \cong \int^m Gm \otimes \mathcal{P}(m, \_) \cdot I \cong G. \quad (6.14)$$

The substitution product is highly non commutative; as an example take two representable presheaves and compose them in different order with respect to $\odot$. It is however closed on one side.

Theorem 6.3.3. The $\odot$-monoidal structure is *left closed* in the sense that each $\_ \odot G$ has a right adjoint, but not right closed: not each $F \odot \_$ has a right adjoint.

Proof To prove left closure, we compute:
$$\begin{aligned} \mathcal{V}\text{-Cat}(\mathcal{C}, \mathcal{V})(F \odot G, H) &\cong \mathcal{V}\text{-Cat}(\mathcal{C}, \mathcal{V})\Big(\int^m Fm \otimes G^{*m}, H\Big) \\ &\cong \int_k \mathcal{V}\Big(\int^m Fm \otimes G^{*m}, H\Big) \\ &\cong \int_{km} \mathcal{V}(Fm, [G^{*m}k, Hk]) \\ &\cong \int_m \mathcal{V}\Big(Fm, \int_k [G^{*m}k, Hk]\Big) \\ &= \mathcal{V}\text{-Cat}(F, \{G, H\}) \end{aligned}$$

if we define $\{G,H\}m = \int_k [G^{*m}k, Hk]$. Hence the functor $(\_)\odot G$ has a right adjoint for any G.

The functor $F \odot (\_)$ cannot have such an adjoint. (Incidentally, this shows also that the substitution product cannot come from a convolution product with respect to a *promonoidal structure* in the sense of 5.5.3.) We leave the reader to find a counterexample (find a colimit that is not preserved by $F \odot \_$). □

Definition 6.3.4. An *operad* in $\mathcal{V}$ consists of a $\odot$-monoid object in $\mathcal{V}$-Cat$(\mathcal{P},\mathcal{V})$.

In more explicit terms, an operad is a functor $T \in \mathcal{V}$-Cat$(\mathcal{P},\mathcal{V})$ endowed with a natural transformation called *multiplication*, $\mu : T \odot T \to T$ and a *unit* $\eta : J \to T$ such that

$$\begin{array}{ccc} T\odot T\odot T & \xrightarrow{T\odot\mu} & T\odot T \\ {\scriptstyle \mu\odot T}\downarrow & & \downarrow{\scriptstyle \mu} \\ T\odot T & \xrightarrow[\mu]{} & T \end{array} \qquad \begin{array}{ccccc} J\odot T & \xrightarrow{\eta\odot T} & T\odot T & \xleftarrow{T\odot\eta} & T\odot J \\ & {\scriptstyle\cong}\searrow & \downarrow{\scriptstyle\mu} & \swarrow{\scriptstyle\cong} & \\ & & T & & \end{array} \tag{6.15}$$

are commutative diagrams.

Remark 6.3.5. Unravelling 6.3.4, we notice that an operad in $\mathcal{V}$ consists of the following.

- A natural transformation $\eta : J \Rightarrow T$ that amounts to a map $\eta_1 : I \to T(1)$, since $J(1) = I$, $J(n) = \varnothing$ for $n \neq 1$.
- A natural transformation $\mu : T \odot T \Rightarrow T$ that, in view of the universal property of the two coends involved, amounts to a cowedge

$$Tm \otimes \mathcal{P}(\vec{n},k)\cdot Tn_1 \otimes \cdots \otimes Tn_m \xrightarrow{\quad\tau\quad} Tk \tag{6.16}$$

for any $m, n_1, \dots, n_m, k \in \mathbb{N}$, natural in k and the n_i and such that the following diagram commutes:

$$\begin{array}{ccc} Tm\otimes\mathcal{P}(\sum n_i,k)\cdot T^{\otimes\vec{n}} & \xrightarrow{\sigma^*} & Tm\otimes\mathcal{P}(\sum n_{\sigma i},k)\cdot T^{\otimes\vec{n}} \\ {\scriptstyle\sigma_*}\downarrow & & \downarrow \\ Tm\otimes\mathcal{P}(\sum n_i,k)\cdot T^{\otimes\sigma\vec{n}} & \longrightarrow & Tk \end{array} \tag{6.17}$$

(the notation is self-evident) for every morphism $\sigma \in \mathcal{P}$. This is equivalent to a transformation

$$Tm \otimes Tn_1 \otimes \cdots \otimes Tn_m \xrightarrow{\hat{\tau}} [\mathcal{P}(\textstyle\sum n_i, \_), T(\_)] \tag{6.18}$$

(considering the n_i fixed and the first functor constant in k), i.e. by the Yoneda lemma a natural transformation from $Tm \otimes Tn_1 \otimes \cdots \otimes Tn_m$ to $T(\sum n_i)$.

Example 6.3.6 (Examples of operads).

EO1. For any $F \in \mathcal{V}\text{-Cat}(\mathcal{P}, \mathcal{V})$ the object $\{F, F\}$ is an operad whose multiplication is the adjunct of the arrow

$$\{F,F\} \odot \{F,F\} \odot F \xrightarrow{\{F,F\}\odot\epsilon} \{F,F\} \odot F \xrightarrow{\text{ev}} F \tag{6.19}$$

(ϵ is the counit of the $\_ \odot F \dashv \{F, \_\}$ adjunction) and whose unit is the adjunct of the isomorphism $J \odot F \cong F$. This is called the *endomorphism operad.*

EO2. (See [Lei04, 2.2.11].) Let T : Set $\to$ Set be a monad; then, the collection $(Tn \mid n \in \text{Set})$ (i.e. the restriction of T on *finite* sets) defines an operad if we take

- the unit $\eta_1 : 1 \to T1$ to be the component of the unit of T at the singleton set;
- the multiplication maps (see 6.3.5 above)

$$\gamma_{m,\vec{n}} : Tm \times Tn_1 \times \cdots \times Tn_m \to T\left(\sum n_i\right) \tag{6.20}$$

via the following rule. Let $n = \sum n_i$, and let us consider the image under T of the coproduct inclusion $u_i : n_i \to n = n_1 \sqcup \cdots \sqcup n_m$, i.e. the map $Tu_i : Tn_i \to Tn$. This defines a map $\prod_{i=1}^m Tn_i \to \prod_{i=1}^m Tn = (Tn)^m = \text{Set}(m, Tn)$ that can be post-composed with the action of T on arrows, and with the multiplication of the monad, obtaining

$$\begin{array}{c} Tn_1 \times \cdots \times Tn_m \to \text{Set}(m, Tn) \\ \downarrow \\ \text{Set}(Tm, TTn) \\ \downarrow {\scriptstyle \text{Set}(Tm,\mu_n)} \\ \text{Set}(Tm, Tn); \end{array} \tag{6.21}$$

the transpose of this map is the desired $\gamma_{m,\vec{n}}$.

As a motivation for this, we recall that in an algebraic theory the free algebra $T(A)$ is the set of all terms (inductively) built by applying the operations of the theory T to the 'variables' in A. In this interpretation, there is a substitution map $\gamma_{m,\vec{n}}$ defined by sending a term $t \in Tm$ in the variables $x_1, \dots, x_m$, and m terms $g_1, \dots, g_m$ in the variables $(x_1^j, \dots, x_{n_j}^j)$ with $j = 1, \dots, m$; $\gamma_{m,\vec{n}}(t; g_1, \dots, g_m)$ is then defined as $t[x_j := g_j]$ (a moderate amount of knowledge of λ-calculus will allow an appreciation of the notation).

EO3. As a particular case, the monad 'free commutative k-algebra' yields that the family of polynomial algebras $(k[X_1, \dots, X_n] \mid n \in \mathbb{N})$ defines an operad if the unit $k \to k[X]$ chooses the multiplicative identity, and the multiplication (if $n = \sum n_i$)

$$\begin{array}{c} k[T_1, \dots, T_m] \otimes k[X_1^1, \dots, X_{n_1}^1] \otimes \cdots \otimes k[X_1^m, \dots, X_{n_m}^m] \\ \Big\downarrow \\ k[X_1, \dots, X_n] \end{array} \tag{6.22}$$

is defined by the rule

$$(q; g_1, \dots, g_m) \mapsto q(g_1, \dots, g_m) \tag{6.23}$$

(evaluation of an m-variable polynomial into an m-tuple of polynomials in $n_1, n_2, \dots, n_m$ variables each). This is evidently an associative and unital composition law, compatible with the action of the permutation group on n elements.

As already said, every attempt to propose a self-contained treatment of operads in a single chapter would turn into a complete failure: the theory is too large to be reduced to a bunch of not-too-technical concepts. In some sense, this is to be expected: an operad is nothing but a multicategory (see [Lei04]) with a single object; the resulting theory must then be at least as expressive as the theory of categories.

Thus, we content ourselves with recording a few of the less-than-elementary results (not without a certain dose of cherry-picking towards the ones that can be easily expressed using (co)end calculus). This should be sufficient to clarify how operads are a fundamental tool to describe 'stuff endowed with n-ary operations' that are coherent with the environment.

6.3.1 Substitution and Algebraic Theories

Once one becomes familiar with their definition, one might find that operad-like structures are way more common than expected. The scope of the present subsection is to convey an intuitive explanation for their ubiquity, while at the same time showing, as it should be, that most of the gadgets devised to axiomatise 'structures borne by operations and properties thereof' form equivalent categories. The non-trivial connection between combinatorics and the theory of structures defined by operations and equations is still far from being perfectly understood, despite the enormous effort spent in studying its main features.

The subsection builds on various classical presentations of the subject, such as [Law63, LR11, HP07], and contains nothing essentially new.

Let Fin be the category of finite sets and functions. As we have already seen, the opposite category Fin^{op} exhibits the universal property of the free category with products generated by the point; namely, if $\mathcal{C}$ is any category with finite products, the subcategory of product-preserving functors $F : \text{Fin}^{\text{op}} \to \mathcal{C}$ is equivalent to $\mathcal{C}$:

$$\mathcal{C} \cong \text{Cat}(*, \mathcal{C}) \cong \text{Cat}_\times(\text{Fin}^{\text{op}}, \mathcal{C}). \tag{6.24}$$

In more concrete terms, this means that a product-preserving functor $p : \text{Fin}^{\text{op}} \to \mathcal{C}$ is uniquely determined by the image of $[1] \in \text{Fin}$, because it must send every other $[n]$ to $p[n] = p[1]^n$.

An *identity on objects* functor ('idonob' functor, for short) is just a functor $F : \mathcal{C} \to \mathcal{D}$ whose function on objects $F_o : \mathcal{C}_o \to \mathcal{D}_o$ is the identity of the set/class $\mathcal{C}_o = \mathcal{D}_o$.

Definition 6.3.7. We call a *Lawvere theory* an idonob functor $p : \text{Fin}^{\text{op}} \to \mathcal{L}$ that strictly preserves products.

The category Law of Lawvere theories is thus defined as having objects the pairs $(p, \mathcal{L})$ as above, and morphisms $(p, \mathcal{L}) \to (p', \mathcal{L}')$ the functors $h : \mathcal{L} \to \mathcal{L}'$ such that $h \circ p = p'$; note that this also says that h is bijective on objects and product-preserving.

When regarded as a subcategory of the coslice $\text{Fin}^{\text{op}}/\text{Cat}$, the category of Lawvere theory seems to be a 2-category; the requirement that a Lawvere theory is bijective on objects, however, forces a natural transformation $\alpha : h \Rightarrow h'$ to be the identity, thus proving that Law is a *2-discrete* 2-category.

Definition 6.3.8 (Finitary monad). A monad (see A.6.3) on Set is *finitary* if it preserves colimits over a filtered category (see A.9.2); finitary monads form a full subcategory of the category of monads, where a morphism of monads has been defined in A.6.15.

Definition 6.3.9. Let us consider the category Cat(Fin, Set) of all functors $F : \text{Fin} \to \text{Set}$, endowed with its cartesian monoidal structure. By mimicking the construction of operads in 6.3.1 and 6.3.4, we can define a *substitution* monoidal structure on Cat(Fin, Set) as follows (evidently, $\text{Fin}(n, m)$ is a much larger set of morphisms than $\mathcal{P}(n, m)$, so the equaliser defining the following coend performs a much smaller quotient):

$$F \odot G : n \mapsto \int^m Fm \times G^m n \tag{6.25}$$

where the mth power of G, $G^m n = Gn \times \cdots \times Gn$, is the product of Gn with itself repeated m times, and it coincides with the m-fold convolution of G with itself when both Fin and Set are given their cartesian monoidal structure.

The monoids with respect to this monoidal structure are called *cartesian operads* or *clones* in [Cur12] and in our 6.4.5 below.

When the obvious notion of homomorphism of internal monoids is specialised to this particular case, we obtain a category Clo of clones.

Lemma 6.3.10. There is an equivalence between the category of functors $\text{Fin} \to \text{Set}$ and the category $\text{Cat}_\omega(\text{Set}, \text{Set})$ of finitary (see A.9.3) endofunctors of Set; this equivalence is monoidal, when Cat(Fin, Set) is endowed with the substitution product defined above.

Proof The equivalence $\text{Cat}(\text{Fin}, \text{Set}) \cong \text{Cat}_\omega(\text{Set}, \text{Set})$ is induced by the adjunction $\text{Lan}_J \dashv J^*$, if $J : \text{Fin} \subset \text{Set}$, and the category of finitary endofunctors of Set is clearly monoidal with respect to composition.

It is a general fact that given an equivalence of categories $F : \mathcal{C} \leftrightarrows \mathcal{D} : G$, if $(\mathcal{D}, \otimes, I)$ is monoidal we can define a monoidal structure on $\mathcal{C}$ that turns the adjunction (F, G) into a monoidal equivalence, by setting $C \diamond C' := G(FC \otimes FC')$ and $J := GI$.

It remains to show that the composition on $\text{Cat}_\omega(\text{Set}, \text{Set})$, when transported to Cat(Fin, Set) along the equivalence, coincides with the substitution product, i.e. that

$$\text{Lan}_J F \circ \text{Lan}_J G \cong \text{Lan}_J(F \odot G) \tag{6.26}$$

for every $F, G : \text{Fin} \to \text{Set}$ if $J : \text{Fin} \subset \text{Set}$ is the inclusion, and vice versa that

$$UJ \odot VJ \cong U \circ V \circ J \tag{6.27}$$

if $U, V : \text{Set} \to \text{Set}$ are finitary endofunctors.

We can easily show both these equations using coend calculus:

$$\begin{aligned}
\text{Lan}_J(F \odot G)A &\cong \int^n \text{Set}(n, A) \times (F \odot G)(n) \\
&\cong \int^n \text{Set}(n, A) \times \int^m Fm \times G^m n \\
&\cong \int^{nm} \text{Set}(n, A) \times Fm \times G^m n \\
&\cong \int^m Fm \times \text{Set}(m, \text{Lan}_J GA) \\
&\cong \text{Lan}_J F(\text{Lan}_J GA),
\end{aligned}$$

where the last passage is motivated by the isomorphism

$$\int^n \text{Set}(n, A) \times G^m n \cong \text{Lan}_J(\text{Set}(m, G_-))(A) \cong \text{Set}(m, \text{Lan}_J GA). \tag{6.28}$$

Conversely,

$$\begin{aligned}
(UJ \odot VJ)n &\cong \int^m UJm \times (Vn)^m \\
&\cong \int^m UJm \times \text{Set}(m, Vn) \\
&\cong \int^m Um \times \text{Set}(Jm, Vn) \\
&\cong \text{Lan}_J(UJ)(Vn) = UVJn.
\end{aligned}$$

This concludes the proof. □

Corollary 6.3.11. There is an equivalence between the category of finitary monads $T : \text{Set} \to \text{Set}$ and the category Clo of clones.

Definition 6.3.12. Let $T : \text{Cat}(\text{Fin}, \text{Set}) \to \text{Cat}(\text{Fin}, \text{Set})$ be a monad; we call it a CMC *functor* if it preserves colimits and finite products (which, in this case, coincide with Day convolution as already said). We denote CMC(Fin, Set) the category of CMC monads on Cat(Fin, Set) and monad morphisms (see A.6.15).

The reader can find a more detailed account of this definition in 6.4.3 below.

Definition 6.3.13. Recall that the category $\mathsf{Prof}(\text{Fin}^{\text{op}}, \text{Fin}^{\text{op}})$ is monoidal with respect to composition of profunctors; a monoid internal to it is a *promonad*, i.e. a profunctor $\mathfrak{t} : \text{Fin}^{\text{op}} \rightsquigarrow \text{Fin}^{\text{op}}$ that in addition is a monoid with respect to profunctor composition.

All these seemingly diverse structures turn out to be the same.

Theorem 6.3.14. There is an equivalence between the categories of 6.3.7, 6.3.8, 6.3.9, 6.3.12, 6.3.13.

The equivalence between 6.3.8 and 6.3.9 follows from 6.3.10 and in particular from the fact that the equivalence is monoidal: it has been said already in 6.3.11.

The equivalence between 6.3.12 and 6.3.13 follows from the fact that the equivalence in 6.3.10 can be promoted to a monoidal equivalence as well, when both hom categories are endowed with the composition monoidal structure; in our case, this boils down to the statement that a promonad $\mathfrak{t}$ corresponds to a cocontinuous monad T on $\text{Cat}(\text{Fin}, \text{Set})$ under the equivalence of 6.3.10 and it is the content of Exercise 5.11.

The remaining implications need an intricate series of definitions and preliminary results, thus their proof occupies the following two subsections.

Algebraic Theories are Finitary Monads

We will prove the equivalence between 6.3.7 and 6.3.8 building mutually functors in opposite directions between the category Law and the category of finitary monads on Set.

Remark 6.3.15. A preliminary remark is in order: the notion of Lawvere theory abstracts the notion of algebraic theory in the sense that one can define a category $\mathcal{L}_M$ with the following properties.

- $\mathcal{L}_M$ has objects the natural numbers $[0], [1], \dots$, and the operation $[n], [m] \mapsto [n+m]$ is a functor $\mathcal{L}_M \times \mathcal{L}_M \to \mathcal{L}_M$ endowing $\mathcal{L}_M$ with (strictly associative) finite products.
- $\mathcal{L}_M$ contains morphisms $\mu : [2] \to [1], \eta : [0] \to [1]$ subject to equations given by the commutative diagrams

$$
\begin{array}{ccc}
[2+1] = [1+2] & \xrightarrow{1+\mu} & [2] \\
{\scriptstyle \mu+1}\downarrow & & \downarrow{\scriptstyle \mu} \\
[2] & \xrightarrow[\mu]{} & [1]
\end{array}
\qquad
\begin{array}{ccc}
[0+1] = [1+0] & \xrightarrow{1+\eta} & [1+1] \\
{\scriptstyle \eta+1}\downarrow & \searrow^{=} & \downarrow{\scriptstyle \mu} \\
[1+1] & \xrightarrow[\mu]{} & [1]
\end{array}
\tag{6.29}
$$

- $\mathcal{L}_M$ has a category of *models*, i.e. functors $F : \mathcal{L}_M \to \text{Set}$ such that $F[n] = F[1]^n$, and thus the functions $F\mu : A^2 \to A$, $F\eta : A^0 = * \to A$ endow the set $A = F[1]$ with an associative, unital binary operation; in short, A is a *monoid.* A natural transformation between models turns out to be just a homomorphism between the monoids.

$\mathcal{L}_M$ can thus be thought of as the 'theory' whose models $F : \mathcal{L}_M \to \text{Set}$ are monoids; it is a category harbouring the most abstract shape a monoid can possibly have, and a concrete realisation of such an abstraction inside set theory turns out to be just a functor $F : \mathcal{L}_M \to \text{Set}$.

Remark 6.3.16. At this point, the reader might want to embark on an easy exercise: are natural transformations $F \Rightarrow G : \mathcal{L}_M \to \text{Set}$ of models really in bijection with monoid homomorphisms? Meaning: should not we ask for a natural transformation to preserve the cartesian product itself? It turns out that this is not needed, as such an $\alpha : F \Rightarrow G$ *must* preserve products, in the sense that the map $\alpha_{[n]} : F[1]^n \to G[1]^n$ must be $\alpha^n_{[1]}$.

Given a general Lawvere theory $p : \text{Fin}^{\text{op}} \to \mathcal{L}$ we can form the category of its models as the functors

$$F : \mathcal{L} \to \text{Set} \tag{6.30}$$

that preserve products. Such models are uniquely determined by their action on the object $[1] \in \mathcal{L}$, in the sense that if F is such a model, $F[n] \cong F[1]^n$ for every $[n] \in \text{Fin}^{\text{op}}$, so we can characterise the subcategory $\text{Mod}(p, \mathcal{L}) \subset \text{Cat}(\mathcal{L}, \text{Set})$ as the upper left corner in the pullback

$$\begin{array}{ccc} \text{Mod}(p,\mathcal{L}) & \xrightarrow{j} & \text{Cat}(\mathcal{L},\text{Set}) \\ {\scriptstyle U}\downarrow & \lrcorner & \downarrow{\scriptstyle p^*} \\ \text{Set} & \xrightarrow[N_J]{} & \text{Cat}(\text{Fin}^{\text{op}},\text{Set}) \end{array} \tag{6.31}$$

where $N_J : A \mapsto (\lambda n.\text{Set}(n, A))$ is the nerve (see 3.2.2) associated to the functor $J : \text{Fin} \subset \text{Set}$, and U the functor that evaluates a model $F : \mathcal{L} \to \text{Set}$ on the object $[1]$.

Remark 6.3.17. The functor U commutes with all limits and with filtered colimits.

Proof Every limit of a diagram (F_i) of product-preserving functors is still product-preserving, since limits commute with limits.[5] Now, the 'evaluation at [1]' functor U is easily seen to be isomorphic to the functor

$$F \mapsto \mathrm{Cat}(\mathcal{L}, \mathrm{Set})(よ[1], F) \tag{6.32}$$

by virtue of the Yoneda lemma; thus, it preserves all limits (because every covariant representable $\mathcal{C}(X, \_)$ does), and all filtered colimits, since the representables are finitely presentable (see A.9.3) objects of $\mathrm{Mod}(p, \mathcal{L})$. □

The functor U does not, however, preserve all colimits: for example, it does not always preserve coproducts or initial objects (consider, for example, the category of monoids $\mathrm{Mod}(\mathcal{L}_M)$).[6]

Lemma 6.3.18. The functor U admits a left adjoint.

Proof Given a set A, consider the functor $N_J(A) : \mathrm{Fin}^{\mathrm{op}} \to \mathrm{Set}$, and the functor $p : \mathrm{Fin}^{\mathrm{op}} \to \mathcal{L}$ defining the theory; we shall prove that $A \mapsto \mathrm{Lan}_p(N_J(A))$ in the diagram

$$\begin{array}{ccc} & \mathrm{Fin} & \\ {}^{p}\swarrow & & \searrow^{N_J A} \\ \mathcal{L} & \xrightarrow[\mathrm{Lan}_p N_J A]{} & \mathrm{Set} \end{array} \tag{6.33}$$

defines a functor that is a left adjoint to U: to do this, we call $F_p A := \int^n \mathcal{L}(m, n) \times A^n$ and prove that there is an adjunction

$$\mathrm{Mod}(p, \mathcal{L})(F_p A, H) \cong \mathrm{Set}(A, H[1]) \tag{6.34}$$

for every model H of p. In order to see that, one can expand the definition of the left hand side:

$$\begin{aligned} \mathrm{Mod}(p, \mathcal{L})(F_p A, H) &\cong \int_n \mathrm{Set}(F_p A n, H n) \\ &\cong \int_n \mathrm{Set}\left(\int^m \mathcal{L}(n, m) \times A^m, H n\right) \end{aligned}$$

[5] More formally, if $\left[\begin{smallmatrix} n \\ m \end{smallmatrix}\right]$ is the functor $\{0,1\} \to \mathcal{C}$ choosing two objects $n, m \in \mathcal{L}$, then $[m] \times [n] \cong \lim_{\{0,1\}} \left[\begin{smallmatrix} n \\ m \end{smallmatrix}\right]$ and then one has

$$\lim_{\mathcal{I}} \lim_{\{0,1\}} F_i \left[\begin{smallmatrix} n \\ m \end{smallmatrix}\right] \cong \lim_{\{0,1\}} \lim_{\mathcal{I}} F_i \left[\begin{smallmatrix} n \\ m \end{smallmatrix}\right],$$

by virtue of the Fubini rule for limits.

[6] More formally, the representable object $y[1]$ on $[1]$ is tiny (i.e. $\hom(y[1], \_)$ is cocontinuous) as an object of $\mathrm{Cat}(\mathcal{L}, \mathrm{Set})$ by virtue of the Yoneda lemma, *but not* as an object of $\mathrm{Mod}(p, \mathcal{L})$, as there are colimits in $\mathrm{Mod}(p, \mathcal{L})$ that are constructed differently from the way they are constructed in $\mathrm{Cat}(\mathcal{L}, \mathrm{Set})$.

$$
\begin{aligned}
&\cong \int_{nm} \mathrm{Set}\,(\mathcal{L}(n,m) \times A^m, Hn) \\
&\cong \int_{nm} \mathrm{Set}(A^m, \mathrm{Set}(\mathcal{L}(n,m), Hn)) \\
&\cong \int_{m} \mathrm{Set}(A^m, H[1]^m) \cong \mathrm{Set}(A, H[1])
\end{aligned}
$$

where in the last steps we used the ninja Yoneda lemma 2.2.1 twice. □

So, given a Lawvere theory $(p, \mathcal{L})$ we can associate a monad to it, precisely the monad induced by the adjunction $F_p \dashv U$, i.e. to the functor $UF_p : \mathrm{Set} \to \mathrm{Set}$.

Proposition 6.3.19. The monad obtained in this way is finitary.

Proof The functor U commutes with finitely filtered colimits, and F_p commutes with all colimits because it is a left adjoint; thus, the composition UF_p is finitary. □

Finally, let us verify that the adjunction $F_p \dashv U$ is *monadic* in the sense of A.6.10: this means that the category of UF_p-algebras coincides up to equivalence with the category of $(p, \mathcal{L})$-modules.

It is rather easy to prove that U is conservative having in mind the pullback in (6.31). It remains to show that U creates the coequalisers of U-split pairs, in order to fulfil all the requirements of A.6.12. This can be checked directly, and in fact it is a general result about pulling back a monadic functor along another functor; we leave the details as an Exercise 6.10; note that it is not possible to remove the assumption that the lower horizontal functor in the pullback (6.31) is fully faithful.

We now construct a correspondence in the opposite direction: given a finitary monad T on Set, we consider the composition

$$
\mathrm{Fin} \xrightarrow{J} \mathrm{Set} \xrightarrow{F^T} \mathrm{Alg}(T) \tag{6.35}
$$

where the functor F^T is the free functor of A.6.7, and its factorisation as an idonob functor $p : \mathrm{Fin} \to \mathcal{L}^{\mathrm{op}}$ followed by a fully faithful functor $\mathcal{L}^{\mathrm{op}} \to \mathrm{Alg}(T)$. The functor p now is a Lawvere theory, and its category of models coincides with the category of T-algebras (see A.6.5).

These two correspondences in opposite directions extend to functors as follows. On the category of finitary monads we take only *restrained* morphisms in the sense of A.6.16.

- Given a morphism of Lawvere theories $h : (p, \mathcal{L}) \to (q, \mathcal{M})$, i.e. a

commutative triangle

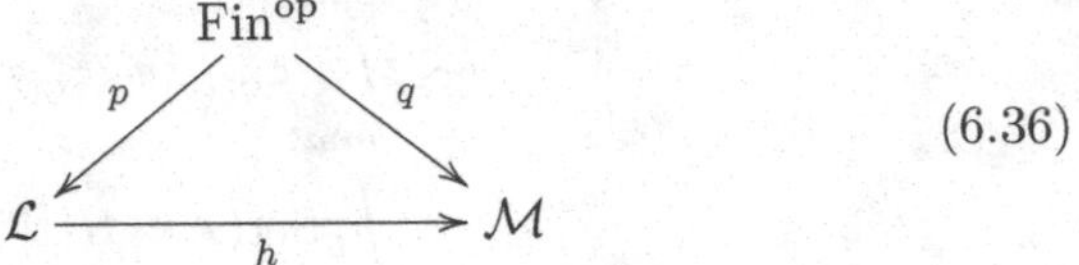

(6.36)

Let us say p gives rise to the monad UF and q to the monad $U'F'$; then, we define a natural transformation

$$\lambda : UF \xrightarrow{UF*\eta'} UFU'F' \to UFU\bar{k}F' \xrightarrow{U*\epsilon*\bar{k}F'} U\bar{k}F' = U'F' \quad (6.37)$$

using the unit of $F' \dashv U'$, the counit of $F \dashv U$ and the functor $\bar{h} : \mathrm{Mod}(q, \mathcal{M}) \to \mathrm{Mod}(p, \mathcal{L})$ obtained from the universal property of the pullback, from $k^* : \mathrm{Cat}(\mathcal{M}, \mathrm{Set}) \to \mathrm{Cat}(\mathcal{L}, \mathrm{Set})$.

It is now a matter of using the zig-zag identities of the two adjunctions, to show that λ satisfies the axioms of A.6.15; we leave this as an unenlightening exercise for the reader.

- Given finitary monads T, S on Set, and a restrained morphism of monads $\lambda : S \Rightarrow T$, we shall obtain a morphism of factorisations to fill in the centre of the diagram

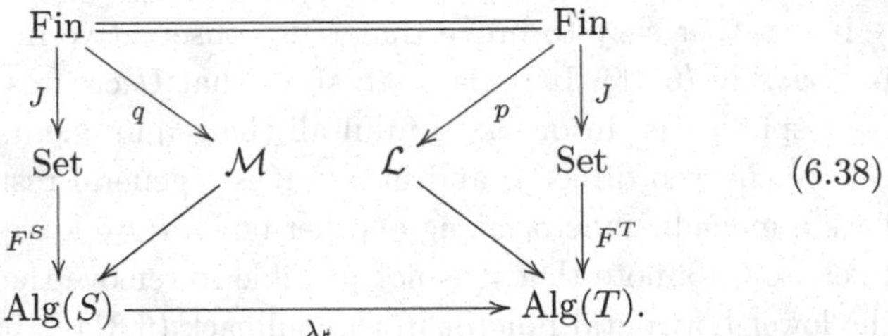

(6.38)

It is evident that λ induces a morphism between T- and S-algebras

$$\lambda^\sharp : \mathrm{Alg}(S) \to \mathrm{Alg}(T) \quad (6.39)$$

defined as sending a T-algebra $\begin{bmatrix} TA \\ \downarrow \\ A \end{bmatrix}$ into $SA \xrightarrow{\lambda_A} TA \to A$; the axioms satisfied by a monad morphism now entail that this composition is an S-algebra. However, the outer rectangle in (6.38) does not commute, because $\lambda^\sharp$ does not restrict to a morphisms of *free* algebras; in fact, the triangle

$$\begin{array}{ccc} & \mathrm{Set} & \\ {}^{F^S}\swarrow & & \searrow^{F^T} \\ \mathrm{Alg}(S) & \xrightarrow[\lambda_\sharp]{} & \mathrm{Alg}(T) \end{array} \quad (6.40)$$

has no reason to commute.

Instead, *if* $\lambda^\sharp$ had a left adjoint $\lambda_!$, it would be true that $\lambda_! \circ F^T = F^S$, and we could use it to define a morphism between the factorisations $\mathcal{L}, \mathcal{M}$ in (6.38) above, and thus a morphism of theories.

It turns out that $\lambda^\sharp$ is indeed a right adjoint: both $\text{Alg}(S), \text{Alg}(T)$ are complete and accessible categories, $\lambda^\sharp$ commutes with limits and filtered colimits because S, T are finitary monads, and thus for the adjoint functor theorem [AR94, 2.45] it must admit a left adjoint $\lambda_!$. An explicit description of such a left adjoint as a sequential colimit that exploits finite accessibility of $\text{Alg}(S), \text{Alg}(T)$ is left as Exercise 6.15.

Substitution Monoids are Cocontinuous Monads

To conclude the section, we prove the equivalence between 6.3.9 and 6.3.12. This is a slicker argument, thanks to coend calculus!

Let T be an $\odot$-monoid; it is a general fact that tensoring with an internal monoid in a monoidal category gives a monad $\_ \odot T : \text{Cat}(\text{Fin}, \text{Set}) \to \text{Cat}(\text{Fin}, \text{Set})$;. This gives a way to construct a monad out of T; this correspondence is of course functorial, as $\odot$ is a bifunctor. The resulting monad is furthermore cocontinuous, since the substitution monoidal structure is left closed (see 6.2.4). We leave to the reader the proof that the monad is convolution preserving.

On the other hand, given a cocontinuous monad S on $\text{Cat}(\text{Fin}, \text{Set})$, we evaluate S on the representable $J = y[1]$, i.e. on the substitution monoidal unit, and it is now a matter of coend calculus to show that

$$S(A) \cong A \odot SJ, \tag{6.41}$$

so that every cocontinuous monad arises this way:

$$\begin{aligned}
(A \odot SJ) &= \int^m Am \times Sy[1]^m \\
&\cong \int^m Am \times \text{Set}(m, S(y[1])_) \\
&\cong \int^m S\left(Am \times \text{Set}(m, y[1]_)\right) \\
&\cong S\left(\int^m Am \times \text{Set}(m, _)\right) = (SA)n.
\end{aligned}$$

This concludes the proof of the equivalence between CMC monads and substitution monoids.

6.4 Some More Advanced Results

We expand a little bit more on the theory of operads, also putting in a broader perspective the results we have already encountered in the previous section (notably, the notion of cartesian operad or *clone*).

We begin with a theorem linking operads and monads: every $\mathcal{V}$-operad induces a monad on $\mathcal{V}$.

Theorem 6.4.1. Let $T : \mathcal{P} \to \mathcal{V}$ be an operad; then we can define a monad on $\mathcal{V}$ by the rule

$$M_T : A \mapsto \int^{n\in\mathcal{P}} Tn \otimes A^n. \tag{6.42}$$

Proof We shall first define the unit and multiplication.

- To define the unit, we employ Exercise 6.6. There is a functor $\Psi : \mathcal{V} \to \mathcal{V}\text{-Cat}(\mathcal{P}, \mathcal{V})$, precisely the left adjoint to evaluation at $0 \in \mathcal{P}$, acting as follows: $\Psi(A)$ is the functor that sends 0 to A, and all other $n \geq 1$ to the initial object of $\mathcal{V}$. Unwinding this definition, we see that $A \cong \mathcal{V}(I, \_) \odot \Psi A$ and that $T \odot \Psi A = M_T A$. This means that we can define the components of the unit as
$$C \cong \mathcal{V}(I, \_) \odot \Psi C \xrightarrow{\eta\odot\Psi C} T \odot \Psi C = M_T(C). \tag{6.43}$$
- To define the multiplication, let us follow the string of equivalences
$$\begin{aligned} \int^m Tm \otimes (M_T(C))^m &\cong \int^m Tm \otimes \left[\int^n Tn \otimes C^n\right]^m \\ &\cong \int^m Tm \otimes \int^{\vec{n}} Tn_1 \otimes \cdots \otimes Tn_m \otimes C^{n_1+\cdots+n_m} \\ &\cong \int^{m,\vec{n}} Tm \otimes Tn_1 \otimes \cdots \otimes Tn_m \otimes C^{n_1+\cdots+n_m}. \end{aligned}$$
Now, using the multiplication of the operad we get that this object maps canonically to $\int^k Tk \otimes C^k$.

With a certain amount of work, associativity and unitality for T now prove the associativity and unitality for the endofunctor M_T, thus showing that it is a monad (see A.6.3). □

Remark 6.4.2. The correspondence described in 6.3.6.EO2 can be promoted to a functor from the category of monads to the category of *Set*-operads; it pairs with the correspondence described above, yielding an adjunction between the category of monads and the category of operads. This functor is however not an equivalence, because there are non-isomorphic operads that generate the same monad (see [Lei06]).

Our analysis continues with the main theorem in [Cur12]:

> There exist 2-comonads on the bicategory Prof [*author: see our 5.1.2*] whose coKleisli categories have objects respectively operads, symmetric operads and clones.

We begin by introducing a special class of operads of particular interest in universal algebra, called *clones*. In short, a clone is an operad with respect to a cartesian monoidal structure.

Definition 6.4.3. A category $\mathcal{C}$ is called cartesian monoidally cocomplete (CMC for short) if it admits finite products, all colimits, and each functor $A \times \_ : \mathcal{C} \to \mathcal{C}$ preserves these colimits.

Recall from Exercise A.17 that we denote by Fin_* the category $1/\text{Set}_{<\omega}$ of pointed finite sets; in the following, Fin will denote the category of unpointed finite sets.

Proposition 6.4.4. Let $\mathcal{C}$ be a CMC category; then there are equivalences of categories

$$\mathcal{C} \cong \text{Cat}_\times(\text{Fin}^{\text{op}}, \mathcal{C}) \cong \text{Cat}_{\times,!}(\text{Cat}(\text{Fin}, \text{Set}), \mathcal{C}) \tag{6.44}$$

of $\mathcal{C}$ with the category of product-preserving functors $\text{Fin} \to \mathcal{C}$, and with the category of colimit and product-preserving functors $\text{Fin} \to \text{Set}$ (we call them *CMC functors* for short: see 6.4.3 for the motivation behind this notation), induced by evaluation at the inclusion functor $\text{Fin} \subset \text{Set}$ (this is colimit- and product-preserving).

Proof The second equivalence is just the universal property of the Yoneda embedding described in 3.1.1 and specialised to this context.

As for the first equivalence, an object $C \in \mathcal{C}$ goes to the functor $W : [n] \mapsto C^n$. Each such functor is uniquely determined by its action on $[1]$, since $[n] \cong [1] \times \cdots \times [1]$ in Fin^{op} (or equivalently, $[n] = [1] \sqcup \cdots \sqcup [1]$ in Fin). □

We can actually trace the image of an object C into the category of CMC functors: C goes first to the functor $[n] \mapsto C^n$, and then to the functor that sends $F : \text{Fin} \to \text{Set}$ to its W-weighted colimit

$$\int^{[n] \in \text{Fin}} Fn \otimes C^n \tag{6.45}$$

where $\otimes$ is the usual tensor of a cocomplete category over Set (see 2.2.3).

An immediate corollary of our 6.4.4 is that we have an equivalence of categories

$$\mathrm{Cat}_{\times,!}(\mathrm{Cat}(\mathrm{Fin},\mathrm{Set}),\mathrm{Cat}(\mathrm{Fin},\mathrm{Set}))\cong\mathrm{Cat}(\mathrm{Fin},\mathrm{Set}) \tag{6.46}$$

between the CMC-endofunctors of Cat(Fin, Set) and itself. In fact, something more general is true (see [Tri]): the monoidal product on endofunctors given by composition transfers across the equivalence to a monoidal product on Cat(Fin, Set), and this monoidal product is exactly the non-symmetric substitution

$$F\odot G=\int^{[n]\in\mathrm{Fin}} Fn\otimes G^n \tag{6.47}$$

where $G^n = G\times\cdots\times G$.

As a consequence, we can define clones.

Definition 6.4.5. A *cartesian operad*, also called a *clone*, is a monoid object in $(\mathrm{Cat}(\mathrm{Fin},\mathrm{Set}),\odot)$.

Now, the main theorem in [Cur12] that characterises operads as the coKleisli category of a certain comonad on Prof involves the choice of a monad on Cat, from among three options.

- The *free monoidal category* $M(\mathcal{A})$: given a category $\mathcal{A}$, the free strict monoidal category $M(\mathcal{A})$ has objects the finite sequences of objects of $\mathcal{A}$, and the hom-set between two tuples $\underline{A},\underline{A}'$ is defined to be

$$M(\mathcal{A})(\underline{A},\underline{A}')=\begin{cases}\prod_{i=1}^n\mathcal{A}(A_i,A_i') & \text{if } \ell(\underline{A})=\ell(\underline{A}')\\ \varnothing & \text{otherwise}\end{cases} \tag{6.48}$$

  where $\ell : M(\mathcal{A})_o=\coprod_{n\geq 0}\mathcal{A}_o^n\to\mathbb{N}$ sends a tuple $\underline{A}=(A_1,\dots,A_n)$ to its *length* n.
- The *free symmetric monoidal category* $S(\mathcal{A})$: given a category $\mathcal{A}$, the objects of $S(\mathcal{A})$ are finite tuples $\vec{A}=(A_1,\dots,A_n)$ of objects of $\mathcal{A}$; morphisms between two tuples $\vec{A}$ and $\vec{B}$ exist only if the tuples have the same length; the monoidal product $\uplus$ is given by juxtaposition of tuples. Morphisms $f\uplus g$ are defined accordingly.

  The category $S(\mathcal{A})$ thus splits as a disjoint union of categories $S^n(\mathcal{A})$ whose objects are n-tuples of objects $\vec{A}=(A_1,\dots,A_n)$ and the hom-objects are

$$S^n(\mathcal{A})(\vec{X},\vec{Y}):=\coprod_{\sigma\in S_n}\mathcal{A}(X_1,Y_{\sigma 1})\times\cdots\times\mathcal{A}(X_n,Y_{\sigma n}). \tag{6.49}$$

The symmetry of the monoidal structure has components $\vec{X} \uplus \vec{Y} \to \vec{Y} \uplus \vec{X}$, the shuffle permutation swapping the first m-elements of the first sequence with the last n-elements of the second. The unit is the empty sequence () and $S^0(\mathcal{A}) = \{()\}$ is the terminal category. The inclusion functor $\mathcal{A} \hookrightarrow S(\mathcal{A})$ takes an object x to the one-element sequence $(x) \in S^1(\mathcal{A})$.
- The *free cartesian category* $K(\mathcal{A})$: defined as adjoining to a category $\mathcal{A}$ all finite products, of course, K is a 'submonad' of S, in that every cartesian category is (non-strictly) symmetric monoidal.

Each functor $T \in \{M, S, K\}$ induces a monad $\boldsymbol{P} \circ T$, and by self-duality a *comonad*, on the bicategory of profunctors, whose coKleisli category is

- the category of operads, if $T = M$,
- the category of symmetric operads, if $T = S$,
- the category of clones, as defined in 6.4.5, if $T = K$.

The whole proof boils down to the existence of a (pseudo)distributive[7] law (see A.6.14) $\gamma : T \circ \boldsymbol{P} \Rightarrow \boldsymbol{P} \circ T$, whose definition and well-posedness can be proved by means of (co)end calculus.

Lemma 6.4.6. Let T be any one of the three monads M, S, K; then there is a distributive law $\gamma : T \circ \boldsymbol{P} \Rightarrow \boldsymbol{P} \circ T$ between T and the presheaf construction $\boldsymbol{P}$.

Proof We define γ on components, as [Cur12, §9] does: let T be for example the monad M and let us take $\mathcal{V} = \text{Set}$ (in all other cases, a similar discussion can be carried over almost unchanged). We define the $\mathcal{A}$-component of γ to be

$$\gamma_A : M(\boldsymbol{P}\mathcal{A}) \to \boldsymbol{P}(M\mathcal{A})$$
$$(F_1, \dots, F_n) \longmapsto \int^{\underline{A}} F_1 A_i \times \dots \times F_n A_n \times M(\mathcal{A})(\_, \underline{A}) \qquad (6.50)$$

where $\underline{A} = A_1, \dots, A_n$ (recall that by its very definition γ must turn 'tuples of presheaves' into 'presheaves over tuples' in each degree $n \geq 0$).

We shall now show that the constraints listed in our A.6.14 hold, thus defining a distributive law.

[7] We decide to hide the coherence constraints of the presheaf construction functor $\boldsymbol{P} : A \mapsto \text{Cat}(A^{\text{op}}, \text{Set})$, like [Cur12] does; this is a hairy matter that we have no room or reason to expand as it would deserve; the interested reader can consult [FGHW18]. We also set aside the problem given by the fact that, strictly speaking, $\boldsymbol{P}$ is not an endofunctor. [FGHW18] addresses this matter in the best possible way. See also [ACU10, DLL19].

- The unit constraint of A.6.14 means that the triangle

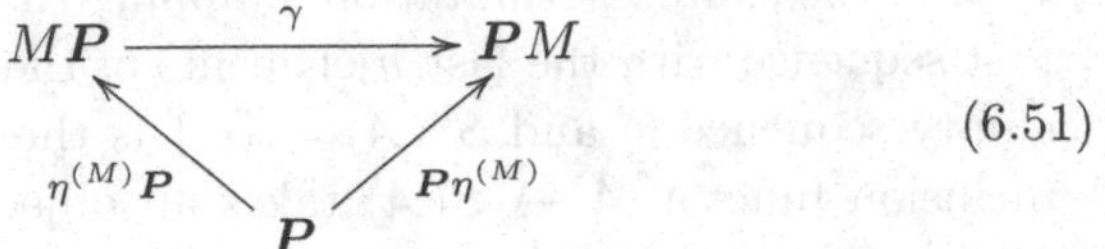

(6.51)

commutes; this is evident, as the composition $\gamma \circ \eta^{(M)}\boldsymbol{P}$ sends a presheaf $F \in \boldsymbol{P}\mathcal{A}$ into $\int^A FA \times M(\mathcal{A})(\_, A)$, i.e. into the left Kan extension of F along the unit $\eta^{(M)}_{\mathcal{A}} : \mathcal{A} \to M\mathcal{A}$, i.e. exactly into the image of F along $\boldsymbol{P}\eta^{(M)}$.

- The second constraint of A.6.14 involves the diagram

$$\begin{array}{ccccc} MM\boldsymbol{P} & \xrightarrow{M\gamma} & M\boldsymbol{P}M & \xrightarrow{\gamma M} & \boldsymbol{P}MM \\ {\scriptstyle \mu^{(M)}\boldsymbol{P}}\downarrow & & & & \downarrow{\scriptstyle \boldsymbol{P}\mu^{(M)}} \\ M\boldsymbol{P} & & \xrightarrow[\gamma]{\qquad\qquad} & & \boldsymbol{P}M \end{array} \tag{6.52}$$

Establishing its commutativity is a bit harder: following [Cur12, §9] very closely, we will keep track of the position of the diagram we are in, as the coend computation proceeds.

First, let us fix a component $\mathcal{A}$: the generic element of the upper left corner of (6.52) is thus a tuple of tuples of presheaves, and has the form

$$\lambda\underline{C}.\int^{\underline{D_1},\dots,\underline{D_n}} F_1^1 D_1^1 \times \cdots \times F_n^{i_n} D_n^{i_n} \times M(\mathcal{A})\Big(\underline{C}, \underline{(\underline{D_1},\dots,\underline{D_n})}\Big) \tag{6.53}$$

where the tuple of tuples $F_1^1 D_1^1 \times \cdots \times F_n^{i_n} D_n^{i_n}$ has not been flattened, and the tuple of tuples $\underline{\underline{D}} = \underline{(\underline{D_1},\dots,\underline{D_n})}$ has been flattened to a long tuple $\underline{D}$ using the multiplication of the monad. This is sent to the lower left corner, to an identically written object where said tuple has been flattened by $\mu^{(M)}$ (formally, we consider the product $F_1^1 D_1^1 \times \cdots \times F_n^{i_n} D_n^{i_n}$ parenthesised in two different, but equivalent, ways). Thus, the object

$$\lambda\underline{C}.\int^{\underline{D_1},\dots,\underline{D_n}} F_1^1 D_1^1 \times \cdots \times F_n^{i_n} D_n^{i_n} \times M(\mathcal{A})\Big(\underline{C}, \underline{(\underline{D_1},\dots,\underline{D_n})}\Big)$$

lies in the lower left corner of the diagram. Applying the distributive law γ now yields the functor

$$\lambda\underline{C}.\int^{\underline{A_1},\dots,\underline{A_n}} \int^{\underline{D_1},\dots,\underline{D_n}} F_1^1 D_1^1 \times \cdots \times F_n^{i_n} D_n^{i_n}$$

$$\times MM(\mathcal{A})(\underline{\underline{A}}, \underline{\underline{D}}) \times M(\mathcal{A})(\underline{C}, (A_1^1, \ldots, A_p^{k_p})). \quad (6.54)$$

We shall now transport the object of (6.53) to the upper right corner, by application of $\gamma M \circ M\gamma$: the result is equal to

$$\lambda(\underline{A_1}, \ldots, \underline{A_p}). \int^{\underline{B_1}, \ldots, \underline{B_n}} \int^{\underline{D_1}, \ldots, \underline{D_n}} F_1^1 D_1^1 \times \cdots \times F_n^{i_n} D_n^{i_n}$$
$$\times M(\mathcal{A})(\underline{B_1}, \underline{D_1}) \times \cdots \times M(\mathcal{A})(\underline{B_n}, \underline{D_n}) \times MM(\mathcal{A})(\underline{\underline{A}}, \underline{\underline{B}}).$$

This is in turn equal to

$$\lambda(\underline{A_1}, \ldots, \underline{A_p}). \int^{\underline{D_1}, \ldots, \underline{D_n}} F_1^1 D_1^1 \times \cdots \times F_n^{i_n} D_n^{i_n}$$
$$\times \int^{\underline{B_1}, \ldots, \underline{B_n}} M(\mathcal{A})(\underline{B_1}, \underline{D_1}) \times \cdots \times M(\mathcal{A})(\underline{B_n}, \underline{D_n}) \times MM(\mathcal{A})(\underline{\underline{A}}, \underline{\underline{B}})$$

and by application of the ninja Yoneda lemma 2.2.1 this is equal to

$$\lambda(\underline{A_1}, \ldots, \underline{A_p}). \int^{\underline{D_1}, \ldots, \underline{D_n}} F_1^1 D_1^1 \times \cdots \times F_n^{i_n} D_n^{i_n} \times MM(\mathcal{A})(\underline{\underline{A}}, \underline{\underline{D}}).$$

Flattening, i.e. applying the functor $\boldsymbol{P}\mu^{(M)}$, falls in the lower right corner, and results in the same functor as in (6.54).

This concludes the proof. □

Theorem 6.4.7. The monad $\boldsymbol{P} \circ T$ can be turned into a comonad under the equivalence $\mathsf{Prof} \cong \mathsf{Prof}^{\mathrm{op}}$, and the coKleisli category of such a comonad corresponds to: the category of operads if $T = M$, the category of symmetric operads if $T = S$, and the category of clones if $T = K$.

This means, in more detail, that

- a functor $F : \mathcal{C} \to \mathcal{V}$ is a free $(\boldsymbol{P} \circ M)$-coalgebra if and only if it is an operad (and similarly for symmetric operads, and clones if $\mathcal{V}$ is cartesian);
- the composition in the coKleisli category of $\boldsymbol{P} \circ T$ coincides with the substitution product for $T = M$, with the symmetric substitution for $T = S$, and with the cartesian substitution if $T = K$.

Exercises

6.1 Let $\mathcal{A}$ be an ordinary (i.e. Set-enriched) category, and let $\mathcal{V}$ be a cosmos; define the *free* $\mathcal{V}$-category $\overline{\mathcal{A}}^{\mathcal{V}}$ on $\mathcal{A}$ as the $\mathcal{V}$-category with the same objects of $\mathcal{A}$, and where

$$\overline{\mathcal{A}}^{\mathcal{V}}(A, A') := \mathcal{A}(A, A') \cdot I$$

is an $\mathcal{A}(A, A')$-fold tensor of the monoidal unit I of $\mathcal{V}$ (i.e. a coproduct of as many copies of I as there are elements in $\mathcal{A}(A, A')$).

- Prove that there is an isomorphism of categories $\overline{(\mathcal{A} \times \mathcal{B})}^{\mathcal{V}} \cong \overline{\mathcal{A}}^{\mathcal{V}} \times \overline{\mathcal{B}}^{\mathcal{V}}$; deduce that if $\mathcal{A}$ is a monoidal category, $\overline{\mathcal{A}}^{\mathcal{V}}$ is a monoidal $\mathcal{V}$-category.
- Does the 'free' $\mathcal{V}$-category $\overline{\mathcal{A}}^{\mathcal{V}}$ deserve its name? Is is true that $\mathcal{V}$-enriched functors

$$F : \overline{\mathcal{A}}^{\mathcal{V}} \to \mathcal{C}$$

to a $\mathcal{V}$-category $\mathcal{C}$ correspond to unenriched functors

$$F_0 : \mathcal{A} \to |\mathcal{C}|$$

to the underlying ordinary category of $\mathcal{C}$?

6.2 Show that the convolution product on $[\mathcal{C}, \mathcal{V}]$ results from the following left Kan extension:

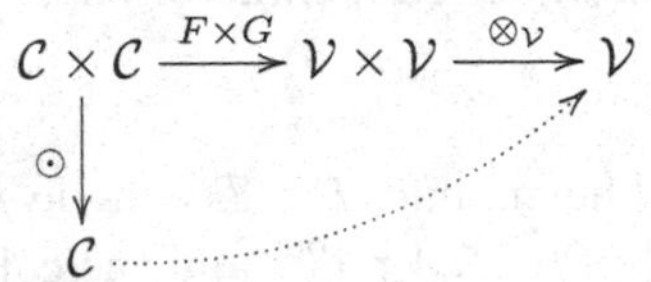

6.3 Show that the Isbell duality

$$\mathcal{V}\text{-Cat}(\mathcal{C}, \mathcal{V})^{\text{op}} \underset{\text{Spec}}{\overset{\mathsf{O}}{\rightleftarrows}} \mathcal{V}\text{-Cat}(\mathcal{C}^{\text{op}}, \mathcal{V})$$

of 3.2.18 is a pair of adjoint strong monoidal functors, when (the category $\mathcal{C}$ is monoidal and) their domains are endowed with the convolution product.

6.4 Let Set_* be the category of pointed sets; such a category is monoidal with respect to the *smash product*, where a pointed set (X, x_0) and a pointed set (Y, y_0) are smashed into the set $X \wedge Y$ defined as the pushout

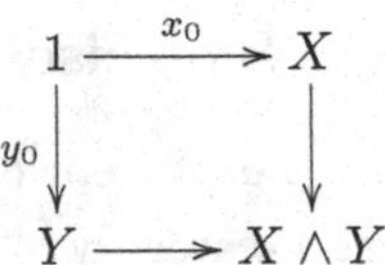

Let Seq be the discrete category whose objects are natural numbers $\{0, 1, 2, \dots\}$. Endow the category $\mathrm{Cat}(\mathrm{Seq}, \mathrm{Set}_*)$ with the convolution product, and show that

$$F * G = \lambda n. \coprod_{i+j=n} Fi \wedge Gj.$$

6.5 Describe the convolution product in the category $\mathrm{Cat}(\mathbb{N}, \mathrm{Set}_*)$, where $\mathbb{N} = \{0 < 1 < 2 < \cdots\}$ (monoidal with respect to ordinal sum, cf. 6.2.2), and in the category $\mathrm{Cat}(\mathcal{N}, \mathrm{Set}_*)$, where $\mathcal{N}$ is the monoid $(\mathbb{N}, \cdot)$ regarded as a category with a single object (the category of pointed sets is always endowed with the smash product). Compare the latter with the convolution product on the category $\mathrm{Cat}(\mathcal{N}, \mathrm{Set})$ (cartesian monoidal structure).

6.6 Define two functors

$$\Phi : \mathcal{V}\text{-}\mathrm{Cat}(\mathcal{P}, \mathcal{V}) \to \mathcal{V} \qquad \text{(evaluation at 0)}$$
$$\Psi : \mathcal{V} \to \mathcal{V}\text{-}\mathrm{Cat}(\mathcal{P}, \mathcal{V}) \qquad \text{(the left adjoint to } \Phi\text{)}.$$

Prove that $\Psi(A \oplus B) \cong \Psi A \otimes \Psi B$ and $\Phi \circ \Psi \cong 1$. Finally, for every object $V \in \mathcal{V}$, if ょV is the representable functor on V, then prove that $\mathcal{V}\text{-}\mathrm{Cat}(\mathcal{P}, \mathcal{V})(V * F, G) \cong \mathcal{V}(V, [F, G])$, where

$$[F, G] := \int_n \mathcal{V}(Fn, Gn)$$

has been defined in 4.3.3.

6.7 Fill in the details of the proof that (6.42) defines a monad, by showing that the associativity and unitality of an operad yield associativity and unitality of the monad M_T.

6.8 Let $\mathcal{A}, \mathcal{B}$ be small categories. An *S-profunctor* $P : \mathcal{A} \overset{S}{\rightsquigarrow} \mathcal{B}$ is a profunctor $P : \mathcal{A} \rightsquigarrow S\mathcal{B}$, i.e. a functor $P : S\mathcal{B}^{\mathrm{op}} \times \mathcal{A} \to \mathrm{Set}$, where S is the free symmetric monoidal category functor. Given $F : \mathcal{X} \overset{S}{\rightsquigarrow} \mathcal{Y}$ and $G : \mathcal{Y} \overset{S}{\rightsquigarrow} \mathcal{Z}$, we define the *composition* $G \circ F : \mathcal{X} \overset{S}{\rightsquigarrow} \mathcal{Z}$ as the coend

$$(G \circ F)[\underline{Z}; X] = \int^{\vec{Y}} G^e[\underline{Z}; \vec{Y}] \otimes F[\vec{Y}; X]$$

where

$$G^e[\underline{Z}; \vec{Y}] := \int^{\underline{Z}_1, \dots, \underline{Z}_n} G[\underline{Z}_1, Y_1] \times \cdots \times G[\underline{Z}_n, Y_n] \times S(\mathcal{Z})(\underline{Y}, \underline{Z}_1 \uplus \cdots \uplus \underline{Z}_n)$$

where $\uplus$ is concatenation.

Show that this is indeed an associative composition rule for a bicategory of S-profunctors; find the identity 1-cell of an object $\mathcal{A}$.

6.9 Fill in the details in the proof of 6.4.7, and in particular prove that the composition in the coKleisli category of $\boldsymbol{P} \circ T$ coincides with substitution products.

- Show that the monad $\boldsymbol{P} \circ T$, for $T \in \{M, S, K\}$ induces a monad $\hat{T}$ on Prof, and thus a comonad W by posing $W\mathcal{A} := T(\mathcal{A}^{\mathrm{op}})^{\mathrm{op}}$. (Hint: use the self-duality of 5.3.1.)
- Show that $\hat{T}$ has the following expression on 1-cells: given $\mathfrak{p} : \mathcal{A} \rightsquigarrow \mathcal{B}$,

$$\hat{T}\mathfrak{p} : (\underline{A}, \underline{B}') \mapsto \int^{\underline{B}} \mathfrak{p}(A_1, B_1) \times \cdots \times \mathfrak{p}(A_n, B_n) \times T(\mathcal{B})(\underline{B}, \underline{B}').$$

- Show that given a 2-comonad W on Prof, if we denote Prof_W its coKleisli bicategory, then the coKleisli composition is defined as soon as every map has a *coKleisli lifting*: the composition $g \bullet f = g \circ Wf \circ \sigma_C$ of $f : TC \to C'$ and $g : TC' \to C''$ equals the composition $g \circ R(f)$ in the diagram

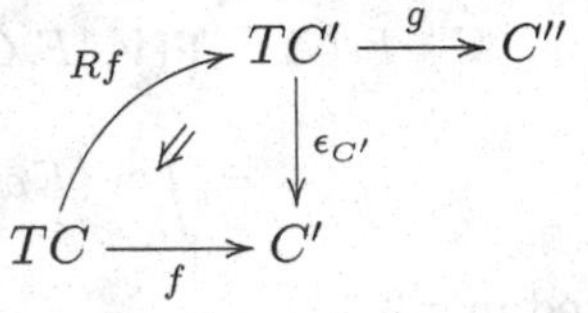

$$\begin{array}{ccc} & \overset{Rf}{\nearrow} TC' & \xrightarrow{g} C'' \\ & \swarrow & \downarrow{\scriptstyle \epsilon_{C'}} \\ TC & \xrightarrow[f]{} & C' \end{array}$$

where Rf is the right lifting of f along $\epsilon_{C'} : TC' \to C'$.
- Use this together with Exercise 5.2 to prove that the coKleisli composition coincides with the substitution product.

6.10 Let

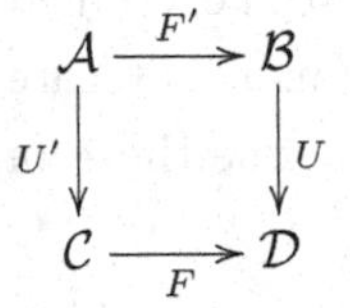

$$\begin{array}{ccc} \mathcal{A} & \xrightarrow{F'} & \mathcal{B} \\ {\scriptstyle U'}\downarrow & & \downarrow{\scriptstyle U} \\ \mathcal{C} & \xrightarrow[F]{} & \mathcal{D} \end{array}$$

be a strict pullback square of categories and functors, and assume that F is fully faithful; prove that if U is monadic, and U' has a left adjoint, then U' is monadic too.

6.11 Show that if $F : \mathcal{E} \to \mathcal{A}$ is a discrete fibration (see A.5.12) that is moreover a strong monoidal functor, then the associated presheaf $\hat{F} : \mathcal{A}^{\mathrm{op}} \to \mathrm{Set}$ is a monoid with respect to the Day convolution product (see 6.2.1) on $\mathrm{Cat}(\mathcal{A}^{\mathrm{op}}, \mathrm{Set})$. Show that this sets up a bijection under the equivalence of A.5.14.

6.12 ◉◉ Let $\mathrm{Cat}(\mathrm{Set}, \mathrm{Set})_s$ be the category of *small* functors[8] $F : \mathrm{Set} \to \mathrm{Set}$ and let F, G be two comonads; show that the Day convolution $F * G$ is itself a comonad. (Hint: there is a neat argument that involves the theory of *duoidal* categories, see [GF16, §8.1].)

6.13 ◉◉ Is it possible to dualise the Day convolution of 6.2.1 to define a *Day involution* operation on $\mathrm{Cat}(\mathcal{A}^{\mathrm{op}}, \mathrm{Set})$? For a suitable operation $A, B \mapsto \langle A, B\rangle$ and presheaves $F, G : \mathcal{A}^{\mathrm{op}} \to \mathrm{Set}$, we define

$$\lceil F, F\rfloor := \int_{AB} \mathcal{A}(\_, \langle A, B\rangle) \pitchfork [FA, GB].$$

How does this operation behave?

6.14 ◉◉ Prove that the operad associated to a monad on Set (see 6.3.6.EO2) really is an operad, by showing the commutativity of the following diagrams:

$$\begin{array}{ccc} Tm \times T1 \times \cdots \times T1 & \xrightarrow{\gamma} & Tm \\ {\scriptstyle Tm\times\eta\times\cdots\times\eta}\uparrow & & \| \\ Tm \times 1 \times \cdots \times 1 & \xrightarrow[\sim]{} & Tm \end{array} \qquad \begin{array}{ccc} T1 \times T1 & \xrightarrow{\gamma} & T1 \\ {\scriptstyle \eta\times T1}\uparrow & & \| \\ 1 \times T1 & \xrightarrow[\sim]{} & T1 \end{array}$$

where the lower horizontal isomorphism comes from the unitor of the cartesian monoidal structure on Set. (Hint: you might want to recall the commutativities given by naturality of the monad structure, A.6.3, naturality of the counit ϵ of the cartesian closed structure of Set, A.4.3.AD4, as well as the fact that ϵ is a cowedge.)

Good luck proving associativity: the axiom is stated as follows. Given a positive integer m, positive integers $p_1, \dots, p_m$, and a matrix of positive integers $\begin{bmatrix} q_{1,1} & \cdots & q_{1,p_1} \\ \vdots & \ddots & \vdots \\ q_{m,1} & \cdots & q_{m,p_m} \end{bmatrix}$, consider the diagram

$$\begin{array}{ccc} T\left(\sum_{i=1}^m p_i\right) & \xleftarrow{\gamma_{m,\vec{p}}\times \mathrm{id}} & Tm \times T\vec{p} \times T\vec{q}_1 \times \cdots \times T\vec{q}_m \\ {\scriptstyle \gamma_{\Sigma p_i, \vec{q}}}\downarrow & & \downarrow{\scriptstyle \mathrm{id}\times\gamma_{p_1,\vec{q}_1}\times\cdots\times\gamma_{p_m,\vec{q}_m}} \\ T\left(\sum_{i=1}^m \sum_{j=1}^{p_i} q_{i,j}\right) & \xleftarrow[\gamma_{m,\vec{q}}]{} & Tm \times T\left(\sum_{j=1}^{p_1} q_{1,j}\right) \times \cdots \times T\left(\sum_{j=1}^{p_m} q_{m,j}\right) \end{array}$$

[8] An endofunctor of Set is called small if it results as the left Kan extension of a functor $\mathcal{A} \to \mathrm{Set}$, where $\mathcal{A} \subseteq \mathrm{Set}$ is a small subcategory, along said inclusion. This restriction is needed in order for $\mathrm{Cat}(\mathrm{Set}, \mathrm{Set})$ to (exist and to) form a locally small category.

(there is an implicit identification between reshuffled copies of the factors in the upper left cartesian product, and we shorten the product $Tq_{j,1} \times \cdots \times Tq_{j,p_j}$ as $T\vec{q}_j$.) The associativity axiom for $\gamma_{-,=}$ amounts to the requirement that this diagram commutes.

6.15 ('The small object argument at work') Let $\lambda : T \Rightarrow S$ be a morphism $S \to T$ between two finitary monads (see A.9.3) on the category of sets; let $\lambda^\sharp : \mathrm{Alg}(S) \to \mathrm{Alg}(T)$ be the functor defined as sending an S-algebra $(A, a : SA \to A)$ into the T-algebra $(A, a \circ \lambda_A)$.

Show that $\lambda^\sharp$ has a left adjoint $\lambda_!$, following this scheme.

- Define the arrow (m_0, t_0) as the pushout

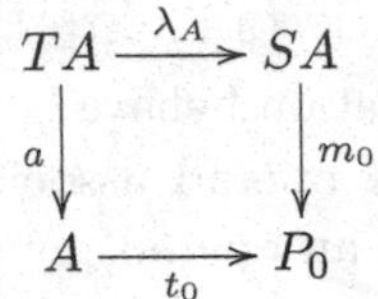

and, by induction, define (m_{i+1}, t_{i+1}) to be the pushout of the pair of arrows St_i and m_i:

$$\begin{array}{ccccccccccc}
TA & \xrightarrow{\lambda_A} & SA & \xrightarrow{St_0} & SP_0 & \xrightarrow{St_1} & SP_1 & \xrightarrow{St_2} & \dots & \longrightarrow & Q \\
{\scriptstyle a}\downarrow & & \downarrow{\scriptstyle m_0} & & \downarrow{\scriptstyle m_1} & & \downarrow & & & & \downarrow{\scriptstyle \alpha = m_\infty} \\
A & \xrightarrow[t_0]{} & P_0 & \xrightarrow[t_1]{} & P_1 & \xrightarrow[t_2]{} & P_2 & \xrightarrow[t_3]{} & \dots & \longrightarrow & P_\infty
\end{array}$$

and let $\alpha : Q \to P_\infty$ be its colimit.

- Show that $SP_\infty \cong Q$; show that α endows P_∞ with the structure of an S-algebra.
- Show that $\lambda_!$ defined as sending (A, a) to (SP_∞, α_a) is a functor. Show that there is an adjunction $\lambda_! \dashv \lambda^\sharp$.

7

Higher Dimensional (Co)ends

SUMMARY. The present chapter studies (co)end calculus in higher category theory; the basic theory of 2-dimensional (co)end calculus is introduced and explored in fair completeness, as well as 'homotopy coherent' versions of (co)ends (in model categories), $(\infty, 1)$-categorical (co)ends, and (co)ends inside a Groth(endieck) derivator [Gro13]. The focus is on showing how (co)end calculus is a paradigm that can be exported in whatever formal context is required to do category theory, rather than merely presenting a set of theorems.

> I Pitagorici raccontano che l'uso di insegnare geometria iniziò così: a uno dei membri della setta, perdute le proprie sostanze in un disastro, fu fatta la concessione di trarre guadagno dall'insegnamento della geometria.
>
> M. Timpanaro Cardini *Vite dei Pitagorici*

We progressively raise the dimension we work in, starting with the theory of (co)ends in 2-categories. We begin the next section by recalling the bare minimum of 2-dimensional category theory needed to appreciate the discussion; the definition of bicategory and 2-category (a strict bicategory) is understood as in A.7.11.

Notation 7.0.1. We freely employ the theory sketched in Section A.7 and in particular A.7.6: as always when dealing with higher dimensional cells and their compositions, there are several 'flavours' in which one can weaken strict commutativity. Besides this strictness (where every diagram commutes with an implicit identity 2-cell filling it), there is a notion of *strong* commutativity and universality, where filling 2-cells are required to be invertible, and *lax* commutativity/universality, where 2-cells are possibly non-invertible.

7.1 2-Dimensional Coends

A *lax functor* $F : \mathcal{C} \dashrightarrow \mathcal{D}$ between two 2-categories $\mathcal{C}, \mathcal{D}$ behaves like a functor, up to the fact that there is a non-invertible 2-cell linking the composition of the images Ff, Fg and the image of a composition $F(gf)$ (see A.7.12 for the precise definition).

A lax natural transformation $\alpha : F \Rightarrow_l G$ is a family of 1-cells $\alpha_C : FC \to GC$ coupled with 2-cells α_f, filling the diagrams

$$\begin{array}{ccc} FC & \xrightarrow{Ff} & FC' \\ {\scriptstyle \alpha_C}\downarrow & \swArrow \alpha_f & \downarrow{\scriptstyle \alpha_{C'}} \\ GC & \xrightarrow[Gf]{} & GC' \end{array} \tag{7.1}$$

These 2-cells are subject to suitable coherence conditions linking α_f, α_g and α_{gf} for composable 1-cells $C \xrightarrow{f} C' \xrightarrow{g} C''$ and characterising α_{id_C}.

In the setting of 2-categories, (co)end calculus has a rather natural interpretation in terms of enriched category theory;[1] even though we strive for clarity, the following discussion has little hope of being a self-contained exposition, and instead heavily relies on the existing literature (see for example [Dub70, Kel05a]).

The definition of a *lax (co)end* of a 2-functor S is given in terms of the notion of a *lax wedge* $\omega : B \Rightarrow S$, and it is the most general and the least symmetric one can give: asking for the diagrams to be filled by isomorphisms, it specialises to the notion of *strong* (co)end, and strict (co)end (of course, we can also define *oplax* (co)ends by reversing all 2-cells); the present subsection is designed in such a way to reduce to the strong and strict cases as particular examples.

This said, we shall warn the reader that 2-category theory exists in many dialects that follow slightly different notational conventions (minor differences that appear innocuous to the expert, but can become annoying when studying a part of category theory still lacking a truly comprehensive monograph).

We try to follow an auto-explicative notation based on our Appendix A and on canonical references like [Kel89, KS74], but we feel free to diverge from it from time to time.

The material on (co)lax (co)ends in the rest of this section comes in

[1] See [Kel89, KS74], for invaluably complete surveys on the matter; the reader should however be aware that 2-category theory is *not* completely subsumed by the theory of Cat-enriched categories.

its entirety from [Boz80]: the original paper, as well as Bozapalides' PhD thesis [Boz76], are very difficult to retrieve, and together they provide useful references and a starting point for developing 2-dimensional coend calculus. We hope this survey can provide a useful and more accessible reference in the future, for a piece of elegant mathematics that is sorely lacking from the mainstream literature.

Borrowing the notation from our Appendix A on basic category theory, we denote by $\mathcal{A}_o$ the class of objects of a category or 2-category $\mathcal{A}$.

Definition 7.1.1 (Lax wedge). Let

$$S : \mathcal{A}^{\mathrm{op}} \times \mathcal{A} \to \mathcal{B} \tag{7.2}$$

be a strict 2-functor between strict 2-categories $\mathcal{A}, \mathcal{B}$. A *lax wedge* ω for S consists of

- a triple $\{B, \omega_o, \omega_h\}$, where $B \in \mathcal{B}_o$ is an object (called the *tip* of the wedge);
- collections of 1-cells $\{\omega_A : B \to S(A,A)\}$, one for each $A \in \mathcal{A}_o$;
- 2-cells $\{\omega_f : S(A,f) \circ \omega_A \Rightarrow S(f,A') \circ \omega_{A'}\}$, in a diagram

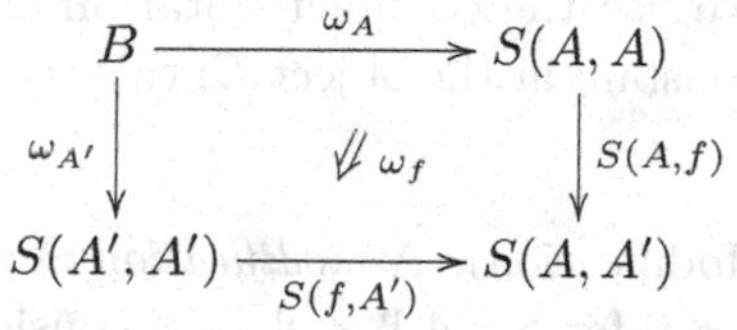

$$\begin{array}{ccc} B & \xrightarrow{\omega_A} & S(A,A) \\ {\scriptstyle \omega_{A'}}\downarrow & \Downarrow \omega_f & \downarrow{\scriptstyle S(A,f)} \\ S(A',A') & \xrightarrow[S(f,A')]{} & S(A,A') \end{array} \tag{7.3}$$

These data must fit together in such a way that the coherence axioms listed below, expressed by the commutativity of the following diagrams of 2-cells, are satisfied:

LW1. The diagram of 2-cells having faces

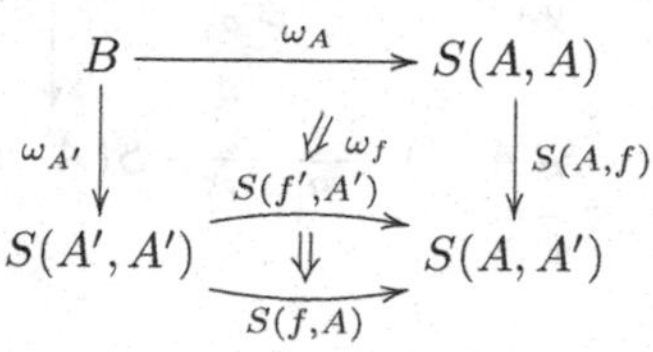

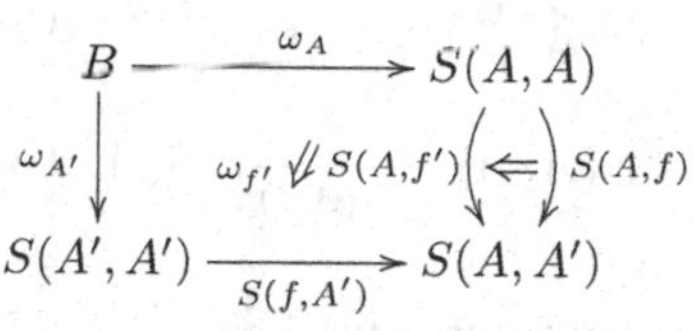

$$\begin{array}{ccc} B & \xrightarrow{\omega_A} & S(A,A) \\ {\scriptstyle \omega_{A'}}\downarrow & \Downarrow \omega_f & \downarrow{\scriptstyle S(A,f)} \\ S(A',A') & \overset{S(f',A')}{\underset{S(f,A)}{\rightrightarrows}}\ \Downarrow & S(A,A') \end{array} \qquad \begin{array}{ccc} B & \xrightarrow{\omega_A} & S(A,A) \\ {\scriptstyle \omega_{A'}}\downarrow & \omega_{f'} \Downarrow & {\scriptstyle S(A,f')}\downarrow \Leftarrow \downarrow{\scriptstyle S(A,f)} \\ S(A',A') & \xrightarrow[S(f,A')]{} & S(A,A') \end{array}$$

is commutative for any $\lambda : f \Rightarrow f'$, i.e. the equation

$$\omega_{f'} \circ (S(A,\lambda) * \omega_A) = (S(\lambda, A') * \omega_{A'}) \circ \omega_f$$

holds.

LW2. For each pair $A \xrightarrow{f} A' \xrightarrow{f'} A''$ of composable arrows in $\mathcal{A}$, the diagram of 2-cells

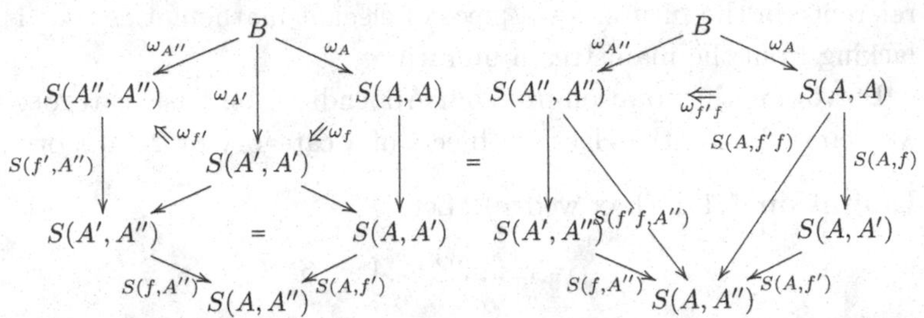

is commutative, i.e. the equation

$$(S(f, A'') * \omega_{f'}) \circ (S(A, f') * \omega_f) = \omega_{f'f}$$

holds.

LW3. For each object $A \in \mathcal{A}$, $\omega_{\mathrm{id}_A} = \mathrm{id}_{\omega_A}$.

Notation 7.1.2. We use the compact notation $\omega : B \Rightarrow S$ for a lax wedge with domain constant at the object B; this is evidently reminiscent of our 1.1.8.

Definition 7.1.3 (Modification). A *modification* $\Theta : \omega \Rrightarrow \sigma$ between two lax wedges $\omega, \sigma : B \Rightarrow S$ for $S : \mathcal{A}^{\mathrm{op}} \times \mathcal{A} \to \mathcal{B}$ consists of a collection of 2-cells $\Theta_A : \omega_A \Rightarrow \sigma_A$ indexed by the objects of $\mathcal{A}$ such that the diagram of 2-cells

$$\begin{array}{ccc} B & \xrightarrow{\omega_{A'}} & S(A', A') \\ \sigma_A \Big\downarrow \overset{\Theta_A}{\Leftarrow} \Big\downarrow \omega_A & \swArrow \omega_f & \Big\downarrow S(f,A') \\ S(A, A) & \xrightarrow[S(A,f)]{} & S(A, A') \end{array} \qquad \begin{array}{ccc} B & \overset{\omega_{A'}}{\underset{\sigma_{A'}}{\rightrightarrows}} \Downarrow \Theta_{A'} & S(A', A') \\ \sigma_A \Big\downarrow & \swArrow \sigma_f & \Big\downarrow S(f,A') \\ S(A, A) & \xrightarrow[S(A,f)]{} & S(A, A') \end{array} \tag{7.4}$$

is commutative, i.e.

$$(S(A, f) * \Theta_A) \circ \omega_f = \sigma_f \circ (S(f, A') * \omega_{A'}).$$

The above definition of a modification is modelled on the definition of modification between (lax) natural transformations of functors.

Remark 7.1.4. There is a more general definition for a modification $\Theta : \omega \Rrightarrow \sigma$ between lax wedges having different domains, say $\{B, \omega\}$ and $\{B', \sigma\}$. It consists of a morphism $\varphi : B \to B'$ and a 2-cell $\lambda_A : \sigma_A \circ \varphi \Rightarrow \omega_A$ such that

$$(\sigma_f * \varphi) \circ (S(A, f) * m_A) = (S(f, A') * m_{A'}) \circ \omega_f \tag{7.5}$$

(draw the corresponding diagram of 2-cells). We are not interested in this alternative definition, and this will not be investigated further. We thus take 7.1.3 as our working definition without further mention.

Such a definition entails that the set $\mathrm{LWd}(B, S)$ of lax wedges $B \Rightarrow S$ is a category having morphisms precisely the modifications $\Theta : \omega \Rrightarrow \sigma$ with the same tip, and the correspondence $\beta_S : B \mapsto \mathrm{LWd}(B, S)$ is functorial. The definition of *lax end* for S relies on the representability of this 2-functor.

Definition 7.1.5 (Lax end of S). Let $S : \mathcal{A}^{\mathrm{op}} \times \mathcal{A} \to \mathcal{B}$ be a 2-functor; an object of $\mathcal{B}$ is called the *lax end* of S, and denoted $\oint_A S(A, A)$, if there is a terminal lax wedge $\oint_A S(A, A) \Rightarrow S$, i.e. such that for any other lax wedge $\sigma : B' \to S$ there exists a unique 1-cell $u : B' \to B$ between the tips of the wedges such that

$$\omega_A \circ u = \sigma_A, \qquad \omega_f * u = \sigma_f. \tag{7.6}$$

(This pair of equations is conveniently depicted by the diagram of 2-cells

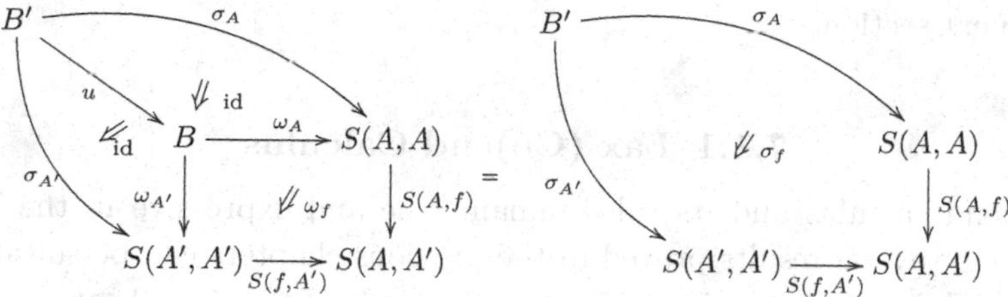

that we require to commute). Moreover, every modification $\Theta : \sigma \Rrightarrow \sigma'$ induces a unique 2-cell $\lambda : u \Rightarrow u'$ (u' is the arrow induced by σ') in such a way that $\lambda * \omega_A = \Theta_A$.

This equivalence sets up an isomorphism of categories between lax wedges with tip B and the category $\mathcal{B}(B, \oint_A S(A, A))$ (if needed, this can be relaxed to an *equivalence* of categories).

Remark 7.1.6. We denote the lax end of S as a 'squared integral'

$$\oint_A S(A, A). \tag{7.7}$$

This notation has a meaning: in the world of n-categories, the n-(co)end operation should be depicted by an integral symbol (with suitable superscripts or subscripts) overlapped by a 2^n-agon; in this way, the polygon for a 2-end has the correct number of edges since it is denoted as a 'square-over-integral' symbol $\boxed{\int}$, and because the circle is a polygon with an infinite number of sides, the notation is consistent for ∞-(co)ends (see 7.3.3, where we denoted the (co)end as an $\oint$ symbol).

Remark 7.1.7. Reversing the direction of 1-cells yields the notion of *lax coend*; but be careful! There is an additional dimension that can be reversed, i.e. the direction of 2-cells. Doing so, we obtain the notion of *oplax* end and coend. We will rarely need to invoke oplax cells; of course every statement involving a lax widget can be properly dualised to get an oplax one.

In compliance with the above 7.1.6, we denote the lax end of S as a squared integral with a superscript,

$$\boxed{\int}^{A} S(A, A). \tag{7.8}$$

Imposing stronger conditions on the 2-cells filling the diagrams of (7.3) and of 7.1.5 we obtain the notion of strong coend (if each component ω_f of a wedge is an isomorphism for every arrow $f : A \to A'$) and strict coend (if each ω_f is the identity). Of course, a similar terminology applies to those cases, as well as the (co)end calculus we are going to develop in the next section.

7.1.1 Lax (Co)end Calculus

(Co)end calculus and its rules remain true and expressive in the lax setting: several results proved in the previous chapters can be suitably 'laxified', justifying the intuition of lax (co)ends as the right 2-categorical generalisation of strict (co)ends.

We collect the most notable examples of this phenomenon in the rest of the section; the content of Section 7.1.2 surely deserves a special mention, as well as other remarks chosen to convey a sense of continuity and analogy. In 7.1.2 we prove that the lax counterpart of the ninja Yoneda lemma 2.2.1 provides a coreflection (using coends) and a reflection (using ends) of the category of strong presheaves into the category of lax presheaves.

Example 7.1.8. The *comma objects* (f/g) [Gra80] of a 2-category $\mathcal{B}$ can be identified with the lax end of functors $[1]^{\mathrm{op}} \times [1] \to \mathcal{B}$ choosing the two 1-cells f, g; this is in perfect analogy with Exercise 1.7, also in view of the characterisation of the comma object (f/g) as a lax limit.

Example 7.1.9. If $F, G : \mathcal{A} \to \mathcal{B}$ are 2-functors, then the lax end of the functor

$$\mathcal{B}(F, G) : \mathcal{A}^{\mathrm{op}} \times \mathcal{A} \to \mathrm{Cat} \tag{7.9}$$

characterises *lax natural transformations between lax functors* $F, G : \mathcal{A} \to \mathcal{B}$:

$$\oint_A \mathcal{B}(FA, GA) \cong \text{2-Cat}_l(\mathcal{A}, \mathcal{B})(F, G) \tag{7.10}$$

(see [Gra80] for more information).

Proof We abbreviate $\text{2-Cat}_l(\mathcal{A}, \mathcal{B})(F, G)$ as $\hom(F, G)$. The reader will notice that the argument is fairly elementary and echoes our proof of 1.4.1.

A lax wedge τ for the 2-functor $(A, A') \mapsto \hom(FA, GA')$ amounts to a diagram

$$\begin{array}{ccc} \mathcal{E} & \xrightarrow{\tau_A} & \hom(FA, GA) \\ {\scriptstyle \tau_{A'}}\downarrow & \Downarrow \tau_f & \downarrow {\scriptstyle Gf \circ \_} \\ \hom(FA', GA') & \xrightarrow[\_ \circ Ff]{} & \hom(FA, GA') \end{array} \tag{7.11}$$

filled by a 2-cell $\tau_f : Gf \circ \tau_A \Rightarrow \tau_{A'} \circ Ff$. Similarly to what happens in 1.4.1, each of the functors $\tau_A : \mathcal{E} \to \hom(FA, GA)$ sends an object $E \in \mathcal{E}$ into an object $\tau_A(E) : FA \to GA$ such that

$$G(f) \circ \tau_A(E) \xRightarrow{\tau_{f,E}} \tau_{A'}(E) \circ F(f) \tag{7.12}$$

is a 2-cell filling the square above. This is precisely the lax naturality condition needed to show that the correspondence $E \mapsto \{\tau_A E\}_{A \in \mathcal{A}}$ factors through $\text{2-Cat}_l(\mathcal{A}, \mathcal{B})(F, G)$ (a moderate amount of work is needed to check that the correct coherence properties hold for $\tau_{f,E}$, but we leave this minor chore to the willing reader). □

Lax natural transformations $\eta : F \dashrightarrow G$, described as the lax end above, can also be characterised as *lax limits* in the enriched sense: this motivates the search for a more general description of lax (co)ends as lax (co)limits, analogue to our 1.2, where instead of strict (co)equalisers we use the notion of *(co)inserter* (see 4.2.7) from [Kel89].

Definition 7.1.10 (Lax (co)limit). Let $F : \mathcal{A} \to \mathcal{B}$ be a 2-functor; a *lax limit* for F consists of a family of 1-cells $p_A : \operatorname{llim} F \to FA$ from an object $\operatorname{llim} F$, together with 2-cells $\pi_f : Ff \circ p_A \Rightarrow p_{A'}$ for every $f : A \to A'$, such that the pair of families (p_A, π_f) is terminal with this property.

This means that every other family of 1-cells $u_A : X \to FA$ and 2-cells $v_f : Ff \circ u_A \Rightarrow u_{A'}$ factors uniquely through a 1-cell $\bar{u} : X \to \operatorname{llim} F$, as in the diagram

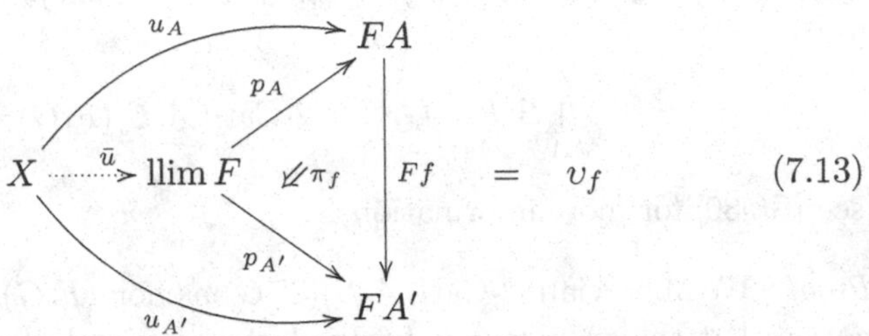

(7.13)

(this takes into account the 1-dimensional universal property of $\operatorname{llim} F$) and moreover, given a 2-cell $\mu : p_A\bar{u} \Rightarrow p_A\bar{v}$, there is a unique 2-cell $\bar{\mu}$ such that

$$\begin{array}{ccccc} & & \operatorname{llim} F & & \\ & \overset{\bar{u}}{\nearrow} & & \overset{p_A}{\searrow} & \\ X & & \Downarrow\mu & & FA \\ & \underset{\bar{v}}{\searrow} & & \underset{p_A}{\nearrow} & \\ & & \operatorname{llim} F & & \end{array} \quad = p_A * \bar{\mu} \tag{7.14}$$

Reversing the direction of 1-cells yields the notion of *lax colimit*; but be careful! There is an additional dimension that can be reversed, i.e. the direction of 2-cells. Doing so, we obtain the notion of *oplax* limit and colimit. Again, we employ without further mention each of these names whenever needed.

Proposition 7.1.11 ((Co)ends of mute functors). Suppose that the 2-functor $S : \mathcal{A}^{\text{op}} \times \mathcal{A} \to \mathcal{B}$ is mute (see 1.1.9) in the contravariant variable, i.e. that there is a factorisation $S = S' \circ p : \mathcal{A}^{\text{op}} \times \mathcal{A} \xrightarrow{p} \mathcal{A} \xrightarrow{S'} \mathcal{B}$; then the lax end of S exists if and only if the lax limit of S does. Moreover, the two objects are canonically isomorphic, sharing the same universal property:

$$\oint_A S(A, A) \cong \operatorname{llim} S'. \tag{7.15}$$

This means that every lax (co)limit is a degenerate form of lax (co)end.

Example 7.1.12. As a particularly simple example of this, if $\mathcal{A}$ is locally discrete (informally, this means that it can be identified with a 1-category) and if the functor $c_B : \mathcal{A} \to \mathcal{B}$ is constant on a single object B, i.e. $c_B(A) \equiv B$ for each $A \in \mathcal{A}$, and $c_B(f) \equiv \mathrm{id}_B$, then $\oint_A c_B$ is canonically identified with the *cotensor* of B by $\mathcal{A}$ and is denoted $\mathcal{A} \pitchfork B$.

Proof It is easily seen that a lax wedge $B' \to \oint c_B$ corresponds to a family of maps $B' \to B$ indexed by the objects of $\mathcal{A}$; the wedge condition now entails that the correspondence $A \mapsto f_A : B' \to B$ can be lifted to a functor $\mathcal{A} \to \mathcal{B}(B', B)$. This concludes the proof of an isomorphism

$$\mathrm{Cat}(\mathcal{A}, \mathcal{B}(B', B)) \cong \mathcal{B}\Big(B', \oint_{\mathcal{B}} c_B\Big). \qquad \square$$

Theorem 7.1.13 (Parametric lax ends). Whenever we have a functor $F : \mathcal{A}^{\mathrm{op}} \times \mathcal{A} \times \mathcal{B} \to \mathcal{C}$, and the lax end $\oint_A F(B, A, A)$ exists for every object $B \in \mathcal{B}$, then $B \mapsto \oint_A F(B, A, A)$ extends to a 2-functor $\mathcal{B} \to \mathcal{C}$ which has the universal property of the lax end of its mate $\hat{F} : \mathcal{A}^{\mathrm{op}} \times \mathcal{A} \to \mathcal{C}^{\mathcal{B}}$ under an obvious cartesian closed adjunction.

We omit the proof (one way to proceed is to show how a functor $F : \mathcal{A}^{\mathrm{op}} \times \mathcal{A} \times \mathcal{B} \to \mathcal{C}$ transposes to $\hat{F} : \mathcal{B} \to 2\text{-}\mathrm{Cat}(\mathcal{A}^{\mathrm{op}} \times \mathcal{A}, \mathcal{C})$ once the functoriality of $\oint : 2\text{-}\mathrm{Cat}(\mathcal{A}^{\mathrm{op}} \times \mathcal{A}, \mathcal{C}) \to \mathcal{C}$ has been proved). We leave the reader to enjoy this exercise, letting them discover how strict the resulting functor $\oint_A F(\_, A, A)$ shall be.

We also offer without proof the statement the 2-dimensional analogue of the *Fubini theorem* in 1.3.1. This is a much more daunting exercise, but the challenge is merely notational, not conceptual.

Theorem 7.1.14 (Fubini rule for lax (co)ends). If one among the following lax ends exists, then so do the others, and the three are canonically isomorphic:

$$\begin{aligned} \oint_{B,C} T(B,C,B,C) &\cong \oint_B \Big(\oint_C T(B,C,B,C)\Big) \\ &\cong \oint_C \Big(\oint_B T(B,C,B,C)\Big). \end{aligned} \tag{7.16}$$

In a similar fashion,

$$\begin{aligned} \oint^{B,C} T(B,C,B,C) &\cong \oint^B \Big(\oint^C T(B,C,B,C)\Big) \\ &\cong \oint^C \Big(\oint^B T(B,C,B,C)\Big). \end{aligned} \tag{7.17}$$

Corollary 7.1.15 (Fubini rule for lax (co)limits)**.** Let $T : \mathcal{B} \times \mathcal{C} \to \mathcal{D}$ be a 2-functor. Then

LC1. Lax limits commute with lax limits: we have canonical isomorphisms in $\mathcal{D}$

$$\operatorname*{llim}_{B\in\mathcal{B}}\Big(\operatorname*{llim}_{C\in\mathcal{C}} T(B,C)\Big) \cong \operatorname*{llim}_{C\in\mathcal{C}}\Big(\operatorname*{llim}_{B\in\mathcal{B}} T(B,C)\Big). \tag{7.18}$$

LC2. Lax colimits commute with lax colimits: we have canonical isomorphisms in $\mathcal{D}$

$$\operatorname*{lcolim}_{B\in\mathcal{B}}\Big(\operatorname*{lcolim}_{C\in\mathcal{C}} T(B,C)\Big) \cong \operatorname*{lcolim}_{C\in\mathcal{C}}\Big(\operatorname*{lcolim}_{B\in\mathcal{B}} T(B,C)\Big). \tag{7.19}$$

The hom functor must preserve all lax ends.

Theorem 7.1.16. Let $S : \mathcal{A} \to \mathcal{B}$ be a 2-functor such that $\oint_A S(A,A)$ exists in $\mathcal{B}$. Then we have a canonical isomorphism of categories

$$\mathcal{B}\Big(B, \oint_A S(A,A)\Big) \cong \oint_A \mathcal{B}(B, S(A,A)). \tag{7.20}$$

A completely dual statement involves the lax coend of $S : \mathcal{A} \to \mathcal{B}$: there is a canonical isomorphism of categories

$$\mathcal{B}\Big(\oint^A S(A,A), B\Big) \cong \oint_A \mathcal{B}(S(A,A), B). \tag{7.21}$$

This can also be seen as an alternative definition: the lax (co)end of S exists if and only if for every object $B \in \mathcal{B}$ the functor $\mathcal{A}^{\mathrm{op}} \times \mathcal{A} \xrightarrow{S} \mathcal{B} \xrightarrow{\hom_B} \mathrm{Cat}$ ($\hom_B$ is understood here as a covariant or contravariant representable over $B \in \mathcal{B}$) has a lax (co)end; indeed if this is the case, there is the isomorphism above.

7.1.2 The Lax Ninja Yoneda lemma

In the world of 2-categories and lax morphisms, the ninja Yoneda lemma 2.2.1 acquires a peculiar flavour, since it is equivalent to the (co)reflectivity of strict presheaves $F : \mathcal{C}^{\mathrm{op}} \to \mathrm{Cat}$ inside lax presheaves $F' : \mathcal{C}^{\mathrm{op}} \dashrightarrow \mathrm{Cat}$.

More precisely, there is a diagram of adjoint 2-functors

$$\mathrm{Cat}(\mathcal{C}^{\mathrm{op}}, \mathrm{Cat}) \underset{(-)^\flat}{\overset{(-)^\sharp}{\leftleftarrows\!\!\rightarrow}} \mathrm{Cat}_l(\mathcal{C}^{\mathrm{op}}, \mathrm{Cat}) \tag{7.22}$$

where the central arrow is the inclusion. This means that for each strict 2-functor $H \in$ 2-Cat$(\mathcal{C}^{\text{op}}, \text{Cat})$ there are two natural isomorphisms

$$\begin{aligned}
\text{2-Cat}(\mathcal{C}^{\text{op}}, \text{Cat})(H, F^\flat) &\cong \text{2-Cat}_l(\mathcal{C}^{\text{op}}, \text{Cat})(H, F),\\
\text{2-Cat}(\mathcal{C}^{\text{op}}, \text{Cat})(F^\sharp, H) &\cong \text{2-Cat}_l(\mathcal{C}^{\text{op}}, \text{Cat})(F, H).
\end{aligned} \tag{7.23}$$

The proof is completely formal. To prove the isomorphism, we will show that the functors $F^\flat$ and $F^\sharp$ defined by means of the lax coends

$$F^\sharp \cong \oint^A \mathcal{C}(\_, A) \times FA \qquad F^\flat \cong \oint_A \mathcal{C}(A, \_) \pitchfork FA \tag{7.24}$$

have the above property.

Proof The proof exploits 7.1.9 as well as the evident fact that strict (co)ends are particular cases of lax (co)ends, and (as a consequence) the fact that any number of (co)ends and lax (co)ends commute with each other by virtue of the lax Fubini rule, plus the preservation of lax (co)ends by the hom functor 7.1.16, and the strict ninja Yoneda lemma 2.2.1.

Now, let us denote $F^\flat = \oint_A \mathcal{C}(A, \_) \pitchfork FA$; we have that

$$\begin{aligned}
\text{Cat}(\mathcal{C}^{\text{op}}, \text{Cat})(H, F^\flat) &= \int_C \text{Cat}(HC, F^\flat C)\\
&= \int_C \text{Cat}\left(HC, \oint_A \text{Cat}(\mathcal{C}(A, C), FA)\right)\\
&\cong \int_C \oint_A \text{Cat}(HC, \text{Cat}(\mathcal{C}(A, C), FA))\\
&\cong \oint_A \int_C \text{Cat}(HC \times \mathcal{C}(A, C), FA)\\
&\cong \oint_A \text{Cat}\left(\int^C HC \times \mathcal{C}(A, C), FA\right)\\
&\cong \oint_A \text{Cat}(HA, FA)\\
(7.10) &= \text{2-Cat}_l(\mathcal{C}^{\text{op}}, \text{Cat})(H, F).
\end{aligned}$$

The proof that $\text{Cat}(\mathcal{C}^{\text{op}}, \text{Cat})(F^\sharp, H) \cong \text{2-Cat}_l(\mathcal{C}^{\text{op}}, \text{Cat})(F, H)$ is similarly formal, thus we leave it to the reader. □

The lax coend defining $F^\sharp$ can take as its argument a very simple functor, and spit out a not-so-simple one: an instructive example is given by the $(\_)^\sharp$ of the terminal functor.

Example 7.1.17. Let $\mathcal{I} : \mathcal{C}^{\mathrm{op}} \to \mathrm{Cat}$ be the constant functor sending $C \in \mathcal{C}$ into the terminal category, regarded as a lax functor. Then the strict functor $\mathcal{I}^\sharp : \mathcal{C}^{\mathrm{op}} \to \mathrm{Cat}$ is the lax coend

$$\mathcal{I}^\sharp C \cong \oint^A \mathcal{I}A \times \mathcal{C}(A, C) \cong \oint^A \mathcal{C}(A, C). \tag{7.25}$$

Hence the category $\oint^A \mathcal{C}(A, C)$ coincides with the lax colimit of the strict presheaf $\mathcal{C}^{\mathrm{op}} \to \mathrm{Cat}$, $A \mapsto \mathcal{C}(A, C)$, which is [Str76, p. 171] the *lax slice category* $\mathcal{C}/\!\!/C$ of commutative diagrams of 2-cells

$$\begin{array}{ccc} A & \xrightarrow{h} & A' \\ & \Downarrow\alpha & \\ {\scriptstyle u}\searrow & & \swarrow{\scriptstyle u'} \\ & C & \end{array} \tag{7.26}$$

Another striking instance of this phenomenon, by which quite complex shapes can arise as lax (co)limits of simpler diagrams, is given by the twisted arrow category of $\mathcal{A}$, whose opposite category is isomorphic to the lax colimit of the 'slice category' diagram.

Proposition 7.1.18 (The twisted arrow category as a lax colimit). Let $\mathcal{A}$ be a category; then it is possible to characterise the (opposite of the) twisted arrow category of $\mathcal{A}$, as defined in 1.2.2, as the lax colimit of the diagram $\mathcal{A} \to \mathrm{Cat} : A \mapsto \mathcal{A}/A$, i.e. as the lax coend $\oint^A \mathcal{A}/A$.

Proof In order to prove the statement, we shall first find a good candidate for a universal cocone. Such a cocone $q : \mathcal{A}/\_ \to \mathrm{TW}(\mathcal{A})^{\mathrm{op}}$ is defined on objects sending every $f : X \to A$ into itself, and a morphism

$$\begin{array}{ccc} X & \xrightarrow{u} & X'' \\ {\scriptstyle f}\searrow & & \swarrow{\scriptstyle f'} \\ & A & \end{array} \tag{7.27}$$

again into itself, regarding the triangle as a square with identity bottom:

$$\begin{array}{ccc} X & \xrightarrow{u} & X' \\ {\scriptstyle f}\downarrow & & \downarrow{\scriptstyle f'} \\ A & = & A \end{array} \tag{7.28}$$

(as an aside comment, this definition makes obvious that the functor we are defining is conservative; we shall however get an indirect proof of invertibility).

Let us denote $t_* : \mathcal{A}/A \to \mathcal{A}/A'$ the functor induced from $t : A \to A'$ by post-composition. Then, we get a tautological laxity cell $q_t : q_A \Rightarrow q_{A'} \circ t_*$, having components

$$\begin{array}{ccc} X & = & X \\ & & \downarrow{\scriptstyle f} \\ {\scriptstyle f}\downarrow & & A \\ & & \downarrow{\scriptstyle t} \\ A & \xrightarrow[t]{} & A' \end{array} \tag{7.29}$$

These components glue together defining a lax cocone $q : \mathcal{A}/\_ \to \text{TW}(\mathcal{A})$.

We shall now prove that this is universal: in order to do so, we have to verify that the diagram

$$\begin{array}{ccccc} \mathcal{A}/A & & \xrightarrow{\quad q_A \quad} & & \\ & \searrow^{q_A} & & & \\ {\scriptstyle t_*}\downarrow & \swarrow q_t & \text{TW}(\mathcal{A})^{\text{op}} & \overset{u}{\dashrightarrow} & \mathcal{K} \\ & \nearrow_{q_{A'}} & & & \\ \mathcal{A}/A' & & \xrightarrow[\quad q_{A'} \quad]{} & & \end{array} \tag{7.30}$$

satisfies the (dual of the) conditions in 7.1.10.

- Each time we are given a lax cocone $h : \mathcal{A}/\_ \to \text{TW}(\mathcal{A})^{\text{op}}$, there is a *unique* induced 1-cell $u : \text{TW}(\mathcal{A})^{\text{op}} \to \mathcal{K}$ such that $u * p = h$.
- Every 2-cell $\sigma : u \Rightarrow u'$ such that the horizontal composition $\sigma \boxminus q_t$ in

$$\begin{array}{ccccc} \mathcal{A}/A & & & & \\ & \searrow^{q_A} & & & \\ {\scriptstyle t_*}\downarrow & \swarrow q_t & \text{TW}(\mathcal{A})^{\text{op}} & \underset{u'}{\overset{u}{\rightrightarrows}}\ {\Downarrow}\sigma & \mathcal{K} \\ & \nearrow_{q_{A'}} & & & \\ \mathcal{A}/A' & & & & \end{array} \tag{7.31}$$

  is invertible for every $t : A \to A'$, is itself invertible.

Given h as above, since for every $t : A \to A'$ we have

$$h_{A'}(t \circ f) = h_A(f), \tag{7.32}$$

we can define $u : \text{TW}(\mathcal{A})^{\text{op}} \to \mathcal{K}$ sending the object $f : X \to A$ and the morphism $(s,t) : f \to f'$ respectively to

- $u\begin{bmatrix} X \\ \downarrow \\ A \end{bmatrix} = h_A\begin{bmatrix} X \\ \downarrow \\ A \end{bmatrix} \in \mathcal{K}$,

- $u\begin{bmatrix} X \xleftarrow{s} Y \\ \downarrow \quad \downarrow \\ A \xrightarrow[t]{} B \end{bmatrix} = h_A\begin{bmatrix} X \\ \downarrow \\ A \end{bmatrix} \leftarrow h_B\begin{bmatrix} Y \\ \downarrow \\ B \end{bmatrix}$ obtained as the composition

$$h_B(t) = h_B(tfs) \to h_A(fs) \to h_A f \tag{7.33}$$

where in the first step we use the laxity cell $h_t : h_B(tfs) \to h_A(fs)$ and in the second we apply h_A to the morphism $f \to fs$ in $\mathcal{A}/A$.

Using these definitions, it is rather easy to see that $u * p = h$ and that these form the components of suitable modifications, since

$$u(q_t) = u\begin{bmatrix} X = X \\ \downarrow \quad \downarrow \\ A \xrightarrow[t]{} A' \end{bmatrix} \tag{7.34}$$

or in other words, $u(q_t)$ equals the action of the laxity cell $q_t : u(tf) \to u(f)$ composed with $q_{\mathrm{id}_A} : q_A(f) \to q_A(f)$.

It is equally easy to see that every morphism in $\mathrm{TW}(\mathcal{A})^{\mathrm{op}}$ can be expressed as a morphism in the image of some q_A, and as a consequence, u is uniquely determined by the properties we just checked.

It remains to check the last condition; however, thanks to the definition of q_t, if the diagram (7.31) satisfies the property that the horizontal composition $\sigma \boxminus q_t$ is invertible for a 2-cell $\sigma : u \Rightarrow u'$, and each component of the composition $\sigma \boxminus q_{\mathrm{id}_A}$ is invertible, then each component $\sigma_{(X \to A)}$: $u\begin{bmatrix} X \\ \downarrow \\ A \end{bmatrix} \to u'\begin{bmatrix} X \\ \downarrow \\ A \end{bmatrix}$ is invertible as well. □

The following result is a partial analogue of 4.1.8: in order to characterise a lax (co)end as a weighted (co)limit, we have to 'twist' the hom functor using the $\sharp$ construction of 7.1.2.

Proposition 7.1.19. [Boz80, §2] There is a canonical isomorphism between the lax end of a 2-functor $T : \mathcal{C}^{\mathrm{op}} \times \mathcal{C} \to \mathcal{B}$ and the limit of T weighted by the bifunctor $\mathcal{C}((\_)^\sharp, =)$, i.e.

$$\lim^{\mathcal{C}((\_)^\sharp, =)} T \cong \oint_A T(A, A) \tag{7.35}$$

where $\mathcal{C}((\_)^\sharp, =) : (C, C') \mapsto \mathcal{C}(\_, C')^\sharp(C) = \oint^A \mathcal{C}(C, A) \times \mathcal{C}(A, C')$. A dual statement holds for lax coends.

Proof The proof is completely formal, by virtue of the results established so far: we can compute

$$\mathcal{B}\left(B, \lim^{\mathcal{C}((\_)^\sharp, =)} T\right) = \lim^{\mathcal{C}((\_)^\sharp, =)} \mathcal{B}(B, T) \qquad \text{(see 4.1.14)}$$

$$
\begin{aligned}
(4.1.5) &\cong \int_{C,D} \mathcal{B}(B,T(C,D))^{\mathcal{C}(C^\sharp,D)} \\
&= \int_{C,D} \mathcal{B}(B,T(C,D))^{\oint^A \mathcal{C}(C,A)\times\mathcal{C}(A,D)} \\
&\cong \oint_A \int_{C,D} \Big(\mathcal{B}(B,T(C,D))^{\mathcal{C}(A,D)}\Big)^{\mathcal{C}(C,A)} \\
(7.1.14) &\cong \oint_A \int_C \mathrm{Cat}\left(\mathcal{C}(C,A), \int_D \mathcal{B}(B,T(C,D))^{\mathcal{C}(A,D)}\right) \\
&\cong \oint_A \int_C \mathrm{Cat}\,(\mathcal{C}(C,A), \mathcal{B}(B,T(C,A))) \\
&\cong \oint_A \mathcal{B}(B,T(A,A)) = \mathcal{B}\left(B, \oint_A T(A,A)\right).
\end{aligned}
$$

This concludes the proof. □

Let $S : \mathcal{A}^{\mathrm{op}} \to \mathrm{Cat}$, $T : \mathcal{A} \to \mathcal{B}$ be two functors and suppose $\mathcal{B}$ has Cat-tensors; then the lax coend of the 2-functor $\mathcal{A}^{\mathrm{op}} \times \mathcal{A} \xrightarrow{S\times T} \mathrm{Cat} \times \mathcal{B} \xrightarrow{\otimes} \mathcal{B}$ is the *tensor product* of S and T, denoted

$$S \overline{\otimes} T =: \oint^A Sa \otimes Ta. \tag{7.36}$$

7.1.3 2-Profunctors, Lax Kan Extensions

The present section is intended to provide an analogue of profunctor theory for lax (co)ends; it can be safely skipped at first reading.

The proof of 7.1.19 above suggests that $\mathcal{C}((\_)^\sharp, =)$ is the *lax composition* of two representable profunctors. The intuition that the theory of Chapter 5 can be suitably adapted to the lax context is correct (unfortunately though, we have few applications in mind for such a theory: Exercise 7.1 tries to suggest a few lines of investigation).

In brief, a *2-profunctor* $\mathfrak{p} : \mathcal{A} \rightsquigarrow \mathcal{B}$ is a 2-functor $\mathfrak{p} : \mathcal{B}^{\mathrm{op}} \times \mathcal{A} \to \mathrm{Cat}$. Lax coends provide a weak composition rule for 2-profunctors. More precisely, let

$$\mathcal{A} \overset{\mathfrak{p}}{\rightsquigarrow} \mathcal{B} \overset{\mathfrak{q}}{\rightsquigarrow} \mathcal{C} \tag{7.37}$$

be a couple of composable 2-profunctors, namely two 2-functors $\mathfrak{p} : \mathcal{B}^{\mathrm{op}} \times \mathcal{A} \to \mathrm{Cat}$ and $\mathfrak{q} : \mathcal{C}^{\mathrm{op}} \times \mathcal{B} \to \mathrm{Cat}$; then the composition $\mathfrak{q} \diamond \mathfrak{p}$ is defined by the coend

$$\mathfrak{q} \diamond_l \mathfrak{p}(C, A) = \oint^B \mathfrak{p}(B, A) \times \mathfrak{q}(C, B). \tag{7.38}$$

The compatibility between lax colimits and products ensures that the expected associativity holds up to a canonical identification:

$$(\mathfrak{r} \diamond_l \mathfrak{q}) \diamond_l \mathfrak{p} \cong \mathfrak{r} \diamond_l (\mathfrak{q} \diamond_l \mathfrak{p}) \tag{7.39}$$

for any three $\mathcal{A} \overset{\mathfrak{p}}{\rightsquigarrow} \mathcal{B} \overset{\mathfrak{q}}{\rightsquigarrow} \mathcal{C} \overset{\mathfrak{r}}{\rightsquigarrow} \mathcal{D}$.

A much more useful laxification involves Kan extensions. Let $T : \mathcal{A} \to \mathcal{B}$ and $F : \mathcal{A} \to \mathcal{C}$ be two 2-functors.

Definition 7.1.20. We define the *left lax Kan extension* of F along T to be a 2-functor $\mathrm{lLan}_T F : \mathcal{B} \to \mathcal{C}$ endowed with a lax natural transformation $\eta : F \Rightarrow_l \mathrm{lLan}_T F \circ T$ (a *unit*) such that, for every pair $S : \mathcal{B} \to \mathcal{C}$ and $\lambda : F \Rightarrow_l S \circ T$ (respectively, a 2-functor and a lax natural transformation) then there exists a unique $\zeta : \mathrm{lLan}_T F \Rightarrow_l S$ such that

$$(\zeta * T) \circ \alpha = \lambda \tag{7.40}$$

and moreover, if $\Sigma : \lambda \Rrightarrow \lambda'$ is a modification between lax natural transformations, there is a unique modification $\Omega : \zeta \Rrightarrow \zeta'$ (where ζ is induced by λ, and ζ' by λ') between Cat-natural transformations such that $(\Omega * T) \circ \alpha = \Sigma$.

This entire list of conditions can be expressed by means of the isomorphism

$$\text{2-Cat}(\mathcal{A}, \mathcal{C})\big(F, S \circ T\big) \cong \text{2-Cat}(\mathcal{B}, \mathcal{C})\big(\mathrm{lLan}_T F, S\big), \tag{7.41}$$

which is natural in S.

Example 7.1.21. Let 1 be the terminal 2-category. Then the left lax Kan extension of a 2-functor $F : \mathcal{A} \to \mathcal{C}$ along the terminal 2-functor $\mathcal{A} \to 1$ is the lax colimit of F.

Remark 7.1.22. We can obtain different flavours of lax Kan extension by reversing the directions of $\alpha, \lambda, \zeta \dots$ and imposing invertibility. The example above, as well as the following theorem, shows that the choice of Cat-natural transformations instead of lax natural transformations is the right one (see also [Boz80] for a dual statement).

Theorem 7.1.23. In the same notation as above, assume $\mathcal{C}$ admits tensors for hom categories $\mathcal{B}(TA, B) \otimes FA'$ for each $A, A' \in \mathcal{A}$, and that the lax coend

$$\oint^{A} \mathcal{B}(TA, B) \otimes FA \tag{7.42}$$

exists; then the lax Kan extension of F along T exists too, and it is canonically isomorphic to the coend above.

We can also mimic Exercise 2.3 to obtain a lax analogue.

Proposition 7.1.24. Let $2\text{-}\mathrm{Cat}_l(\mathcal{C}, \mathrm{Cat})(U, V)$ denote the category of lax natural transformations between two 2-functors $U, V : \mathcal{C} \to \mathrm{Cat}$. Then

$$2\text{-}\mathrm{Cat}_l(\mathcal{C}, \mathrm{Cat})(F \times G, H) \cong 2\text{-}\mathrm{Cat}_l(\mathcal{C}, \mathrm{Cat})(F, H^G), \tag{7.43}$$

where $H^G(x) = 2\text{-}\mathrm{Cat}_l(\mathcal{C}, \mathrm{Cat})(よ(A) \times G, H) = \oint_Y \mathrm{Set}(\hom(Y, X) \times GY, HY)$.

Proof Every step can be motivated by results in the present section:

$$\begin{aligned}
2\text{-}\mathrm{Cat}_l(\mathcal{C}, \mathrm{Cat})(F, H^G) &= \oint_X \mathrm{Set}(FX, 2\text{-}\mathrm{Cat}_l(\mathcal{C}, \mathrm{Cat})(よ(A) \times G, H)) \\
&\cong \oint_X \oint_Y \mathrm{Set}(FX, \mathrm{Set}(\mathcal{C}(Y, X) \times GY, HY)) \\
&\cong \oint_Y \mathrm{Set}\Big(\Big(\oint^X FX \times \mathcal{C}(Y, X)\Big) \times GY, HY\Big) \\
&\cong \oint_Y \mathrm{Set}(FY \times GY, HY) \\
&= 2\text{-}\mathrm{Cat}_l(\mathcal{C}, \mathrm{Cat})(F \times G, H).
\end{aligned}$$

This concludes the proof. □

7.2 Coends in Homotopy Theory

Higher category theory is enjoying a Renaissance, thanks to massive collaboration between many people drawing together from various fields of research, and cooperating to re-analyse every feature of category theory inside, or in terms of, the topos of simplicial sets.

The purpose of the present section is to study what this 'homotopification' process does to (co)end calculus.

The urge to keep this chapter self-contained, forces us to take for granted a certain acquaintance with model categories, simplicial categories, ∞-categories *à la* Joyal–Lurie, and dg-categories. Each of these theories is vast and constitutes enough material for a dedicated monograph. Nonetheless, we strive to offer the best intuition we can, in the little space we have.

We start by presenting the theory of *homotopy (co)ends* 7.2.1 in model category theory; the coend functor $\int^{\mathcal{C}} : \mathrm{Cat}(\mathcal{C}^{\mathrm{op}} \times \mathcal{C}, \mathcal{D}) \to \mathcal{D}$ admits a 'derived' counterpart $\int_{\mathbb{L}}^{C}$ (see [Isa09]) that preserves weak equivalences, in the same sense the colimit functor does.

Subsequently, in Section 7.3, we will explore the theory of *quasicategorical* (co)end calculus (providing a proof of the Fubini rule for ∞-coends: this is not strictly new material, but brings together the pieces present in [GHN15]).

In Section 7.2.2, we address the study of *simplicially coherent* (co)ends (i.e. enriched (co)ends in sSet-categories), and the definition of a (co)end in a *derivator* (this is nothing more than a paragraph; some additional results are presented as Exercises 7.11, 7.13, and 7.14).

The discussion closes the circle of (co)end calculus in each of the most common models for higher category theory (model categories, simplicially enriched categories, simplicial sets, derivators). The reader will notice that most of these flavours of (co)end calculus are far from being full-fledged theories. This testifies how much work there is still to be done in the field!

We hope to have convinced our reader that the endeavour to complete the theory is worth undertaking. Given its utility, (co)end calculus should reach the status of a well-understood and standardised tool for category theorists, so that all its users, no matter their background or purpose, can profit from its conceptual simplicity.

Remark 7.2.1. We set aside a rather important question, that of *model dependence*: the models used to study the theory of $(\infty, 1)$-categories form a complicated web of equivalences. Uniqueness results [BSP11, Toë05] or even synthetic approaches [RV15, RV17a, RV17b] are nowadays very trendy as they offer extremely powerful inter-model comparison theorems, but things can become rather hairy in explicitly proving that (say) a homotopy (co)end corresponds to the same notion of an ∞-(co)end, when we move from one model to another.

This is a subtle issue, derailing us from our primary objective. We thus choose to bluntly put it aside, but we maintain at least an agnostic point of view towards the matter: as no model is privileged, we shall glance at them all.

7.2.1 Coends in Model Categories

A *model category* is a category $\mathcal{C}$ having all small (co)limits (or, in the original definition of Quillen, all finite (co)limits) that is endowed with three distinguished classes of morphisms, the *weak equivalences*, the *fibrations* and the *cofibrations*, such that the following properties are satisfied.

- The initial category $\mathcal{C}[\text{WK}^{-1}]$ with a functor $\mathcal{C} \to \mathcal{C}[\text{WK}^{-1}]$ where all weak equivalences become invertible admits a presentation as a category with the same objects, which is a hom-wise quotient of $\mathcal{C}$. More precisely, there is an equivalence relation R_{XY} on each $\mathcal{C}(X,Y)$ such that $\mathcal{C}[\text{WK}^{-1}](X,Y) \cong \mathcal{C}(X,Y)/R_{XY}$; we call $\mathcal{C}[\text{WK}^{-1}]$ the *homotopy category* of $\mathcal{C}$ with respect to WK.
- The pairs (WK ∩ COF, FIB) and (COF, WK ∩ FIB) form two *weak factorisation systems* on $\mathcal{C}$ (simply put, this means that there is a way to factor every $f :\to Y$ in $\mathcal{C}$ as a composition $X \xrightarrow{\text{COF}} E \xrightarrow{\text{WK}\cap\text{FIB}} Y$ and as a composition $X \xrightarrow{\text{WK}\cap\text{COF}} E' \xrightarrow{\text{FIB}} Y$).

Definition 7.2.2. Let $\mathcal{C}$ be a model category; we say that an object $A \in \mathcal{C}$ is *cofibrant* if its initial arrow $\varnothing \to A$ is a cofibration. Dually, we say that an object X is *fibrant* if its terminal arrow $X \to *$ is a fibration.

One of the most important parts of model category theory is the study of *homotopy (co)limits*. In short, the vastness of homotopy theory and homological algebra arises from the fact that the colimit functor colim is rather ill behaved in terms of its interaction with weak equivalences. The story goes as follows.

It turns out that many common functors between model categories do not send weak equivalences to weak equivalences, so their behaviour must be corrected, either by restricting their domain to subcategories of suitably nice objects, or by changing the shape of the functors themselves. An illustrative example of such a common but ill-behaved functor is the colimit functor $\text{colim}_{\mathcal{J}} : \text{Cat}(\mathcal{J},\mathcal{C}) \to \mathcal{C}$ over a diagram of shape $\mathcal{J}$.

Remark 7.2.3. Given a class of weak equivalences $\mathcal{W}$, every diagram category $\text{Cat}(\mathcal{J},\mathcal{C})$ acquires an analogous structure $\text{Cat}(\mathcal{J},\mathcal{W})$, where $\eta : F \Rightarrow G$ is in $\text{Cat}(\mathcal{J},\mathcal{W})$ if and only if each component $\eta_j : Fj \to Gj$ is itself a weak equivalence.

It turns out that the image of such a natural transformation $\eta : F \Rightarrow G$ under the colimit functor, $\text{colim}\,\eta : \text{colim}\,F \to \text{colim}\,G$ is rarely a weak equivalence in $\mathcal{C}$.[2]

[2] A minimal example of this goes as follows: take $\mathcal{J}$ to be the generic span $1 \leftarrow 0 \to 2$ and $F : \mathcal{J} \to \text{Spc}$ the functor sending it to $\{*\} \leftarrow S^{n-1} \to \{*\}$ (S^k is the k-dimensional sphere); the colimit of F is the one-point space $\{*\}$. We can replace F with the diagram $\tilde{F} : D^2 \leftarrow S^{n-1} \to D^2$, and since disks are contractible there is a homotopy equivalence $\tilde{F} \Rightarrow F$; unfortunately, the induced

One of the main tenets of homotopy theory is, nevertheless, that it does not matter if we replace an object of a model category with another, as soon as the two become (controllably) isomorphic in the homotopy category. There is a spark of hope, then, that the category of functors $\mathrm{Cat}(\mathcal{J},\mathcal{C})$ contains a better behaved representative for the functor colim, and that the two are linked by some sort of natural weak equivalence.

That is what homotopy colimits are: they provide *deformations* $\mathrm{hcolim}_{\mathcal{J}}$ of $\mathrm{colim}_{\mathcal{J}}$ that preserve the natural choice of weak equivalences on a functor category (see 7.2.3), and are linked by object-wise weak equivalence hcolim $\Rightarrow$ colim.

Remark 7.2.4. Let $\boxtimes : \mathcal{A}\times\mathcal{B} \to \mathcal{C}$ be the 'tensor' part of a THC situation (see Exercise 3.6), and let us assume that it is a left Quillen functor (see [Hov99, 1.3.1]); let $\mathcal{J}$ be a Reedy category (see [Hov99, 5.2.1]). Then the coend functor

$$\int^{\mathcal{J}} : \mathrm{Cat}(\mathcal{J}^{\mathrm{op}},\mathcal{A}) \times \mathrm{Cat}(\mathcal{J},\mathcal{B}) \to \mathcal{C} \tag{7.44}$$

is a left Quillen bifunctor if we regard the functor categories $\mathrm{Cat}(\mathcal{J}^{\mathrm{op}},\mathcal{A})$ and $\mathrm{Cat}(\mathcal{J},\mathcal{B})$ to be endowed with the Reedy model structure.

The coend functor remains left Quillen even if $\mathcal{J}$ is not Reedy, but we have to assume the categories $\mathcal{A},\mathcal{B},\mathcal{C}$ are all combinatorial,[3] and we put the projective model structure on $\mathrm{Cat}(\mathcal{J},\mathcal{B})$, and the injective model structure on $\mathrm{Cat}(\mathcal{J}^{\mathrm{op}},\mathcal{A})$.[4]

In the following, $\mathcal{C}$ will be a small category, and $F : \mathcal{C} \to \mathcal{D}$, $G : \mathcal{C}^{\mathrm{op}} \to \mathrm{Set}$ will be functors; when needed, we freely employ the notation of Chapter 4 on weighted (co)limits. The category $\mathcal{D}$ will admit the (co)limits allowing the object we define to exist.

Definition 7.2.5 (Bar and cobar complexes).

- The bar complex of a pair of functors F, G is the simplicial object in

arrow $\mathrm{colim}\,\tilde{F} = S^2 \to *$ is not a weak equivalence (because S^2 is not contractible).

3 A model category is *combinatorial* if it is locally presentable [AR94, 1.17] and cofibrantly generated, i.e. if the cofibrations are generated as the weak orthogonal (see A.9.3) of a small set I.

4 It is not worth entering into detail, but see [Hir03, Hov99] for more information: the *projective* model structure on $\mathrm{Cat}(\mathcal{J},\mathcal{C})$ is determined by the fibrations of $\mathcal{C}$, and the *injective* model structure is determined by cofibrations of $\mathcal{C}$, in a suitable sense.

$\mathcal{D}$ $B(G, \mathcal{C}, F)_\bullet$ whose set of n-simplices is

$$B(G, \mathcal{C}, F)_n := \coprod_{C_0,\dots,C_n} (N(\mathcal{C})_n \times GC_n) \otimes FC_0 \tag{7.45}$$

where $\otimes$: Set $\times \mathcal{D} \to \mathcal{D}$ is the tensor functor of 2.2.3. More explicitly, $B(G, \mathcal{C}, F)_n$ is the disjoint union over $(n+1)$-tuples of objects of $\mathcal{C}$ of sets

$$(GC_n \times \mathcal{C}(C_0, C_1) \times \cdots \times \mathcal{C}(C_{n-1}, C_n)) \otimes FC_n. \tag{7.46}$$

- Dually, the cobar complex of a pair of functors F, G is the cosimplicial object in $\mathcal{D}$ $C(G, \mathcal{C}, F)^\bullet$ whose set of n-simplices is

$$C^n(G, \mathcal{C}, F) := \prod_{C_0,\dots,C_n} (N(\mathcal{C})_n \times GC_n) \pitchfork FC_0 \tag{7.47}$$

where $\pitchfork$: Set$^{\text{op}} \times \mathcal{D} \to \mathcal{D}$ is the cotensor functor of 2.2.3. More explicitly, $C(G, \mathcal{C}, F)^n$ is the product over $(n+1)$-tuples of objects of $\mathcal{C}$ of sets

$$(GC_n \times \mathcal{C}(C_0, C_1) \times \cdots \times \mathcal{C}(C_{n-1}, C_n)) \pitchfork FC_n. \tag{7.48}$$

We leave as Exercise 7.4 the check of some elementary properties of these objects; in particular, we will constantly exploit the fact that $B(G, \mathcal{C}, F)$ and $C(G, \mathcal{C}, F)$ are functorial in F, G (with appropriate variance).

The proof of the following statement is conducted in the cited reference. The bar and cobar constructions allow the computation of every (weighted) (co)limit to be reduced to the computation of a certain (weighted) (co)limit over the simplex category $\mathbf{\Delta}$ (these colimits and limits are called *diagonalisation* and *totalisation* respectively, and will return in 7.2.11 and 7.2.12 to define simplicially coherent (co)ends).

Theorem 7.2.6. [Rie14, 4.4.2] Let $F : \mathcal{J} \to \mathcal{D}$ be a functor, and $W : \mathcal{J} \to$ Set be a weight (covariant, if we compute a limit; contravariant, if we compute a colimit in the formula below). Then we have canonical isomorphisms

$$\begin{aligned} \lim^W F &\cong \operatorname{colim}_{\mathbf{\Delta}} B(G, \mathcal{J}, F)_\bullet \\ &\cong \int^{n\in\mathbf{\Delta}} B(G, \mathcal{J}, F)^n \end{aligned} \tag{7.49}$$

$$\begin{aligned} \operatorname{colim}^W F &\cong \lim_{\mathbf{\Delta}} C(G, \mathcal{J}, F)^\bullet \\ &\cong \int_{n\in\mathbf{\Delta}} C(G, \mathcal{J}, F)^n. \end{aligned} \tag{7.50}$$

The same result holds (and it is much more interesting) over most bases of enrichment (but not all: the proof relies on the presence of an *enriched Grothendieck construction*); the most important instance of such an enrichment base is that of simplicial sets.

Moreover, the bar and cobar constructions provide *replacement* functors for colim and lim respectively, and as a consequence they provide models for the homotopy colimit and limit functors: in the following, if $F : \mathcal{J} \to \mathcal{D}$ is a functor and $\mathcal{D}$ is a simplicial model category, we denote $B(\mathcal{J}, \mathcal{J}, F)_\bullet$ the simplicial set $[n] \mapsto B(よ J, \mathcal{J}, F)_n$, and dually $C(\mathcal{J}, \mathcal{J}, F)^n$ is the cosimplicial set $[n] \mapsto C(よ^\vee J, \mathcal{J}, F)^n$. (The statement found in [Rie14] is more general than the one we introduce here; its essence is, however, unchanged.)

Theorem 7.2.7. [Rie14, 5.1.3] The left derived functor of the colimit functor colim : $\mathrm{Cat}(\mathcal{J}, \mathcal{D}) \to \mathcal{D}$ and the right derived functor of lim : $\mathrm{Cat}(\mathcal{J}, \mathcal{D}) \to \mathcal{D}$ are computed respectively as

$$\mathbb{L}\operatorname{colim} F \cong \mathsf{d}\big(B(\mathcal{J}, \mathcal{J}, \tilde{F})\big) \qquad \mathbb{R}\lim F \cong \mathsf{t}\big(C(\mathcal{J}, \mathcal{J}, \hat{F})\big), \tag{7.51}$$

where $\tilde{F}$ is a cofibrant replacement, and $\hat{F}$ is a fibrant replacement for the diagram F, in suitable model structures on the diagram category $\mathrm{Cat}(\mathcal{J}, \mathcal{D})$ i.e. as the (co)ends

$$\begin{aligned} \mathbb{L}\operatorname{colim} F &\cong \int^n \mathbf{\Delta}(\_, [n]) \otimes B(\Delta[n], \mathcal{J}, \tilde{F}) \\ \mathbb{R}\lim F &\cong \int_n \mathbf{\Delta}(\_, [n]) \pitchfork C(\Delta[n], \mathcal{J}, \hat{F}) \end{aligned} \tag{7.52}$$

The other way to compute homotopy (co)limits involves a resolution of the weight: the equivalence between the two approaches was first proved in [Gam10].

Recall 4.2.5: we can express the Bousfield–Kan construction of [BK72] for the homotopy (co)limit functor using (co)end calculus.

Theorem 7.2.8. Let $F \colon \mathcal{J} \to \mathcal{C}$ be a diagram in a model category $(\mathcal{C}, \text{WK}, \text{COF}, \text{FIB})$ which is tensored and cotensored (see 2.2.3) over simplicial sets. Then the homotopy limit hlim F of F can be computed as the end

$$\int_J N(\mathcal{J}/J) \pitchfork F(J), \tag{7.53}$$

where N is the categorical nerve of 3.2.5; in the same notation, the homotopy colimit hcolim F of F can be computed as the coend

$$\int^{J} N(\mathcal{J}/J) \otimes F(J). \tag{7.54}$$

Remark 7.2.9. These two universal objects are the weighted (co)limit of F with the nerve of a slice category as weight; the idea behind this characterisation is that colim is a weighted colimit over the terminal weight. The problem is that usually the constant terminal weight will not be a cofibrant object in $\mathrm{Cat}(\mathcal{J}, \mathrm{sSet})$; thus, when we want to pass to the homotopically correct version of colim we shall replace the weight W with a homotopy equivalent, but cofibrant, diagram $\tilde{W}$.

The Bousfield–Kan formula arises precisely in this process: $N(J/\mathcal{J})$ and $N(\mathcal{J}/J)$ are both contractible categories, and they are linked to $N(*)$ (the nerve of the terminal category) by a homotopy equivalence induced by the terminal functor. The simplicial presheaves $N(\mathcal{J}/\_), N(\_/\mathcal{J})$ can thus be thought as proper *replacements* for the terminal functor.

7.2.2 Simplicially Coherent (Co)ends

All the material in the present subsection comes from [CP97]. We begin the exposition by establishing a convenient notation and a series of useful shorthands to adapt the discussion to our choice of notation.

We strive to keep this introduction equally self-contained and simple, but the reader shall be warned of two things.

- There are a number of sheerly unavoidable sins of omission in the exposition, essentially due to our inability to master the topic in its entirety; moreover, the price we pay to obtain a self-contained exposition is that we deliberately ignore most of the subtleties of the combinatorics of simplicial sets. The blame is on us if the reader feels that our exposition is flawed or incomplete.
- Since [CP97] was published, newer and more systematic approaches to these topics have been developed; among many, the reader should take the exceptionally clear [Rie14, Shu06]. These references reduce the construction of a simplicially coherent (co)end to an application of the 'unreasonably effective' (co)bar construction [Rie14, 4], and it can be proved using [Shu06, 21.4] that the coherent (co)end of T (see 7.2.15) results as a suitable *derived weighted (co)limit* of the functor T. This last remark is important, in particular in light of the fact that it can be exported to define a homotopy coherent (co)end calculus in

every 2-category $\mathcal{V}$-Cat of categories enriched over a monoidal model category (see [Hov99, Ch. 4] and [BM13]).[5]

It is our sincere hope that this does not affect the outreach of this elegant and neglected piece of mathematics, and our clumsy attempt to popularise an account of [CP97] has to be seen as a reverent act of outreach of the branch of algebraic topology called *categorical homotopy theory*.

Notation 7.2.10. All categories $\mathcal{A}, \mathcal{B}, \dots$ appearing in the present subsection are enriched over the category $\mathrm{sSet} = \mathrm{Cat}(\mathbf{\Delta}^{\mathrm{op}}, \mathrm{Set})$. All such categories possess the (co)tensors (see 2.2.3) needed to state definitions and perform computations.

Moreover, the tensor, internal hom, and cotensor functors assemble into a THC situation (see Exercise 3.6) $(\otimes, \hom, \pitchfork)$ where $\otimes : \mathrm{sSet} \times \mathcal{A} \to \mathcal{B}$ determines the variance of the other two functors. A useful shorthand to denote the functor $\pitchfork(K, A) = K \pitchfork A$ is A^K. We feel free to employ such exponential notation when needed (especially when it is necessary to save space or invoke its behaviour, similar to that of an internal hom).

Let $\mathcal{B}$ be a simplicially enriched category. A *simplicial-cosimplicial object* in $\mathcal{B}$ is a functor $Y : \mathbf{\Delta}^{\mathrm{op}} \times \mathbf{\Delta} \to \mathcal{B}$.

Definition 7.2.11 (Totalisation). Given such a simplicial-cosimplicial object, we define its *totalisation* $\mathsf{t}(Y)$ as the end

$$\int_{n \in \mathbf{\Delta}} \mathbf{\Delta}[n] \pitchfork Y_n \tag{7.55}$$

(note that it is a cosimplicial object $m \mapsto \int_{n \in \mathbf{\Delta}} \mathbf{\Delta}[n] \pitchfork Y_m^n$). The totalisation of Y is also denoted with the shorthand $\mathbf{\Delta}^\bullet \pitchfork Y$ or similar.

Dually, a *bisimplicial object* in $\mathcal{B}$ is a functor $X : \mathbf{\Delta}^{\mathrm{op}} \times \mathbf{\Delta}^{\mathrm{op}} \to \mathcal{B}$.

Definition 7.2.12 (Diagonalisation). Given a bisimplicial object $X : \mathbf{\Delta}^{\mathrm{op}} \times \mathbf{\Delta}^{\mathrm{op}} \to \mathcal{B}$ we define the *diagonalisation* $\mathsf{d}(X)$ of X to be the coend

$$\int^{n \in \mathbf{\Delta}} \mathbf{\Delta}[n] \otimes X_n \tag{7.56}$$

[5] This remark deserves expansion, and yet *hanc marginis exiguitas non caperet*: in short, the category $\mathcal{V}\text{-Cat}(\mathcal{C}^{\mathrm{op}} \times \mathcal{D}, \mathcal{V})$ can be endowed with a model structure that allows the computation of cofibrant resolutions $\delta \hom$ for the identity profunctor $\hom_{\mathcal{C}}$ when $\mathcal{C} = \mathcal{D}$; the coherent (co)end of an endoprofunctor $T : \mathcal{C} \rightsquigarrow \mathcal{C}$ is then computed as the homotopy (co)limit of T weighted by the resolved weight $\delta \hom$, in the exact same way as the incoherent (co)end is the (co)limit weighted by hom (see 4.1.8). At the time of writing, this appears to be an interesting subject of further investigation.

(note that it is a simplicial object $m \mapsto \int^{n\in\mathbf{\Delta}} \mathbf{\Delta}[n] \otimes X_n^m$). The diagonalisation of X is also denoted with the shorthand $\mathbf{\Delta}^\bullet \otimes X$ or similar.

Note that as a consequence of the ninja Yoneda lemma 2.2.1, the diagonalisation $\mathbf{\Delta}^\bullet \otimes X$ is the simplicial object $n \mapsto X_n^n$ (often denoted simply X_{nn}).

Notation 7.2.13 (Chain (co)product). Let $\mathcal{A} \in \mathrm{Cat}_{\mathbf{\Delta}}$; we shall denote as $\vec{X}_n = (X_0, \dots, X_n)$ the 'generic n-tuple of objects' in $\mathcal{A}$; given two other objects $A, B \in \mathcal{A}$, we define a bisimplicial set $\amalg\mathcal{A}[A|\vec{X}_\bullet|B]_\bullet$ whose simplicial set of n-simplices is[6]

$$\amalg\mathcal{A}[A|\vec{X}_n|B]_\bullet := \coprod_{X_0,\dots,X_n\in\mathcal{A}} \mathcal{A}(A, X_0)_\bullet \times \mathcal{A}(X_0, X_1)_\bullet \times \cdots \times \mathcal{A}(X_n, B)_\bullet. \tag{7.57}$$

Faces and degeneracies are induced by composition and identity-insertion respectively (see Exercise 7.5).

Finally we define the simplicial set $\delta\mathcal{A}(A, B)$ to be the diagonalisation

$$\mathsf{d}(\amalg\mathcal{A}[A|\vec{X}_\bullet|B]_\bullet) = n \mapsto \amalg\mathcal{A}[A|\vec{X}_n|B]_n. \tag{7.58}$$

Couched as a coend, the object $\delta\mathcal{A}(A, B)$ is written

$$\begin{aligned}
\delta\mathcal{A}(A, B) &\cong \int^{n\in\mathbf{\Delta}} \mathbf{\Delta}[n] \times \amalg\mathcal{A}[A|\vec{X}_n|B] \\
&= \int^{n\in\mathbf{\Delta}} \mathbf{\Delta}[n] \times \coprod_{\vec{X}_\bullet} \mathcal{A}(A, X_0) \times \mathcal{A}(X_0, X_1) \times \cdots \times \mathcal{A}(X_n, B) \\
&\cong \int^{n\in\mathbf{\Delta}} \coprod_{\vec{X}_\bullet} \mathbf{\Delta}[n] \times \mathcal{A}(A, X_0) \times \mathcal{A}(X_0, X_1) \times \cdots \times \mathcal{A}(X_n, B).
\end{aligned}$$

Example 7.2.14. If we consider $\mathcal{A}$ to be trivially enriched over sSet (i.e. as a *discrete* simplicial category, where each $\mathcal{A}(A, A')$ is a discrete simplicial set), then the object $\mathcal{A}[A|\vec{X}_n|B]_\bullet$ coincides with the nerve of the 'double slice' category $A/\mathcal{A}/B$ of arrows 'under A and above B'.

Definition 7.2.15 (Simplicially coherent (co)end). Let $T : \mathcal{A}^{\mathrm{op}} \times \mathcal{A} \to \mathcal{B}$ be an sSet-functor. We define

$$\oint_A T(A, A) := \int_{A', A''} \delta\mathcal{A}(A', A'') \pitchfork T(A', A'')$$

[6] It is useful to extend this notation in a straightforward way: $\mathcal{A}[A|\vec{X}|B]$ denotes the product $\mathcal{A}(A, X_0) \times \mathcal{A}(X_0, X_1) \times \cdots \times \mathcal{A}(X_n, B)$, and $\amalg\mathcal{A}[A|\vec{X}|B]$, $\mathcal{A}[\vec{X}]$, $\amalg\mathcal{A}[\vec{X}]$, $\amalg\mathcal{A}[\vec{X}]$ are defined similarly. Note that $\amalg\mathcal{A}[A|\vec{X}_n|B]_\bullet$ does not depend on $\vec{X}_n$ since the coproduct is quantified over all such $\vec{X}_n$.

$$\oint^{A} T(A,A) := \int^{A',A''} \delta\mathcal{A}(A',A'') \otimes T(A',A'') \tag{7.59}$$

to be respectively the *simplicially coherent end* and *coend* of T.

Expanding these definitions, we see that

$$\oint_A T(A,A) \cong \int_{A',A'',n} \mathbf{\Delta}[n] \times \amalg\mathcal{A}[A|\vec{X}_n|B] \pitchfork T(A',A'')$$
$$\oint^A T(A,A) \cong \int^{A',A'',n} \mathbf{\Delta}[n] \times \amalg\mathcal{A}[A|\vec{X}_n|B] \otimes T(A',A'').$$

(At this point, the reader will surely understand why performing even elementary computations with these sorts of objects compels us to establish a compact notation.)

Remark 7.2.16. In a few words, the definition of a simplicially coherent (co)end mimics the classical construction, in particular the characterisation of a (co)end as a hom weighted (co)limit (see 4.1.8 and 4.1.13.WC3), replacing hom with (co)tensors for a 'fattened up' mapping space $\mathcal{A}[A|\vec{X}|B]$. The 'deformation' perspective is very useful, since we write that $\oint_A T$ corresponds to the end

$$\int_{(A',A'')\in\mathcal{A}^{\mathrm{op}}\times\mathcal{A}} \hom(A',A'') \pitchfork T(A',A'') \tag{7.60}$$

where we applied a suitable 'deformation' (or 'resolution', or 'replacement') functor δ to the hom functor $\mathcal{A}(\_,=)$, seen as the identity profunctor (Remark 5.1.8). This point of view is rather fruitfully explored in [Gen15] in the particular case of dg-Cat (see subsection 7.2.2 below to draw a connection between simplicially enriched and dg-categories, and for a precise definition of dg-category).

This perspective is of great importance to encompass coherent (co)ends into a general theory 'compatible' with a model structure on $\mathcal{V}$-Cat, for some monoidal model $\mathcal{V}$ and the Bousfield–Kan model structure on $\mathcal{V}$-Cat.

We now embark on the study of what we should call *simplicially coherent coend calculus.* Classical (co)end calculus consists of the triptych Fubini - Yoneda - Kan; we shall now reproduce these steps in full detail.

The authors of [CP97] succeed in the quite ambitious task of rewriting the most important pieces of classical category theory in this coherent model describing (co)limits, mapping spaces, the Yoneda lemma, and Kan extensions. The aim of the rest of this subsection is to sketch some of these original definitions, hopefully helping one of the alternative approaches to $(\infty, 1)$-category theory to escape oblivion.

Definition 7.2.17 (The functors $\boldsymbol{Y}$ and $\boldsymbol{W}$). Let $T : \mathcal{A}^{\mathrm{op}} \times \mathcal{A} \to \mathcal{B}$ be a functor; we define $\boldsymbol{Y}(T)^\bullet$ to be the cosimplicial object (in $\mathcal{B}$)

$$\boldsymbol{Y}(T)^n := \prod_{\vec{X}_n=(X_0,\dots,X_n)} \mathcal{A}[\vec{X}_n] \pitchfork T(X_0, X_n), \tag{7.61}$$

where $\mathcal{A}[\vec{X}_n] = \mathcal{A}(X_0, X_1) \times \cdots \times \mathcal{A}(X_{n-1}, X_n)$.

Dually, given the same T, we define $\boldsymbol{W}(T)_n$ to be the simplicial object (in $\mathcal{B}$)

$$\boldsymbol{W}(T)_n := \coprod_{\vec{X}_n=(X_0,\dots,X_n)} \mathcal{A}[\vec{X}_n] \otimes T(X_0, X_n). \tag{7.62}$$

Proposition 7.2.18. Let $T : \mathcal{A}^{\mathrm{op}} \times \mathcal{A} \to \mathcal{B}$ be an sSet-functor. Then there is a canonical isomorphism

$$\oint_A T(A, A) \cong \mathsf{t}(\boldsymbol{Y}(T)^\bullet). \tag{7.63}$$

Proof We make heavy use of the ninja Yoneda lemma 2.2.1 in its enriched form, i.e. that given an sSet-functor $F : \mathcal{A} \to \mathrm{sSet}$ we have a canonical isomorphism

$$\int_X \mathcal{A}(X, B) \pitchfork F(X) \cong F(B) \tag{7.64}$$

and that $K \pitchfork (H \pitchfork A) \cong (K \otimes H) \pitchfork A$, naturally in all arguments.

Now, let $\vec{X}_n = (X_0, \dots, X_n)$ be a generic tuple of objects of $\mathcal{A}$. To save some space we employ the exponential notation A^K to denote the cotensor $K \pitchfork A$, and compute

$$\begin{aligned}
\oint_A T(A, A) &:= \int_{A',A''} T(A', A'')^{\delta\mathcal{A}(A',A'')} \\
&\cong \int_{A',A''} T(A', A'')^{\int_n \Delta[n] \times \amalg \mathcal{A}[A'|\vec{X}|A'']} \\
&\cong \int_{A',A'',n} T(A', A'')^{\Delta[n] \times \amalg \mathcal{A}[A'|\vec{X}|A'']} \\
&\cong \int_{A',A'',n} \left(T(A', A'')^{\mathcal{A}(X_n,A'')}\right)^{\Delta[n] \times \amalg \mathcal{A}[A'|\vec{X}]} \\
&\cong \int_{A',A'',n} \prod_{X_0,\dots,X_n} \left(T(A', A'')^{\mathcal{A}(X_n,A'')}\right)^{\Delta[n] \times \mathcal{A}[A'|\vec{X}]}
\end{aligned}$$

$$\cong \int_{A',n} \prod_{X_0,\dots,X_n} \left(\int_{A''} T(A',A'')^{\mathcal{A}(X_n,A'')} \right)^{\Delta[n]\times \mathcal{A}[A'|\vec{X}]}$$

$$\cong \int_{A',n} \prod_{X_0,\dots,X_n} T(A',X_n)^{\Delta[n]\times \mathcal{A}[A'|\vec{X}]}$$

$$\cong \int_{n} \prod_{X_0,\dots,X_n} \left(\int_{A'} T(A',X_n)^{\mathcal{A}(A',X_0)} \right)^{\Delta[n]\times \mathcal{A}[\vec{X}]}$$

$$\cong \int_{n} \Big(\prod_{X_0,\dots,X_n} T(X_0,X_n)^{\mathcal{A}[\vec{X}_n]} \Big)^{\Delta[n]} \cong \mathsf{t}(\boldsymbol{Y}(T)).$$

This concludes the proof. □

For the sake of completeness, we notice that the universal wedge testifying that $\oint_A T(A,A) \cong \mathsf{t}(\boldsymbol{Y}(T))$ is induced by the morphisms

$$\begin{array}{c} \oint_A T(A,A) = \displaystyle\int_{A',A''} \delta\mathcal{A}(A',A'') \pitchfork T(A',A'') \\ \downarrow\wr \\ \displaystyle\int_{A',A'',n} (\Delta[n] \times \amalg \mathcal{A}[A'|\vec{X}|A'']) \pitchfork T(A',A'') \\ \downarrow \\ \Delta[n] \pitchfork \boldsymbol{Y}(T)^n. \end{array}$$

The reader is invited to prove the dual statement as an exercise (it is of vital importance that you establish a good notation).

Proposition 7.2.19. Let $T : \mathcal{A}^{\mathrm{op}} \times \mathcal{A} \to \mathcal{B}$ be an sSet-functor. Then there is a canonical isomorphism

$$\oint^A T(A,A) \cong \mathsf{d}(\boldsymbol{W}(T)_\bullet). \tag{7.65}$$

Remark 7.2.20. The homotopy coherent (co)ends admit 'comparison' maps to the classical (co)ends, induced by the fact that the 'fattened hom' $\delta\mathcal{A}(-,=)$ has canonical maps to/from the plain hom $\mathcal{A}(-,=)$; this is an example of a general tenet, where homotopically correct objects result as a *deformation* of classical ones, and the deformations maps in/out of the plain object.

The comparison map $\oint T(A,A) \to \int T(A,A)$ arises, here, as a homotopy equivalence between the simplicial set $\mathcal{A}(A,B)$ (seen as bisimplicial, and constant in one direction) and the bisimplicial set $\delta\mathcal{A}(A,B) = \mathsf{d}\mathcal{A}[A|\bullet|B]_\bullet$. This is [CP97, p. 15].

Note that it is possible to write an explicit contracting homotopy between the two objects. The map

$$d_0 : \coprod_{X_0} \mathcal{A}(A, X_0) \times \mathcal{A}(X_0, B) \to \mathcal{A}(A, B) \quad (7.66)$$

given by composition has a homotopy inverse given by

$$s_{-1} : \mathcal{A}(A, B) \to \mathcal{A}(A, A) \times \mathcal{A}(A, B) : g \mapsto (\mathrm{id}_A, g). \quad (7.67)$$

Indeed, the composition $d_0 s_{-1}$ is the identity on $\mathcal{A}(A, B)$, whereas the composition $s_{-1} d_0$ is homotopic to the identity on $\delta\mathcal{A}(A, B)$ (we use the same name for the maps d_0, s_{-1} and the induced maps $\bar{d}_0 : \delta\mathcal{A}(A, B) \to \mathcal{A}(A, B)$, induced by the universal property, and $\bar{s}_{-1} : \mathcal{A}(A, B) \to \delta\mathcal{A}(A, B)$).

There is an important difference between these two maps, though: while d_0 is natural in both arguments, s_1 is natural in B but not in A. This has an immediate drawback: while d_0 can be obtained canonically, as the universal arrow associated to a certain natural isomorphism (see (7.71) below), s_{-1} cannot (the best we can do is to characterise the natural argument of s_{-1} via [CP97, Example 2, p. 16]).

Simplicially Coherent Natural Transformations

As we have seen in 1.4.1, the set of natural transformations between two functors $F, G : \mathcal{C} \to \mathcal{D}$ coincides with the end $\int_X \mathcal{D}(FX, GX)$, and (see 7.1.9) the category of lax natural transformations between two 2-functors coincides with the lax end $\oint_X \mathcal{D}(FX, GX)$. It comes as no surprise, then, that the following characterisation of *homotopy coherent* natural transformations between two simplicial functors hold.

Definition 7.2.21 (Coherent natural transformations). Let $F, G : \mathcal{C} \to \mathcal{D}$ be two simplicial functors; then the simplicial set of *coherent transformations* between F and G is defined as

$$\mathrm{Cat}_{\mathbf{\Delta}}(\mathcal{C}, \mathcal{D})(\!(F, G)\!) := \oint_A \mathcal{D}(FA, GA). \quad (7.68)$$

We define the following operations of *coherent tensoring* and *cotensoring* a simplicial functor.

Definition 7.2.22 (Mean tensor and cotensor). Let $F : \mathcal{A} \to \mathcal{B}$, $G : \mathcal{A} \to \mathrm{sSet}$, $H : \mathcal{A}^{\mathrm{op}} \to \mathrm{sSet}$. We define $G \,\overline{\pitchfork}\, F$, $H \,\underline{\otimes}\, F$ respectively as

$$G \,\overline{\pitchfork}\, F := \oint_A GA \pitchfork FA, \qquad H \,\underline{\otimes}\, F := \oint^A HA \otimes FA. \quad (7.69)$$

This yields the notion of *standard resolution* of a simplicial functor.

Definition 7.2.23 (Standard resolutions). Let $F : \mathcal{A} \to \mathcal{B}$ be a simplicial functor; we define

$$\overline{F}A := \mathcal{A}(A, \_) \,\overline{\pitchfork}\, F = \oint_X \mathcal{A}(A, X) \pitchfork FX$$

$$\underline{F}A := \mathcal{A}(\_, A) \,\underline{\otimes}\, F = \oint^X \mathcal{A}(X, A) \otimes FX.$$

Example 7.2.24. We specialise the above definition to compute the functors $\overline{\mathrm{hom}}(A, \_)$ and $\underline{\mathrm{hom}}(A, \_)$. We concentrate in particular on the second case, since the first is completely dual:

$$\begin{aligned}\underline{\mathrm{hom}}(A, B) &= \oint^A \mathcal{A}(A, X) \times \mathcal{A}(X, B)\\ &\cong \int^{XY} \delta\mathcal{A}(X, Y) \times \mathcal{A}(A, X) \times \mathcal{A}(Y, B)\\ &\cong \int^{XY} \mathcal{A}(A, X) \times \delta\mathcal{A}(X, Y) \times \mathcal{A}(Y, B)\\ &\cong \int^{XYn} \mathcal{A}[A|\tilde{X}_n|B] \times \mathbf{\Delta}[n] \cong \delta\mathcal{A}(A, B).\end{aligned}$$

We leave the proof of the following result as an easy exercise in (co)end calculus (see Exercise 7.7), which shows that the standard resolutions $\underline{F}, \overline{F}$ of F 'absorb the coherence'.

Proposition 7.2.25. Let $F, G : \mathcal{C} \to \mathcal{D}$ be two simplicial functors; then there are canonical isomorphisms

$$[F, \overline{G}] \cong \mathrm{Cat}_{\mathbf{\Delta}}(\mathcal{C}, \mathcal{D})((F, G)) \cong [\underline{F}, G]. \tag{7.70}$$

This result has a number of pleasant consequences: the simplicially coherent setting is powerful enough to retrieve several classical constructions.

- Example 7.2.24 above shows that $\underline{\mathrm{hom}}(A, \_)(b) \cong \delta\mathcal{A}(A, B)$; this entails that there is an isomorphism

$$\mathrm{Cat}_{\mathbf{\Delta}}(\mathcal{A}, \mathrm{sSet})(\delta\mathcal{A}(A, \_), \mathcal{A}(A, =)) \cong \mathrm{Cat}_{\mathbf{\Delta}}(\mathcal{A}, \mathrm{sSet})((\mathcal{A}(A, \_), \mathcal{A}(A, =))) \tag{7.71}$$

  and it is a matter of verifying some straightforward identities to see that the sSet-natural transformation corresponding to the identity coherent transformation is precisely d_0.
- The map d_0 defines additional universal maps η_F, η^F which 'resolve' a functor $F : \mathcal{A} \to \mathcal{B}$ whenever $\underline{F}, \overline{F}$ exist (it is sufficient that $\mathcal{B}$ admits

all the relevant (co)limits to perform the construction of $\underline{F}, \overline{F}$). From the chain of isomorphisms

$$\begin{aligned}
\eta^F : \overline{F}B &= \oint_A \mathcal{A}(B, A) \pitchfork FA \\
&\cong \int_{A',A''} \delta\mathcal{A}(A', A'') \pitchfork \mathcal{A}(B, A'') \pitchfork FA' \\
&\leftarrow \int_{A',A''} \mathcal{A}(A', A'') \pitchfork \mathcal{A}(B, A'') \pitchfork FA' \\
(2.2.1) &\cong FB, \\
\eta_F : \underline{F}B &= \oint^A FA \otimes \mathcal{A}(A, B) \\
&\cong \int^{A',A''} FA' \otimes \mathcal{A}(A'', B)\delta\mathcal{A}(A', A'') \\
&\to \int^{A',A''} FA' \otimes \mathcal{A}(A'', B)\mathcal{A}(A', A'') \\
&\cong FB,
\end{aligned}$$

we obtain natural transformations corresponding to suitable coherent identities under the isomorphism of 7.2.25.

- The maps η_F, η^F behave like resolutions: [CP97, 3.4] shows that they are level-wise homotopy equivalences (meaning that $\eta_F : FA \to \overline{F}A$ induces homotopy equivalences of simplicial sets $\mathcal{B}(B, FA) \xrightarrow{(\eta_F)_*} \mathcal{B}(B, \overline{F}A)$ for each B, naturally in B).[7]

Simplicially Coherent Kan Extensions

The universal property of a Kan extension is inherently 2-dimensional: uniqueness is stated at the level of 2-cells, and any sensible generalisation to the higher world involves a ‘space’ of 2-cells between 1-cells.

This entails that any reasonable definition of a (left or right) Kan extension ultimately relies on a nice definition of a space of coherent natural transformations between functors, which has been the subject of the previous subsection.

There are, nevertheless, several subtleties, as there are many choices available for a definition: in the words of [CP97],

> Clearly one can replace natural transformations by *coherent* ones [in the

[7] We decide to skip the proof of this proposition, as it is quite long, technical, and even though it relies on (co)end calculus it does not add much to the present discussion.

definition of a Kan extension], but should isomorphism be replaced by homotopy equivalence, should they be natural, in which direction should this go...?

Solving this problem can be tricky; one of the reasons for this is that simplicial combinatorics captures the behaviour of $(\infty, 1)$-categories really well, whereas any satisfactory model for homotopy coherent Kan extensions should speak about $(\infty, 2)$-categories.

In order to define coherent Kan extensions for $\mathcal{B} \xleftarrow{G} \mathcal{A} \xrightarrow{F} \mathcal{C}$ we ask for the isomorphisms

$$\mathrm{Cat}_{\boldsymbol{\Delta}}(\mathcal{B}, \mathcal{C})(\!(H, \mathrm{hoRan}_G F)\!) \cong \mathrm{Cat}_{\boldsymbol{\Delta}}(\mathcal{A}, \mathcal{C})(\!(HG, F)\!) \tag{7.72}$$

$$\mathrm{Cat}_{\boldsymbol{\Delta}}(\mathcal{B}, \mathcal{C})(\!(\mathrm{hoLan}_G F, K)\!) \cong \mathrm{Cat}_{\boldsymbol{\Delta}}(\mathcal{A}, \mathcal{C})(\!(F, KG)\!) \tag{7.73}$$

to hold at the level of *coherent* transformations. This can be achieved as follows.

Definition 7.2.26 (Coherent Kan extensions). Let $F : \mathcal{A} \to \mathcal{C}$ and $G : \mathcal{A} \to \mathcal{B}$ be a span of simplicial functors; we define

$$\mathrm{hoRan}_G F(\_) = \oint_A \mathcal{B}(\_, GA) \pitchfork FA$$

$$\mathrm{hoLan}_G F(\_) = \oint^A \mathcal{B}(GA, \_) \otimes FA.$$

Proving that the isomorphisms (7.72) and (7.73) hold follows from an easy computation with the explicit form of the coherent (co)ends above. We leave the reader either to solve this Exercise 7.8, or to consult [CP97].

A Glance at dg-Coends. The Dold–Kan correspondence (see 3.2.10) establishes an equivalence of categories between $\mathrm{Ch}_{\geq}(\mathbb{Z})$, chain complexes of abelian groups concentrated in positive degree, and simplicial objects in the category of abelian groups, i.e. functors $G : \boldsymbol{\Delta}^{\mathrm{op}} \to \mathrm{Mod}(\mathbb{Z})$. The equivalence of categories is generated by a cosimplicial object

$$DK : \boldsymbol{\Delta} \to \mathrm{Ch}_{\geq}(\mathbb{Z}) \tag{7.74}$$

and the universal property of the Yoneda embedding now yields an adjunction

$$\mathrm{Lan}_{よ}(DK) : \mathrm{sSet} \leftrightarrows \mathrm{Ch}_{\geq}(\mathbb{Z}) : \mathrm{Lan}_{DK}\, よ. \tag{7.75}$$

It turns out that this is an equivalence of categories.

(This result can be restated in fair more generality, but in this paragraph we stick to the R-linear case.)

The Dold–Kan equivalence induces an equivalence of 2-categories between simplicial-abelian-group enriched categories on one side, and categories enriched in (positive) chain complexes on the other, or suitable *dg-categories* (i.e. categories enriched in chain complexes) for short.

The theory of dg-categories is deeply rooted in homological algebra and finds applications in algebraic geometry [Gen15, KL14]. In order to study nicer versions of derived categories, one can attach a *derived dg-category* $\underline{\mathbb{D}}(X)$ to a space/scheme/abelian category, and such a category is way better behaved than the 'incoherent' derived category $\mathbb{D}(X)$ (where the cone construction of 4.2.1 is not functorial).

There is of course a link between the two objects; the incoherent category can be recovered as the homotopy category of $\underline{\mathbb{D}}(X)$ as follows: each cohomology functor H^n extends degree-wise to functors $\underline{H}^n$: dg-Cat $\to$ Ab-Cat (=categories enriched on abelian groups, also called *preadditive*), and there is an isomorphism $\mathbb{D}(X) = \underline{H}^0(\underline{\mathbb{D}}(X))$.

In order to approach derived algebraic geometry with the tools of enriched category theory, it might be interesting to restrict coherent (co)end calculus to $[\mathbf{\Delta}^{\text{op}}, \text{Ab}]$-categories, and think about the results as dg-categories using the Dold–Kan equivalence.

This is an enticing application of (co)end calculus in a homotopical/homological setting, and many questions arise naturally from the expressive power of (co)end calculus. For example, if $\mathcal{A}$ is any dg-category, its identity profunctor $\mathcal{A} \rightsquigarrow \mathcal{A}$ is a functor $\mathcal{A}^{\text{op}} \boxtimes \mathcal{A} \to \text{Ch}(\mathbb{Z})$, so that the coherent end

$$\oint_A \mathcal{A}(A, A), \tag{7.76}$$

i.e. the object of derived natural transformations of the identity functor $\text{id}_{\mathcal{A}}$, recovers the *Hochschild complex* of $\mathcal{A}$. Then, if A is an associative algebra regarded as a one-object dg-category concentrated in degree zero, the object $H^n(\int_* A)$ is the *Hochschild cohomology* of A, understood in the classical sense of, say, [Pie82, Ch. 11].

We will not expand on this interesting topic, but one can find that the bicategory of dg-profunctors, and in particular of the *endoprofunctors* of a single object $\mathcal{X}$, gives rise to plenty of derived invariants of a dg-category, but we leave to the interested reader the endeavour of re-reading the paper [KL14] wearing appropriate (co)end-goggles (among many, that paper seems the most liable to a (co)end-theoretic reformulation).

The same approach can be carried over to the more general setting of a category enriched over a monoidal model category; a perfect starting point accounting for the state of the art on the matter is Shulman's article [Shu06].

7.3 (Co)ends in Quasicategories

As a rule of thumb, the translation procedure from category theory to ∞-category theory is based on the following meta-principle: first you rephrase the old definition in a 'simplicially meaningful' way, so that the ∞-categorical definition specialises to the old one for quasicategories $N(\mathcal{C})$ which arise as nerves of categories. Then you forget about the original gadget and keep the simplicial one; this turns out to be the right definition.

The first victim of this procedure is the twisted arrow category 1.2.2 of an ∞-category.

Definition 7.3.1 (Twisted arrow ∞-category)**.** Let $\varepsilon : \mathbf{\Delta} \to \mathbf{\Delta}$ be the functor $[n] \mapsto [n] \star [n]^{\mathrm{op}}$, where $\star$ is the join of simplicial sets [EP08, Joy08b], and the *opposite* of a simplicial set is defined in [Rez17, §6.19]. Let $\mathcal{C}$ be an ∞-category; the twisted arrow category $\mathrm{TW}(\mathcal{C})$ is defined to be the simplicial set $\varepsilon^*\mathcal{C}$, where $\varepsilon^* = \_ \circ \varepsilon^{\mathrm{op}} : \mathrm{sSet} \to \mathrm{sSet}$ is the induced precomposition functor. More explicitly, and consequently, the n-simplices of $\mathrm{TW}(\mathcal{C})$ are characterised by the relation

$$\mathrm{TW}(\mathcal{C})_n \cong \mathrm{sSet}(\mathbf{\Delta}[n], \mathrm{TW}(\mathcal{C})) \cong \mathrm{sSet}(\mathbf{\Delta}[n] \star \mathbf{\Delta}[n]^{\mathrm{op}}, \mathcal{C}). \tag{7.77}$$

The most important feature of the twisted arrow category is that it admits a fibration over $\mathcal{C}^{\mathrm{op}} \times \mathcal{C}$ (part of its essential properties can be deduced from this); the machinery of left and right fibrations exposed in [Lur09, 2.0.0.3] gives that

- there is a canonical simplicial map $\Sigma : \mathrm{TW}(\mathcal{C}) \to \mathcal{C}^{\mathrm{op}} \times \mathcal{C}$ (induced by the two join inclusions $\mathbf{\Delta}[\_], \mathbf{\Delta}[\_]^{\mathrm{op}} \to \mathbf{\Delta}[\_] \star \mathbf{\Delta}[\_]^{\mathrm{op}}$),
- this ∞-functor is a right fibration in the sense of [Lur09, 2.0.0.3].

Remark 7.3.2. It is rather easy to see that the above definition is reasonable: a 0-simplex in $\mathrm{TW}(\mathcal{C})$ is an edge $f : \mathbf{\Delta}[1] \to \mathcal{C}$, and a 1-simplex of $\mathrm{TW}(\mathcal{C})$ is a 3-simplex thereof, which we can depict as a pair of edges (u, v), such that the square having twisted edges

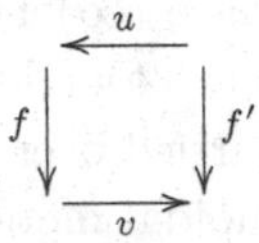

$$\begin{array}{ccc} \cdot & \xleftarrow{u} & \cdot \\ {\scriptstyle f}\downarrow & & \downarrow{\scriptstyle f'} \\ \cdot & \xrightarrow[v]{} & \cdot \end{array} \tag{7.78}$$

commutes. This suggests (as it must be) that the definition of $\text{TW}(\mathcal{C})$ for an ∞-category specialises to the 1-dimensional one and adds higher dimensional information to it.

Definition 7.3.3. Let $\mathcal{C}, \mathcal{D}$ be two ∞-categories; the ∞-*(co)end* of a simplicial map $F : \mathcal{C}^{\text{op}} \times \mathcal{C} \to \mathcal{D}$ is the (co)limit $\oint F$ of the composition

$$\text{TW}(\mathcal{C}) \xrightarrow{\Sigma} \mathcal{C}^{\text{op}} \times \mathcal{C} \xrightarrow{F} \mathcal{D}. \tag{7.79}$$

The main interest of the authors of [GHN15] is to formulate an analogue of A.5.13, which characterises the Grothendieck construction of a Cat-valued functor as a particular weighted colimit (see 4.2.2).

It is rather easy to formulate such an analogue definition: this appears as [GHN15, 2.8].

Definition 7.3.4 (op/lax colimit of F). Let $F : \mathcal{C} \to \text{Cat}_\infty$ be a functor between ∞-categories. We define

- the *slice fibration* for $\mathcal{C} \in \text{QCat}$ to be the functor of quasicategories $\chi_{\mathcal{C}} : \mathcal{C} \to \text{QCat}$ sending $C \in \mathcal{C}$ to $\mathcal{C}/C$, and dually the *coslice fibration* to be $\chi^{\mathcal{C}} : \mathcal{C} \to \text{QCat} : C \mapsto C/\mathcal{C}$;
- the *lax colimit* of F to be the coend

$$\oint^{C} \mathcal{C}/C \times FC; \tag{7.80}$$

- the *oplax colimit* of F to be the coend

$$\oint^{C} \mathcal{C}/C \times FC. \tag{7.81}$$

The Grothendieck construction associated to F, discussed in [Lur09], with the formalism of un/straightening functors is precisely the oplax colimit of F. This is concordant with our 4.2.2 and 4.2.3.

A Fubini Rule for ∞-Coends

This section establishes the analogue of 1.3.1 for ∞-coends; the present proof currently appears as [Lor18].

We freely employ the terminology on ∞-category theory recalled in Section A.8. In particular we denote by Kan the category of *Kan complexes*, i.e. simplicial sets that lie in the orthogonal of all horn inclusions $\Lambda^n_k \hookrightarrow \Delta[n]$, for $0 \leq k \leq n$ and $n \geq 0$. The *nerve* functor $N : \text{Cat} \to \text{sSet}$ establishes a Quillen equivalence between the category of categories and the category of simplicial sets. Fibrant objects in the latter category are the ∞-*categories* of [Lur09].

Lemma 7.3.5. Let $\mathcal{C}$ be a small ∞-category, and $\mathcal{D}$ be a presentable ∞-category; then $\mathcal{D}$ is tensored and cotensored over $\mathcal{S} = N(\mathrm{Kan})$ (the *∞-category of spaces*). This entails that there is a two-variable adjunction

$$\mathcal{D}^{\mathrm{op}} \times \mathcal{D} \xrightarrow{\mathrm{Map}_{\mathcal{D}}} \mathcal{S} \qquad \mathcal{S} \times \mathcal{D} \xrightarrow{\otimes} \mathcal{D} \qquad \mathcal{S}^{\mathrm{op}} \times \mathcal{D} \xrightarrow{\pitchfork} \mathcal{D} \tag{7.82}$$

such that

$$\mathcal{D}(X \otimes D, D') \cong \mathcal{S}(X, \mathrm{Map}_{\mathcal{D}}(D, D')) \cong \mathcal{D}(D, X \pitchfork D'). \tag{7.83}$$

From the existence of these isomorphisms it is clear that

$$V \otimes (W \otimes D) \cong W \otimes (V \otimes D) \cong (V \times W) \otimes D \tag{7.84}$$
$$V \pitchfork (W \pitchfork D) \cong W \pitchfork (V \pitchfork D) \cong (V \times W) \pitchfork D. \tag{7.85}$$

Lemma 7.3.6. Let $F : \mathcal{C}^{\mathrm{op}} \times \mathcal{C} \to \mathcal{D}$ be an ∞-functor and $\mathcal{C}, \mathcal{D}$ ∞-categories as in the assumptions of 7.3.5. Then

- $F \mapsto \int^C F$ is functorial, and it is a left adjoint;
- $F \mapsto \int_C F$ is functorial, and it is a right adjoint.

Proof We only prove the first statement for coends; the second is dual.

Since $\int^C F = \mathrm{colim}_{\mathrm{TW}(\mathcal{C})}(F \circ \Sigma) = \mathrm{colim}_{\mathrm{TW}(\mathcal{C})} \circ \Sigma^*(F)$ results as a composition of ∞-functors, it is clearly functorial; then

$$\int^C : [\mathcal{C}^{\mathrm{op}} \times \mathcal{C}, \mathcal{D}] \underset{\mathrm{Ran}_\Sigma}{\overset{\Sigma^*}{\underset{\perp}{\rightleftarrows}}} [\mathrm{TW}(\mathcal{C}), \mathcal{D}] \underset{c}{\overset{\mathrm{colim}_{\mathrm{TW}(\mathcal{C})}}{\underset{\perp}{\rightleftarrows}}} \mathcal{D} \tag{7.86}$$

is a left adjoint because it is a composition of left adjoints ($c = t^*$ is the constant functor inverse image of the terminal map $\mathrm{TW}(\mathcal{C}) \to *$).

Dually, the left adjoint to $\int_C$ is given by $\mathrm{Lan}_\Sigma \circ c(D)$. $\square$

Now, the Fubini rule asserts that when the domain of a functor $F : \mathcal{A}^{\mathrm{op}} \times \mathcal{A} \to \mathcal{D}$ is of the form $(\mathcal{C} \times \mathcal{E})^{\mathrm{op}} \times (\mathcal{C} \times \mathcal{E})$, then the (co)ends of F can be computed as 'iterated integrals'

$$\int^{(C,E)} F \cong \int^{CE} F \cong \int^{EC} F \tag{7.87}$$
$$\int_{(C,E)} F \cong \int_{CE} F \cong \int_{EC} F. \tag{7.88}$$

These identifications hide a slight abuse of notation, which is worth making explicit in order to avoid confusion: thanks to 7.3.6, the three objects of (7.87) can be thought as images of F along certain functors, and the Fubini rule asserts that they are linked by canonical isomorphisms.

We can easily turn these functors into ones having the same type by means of the cartesian closed structure of sSet (somewhat sloppily, we denote the internal hom of $\mathcal{E},\mathcal{D}$ as $[\mathcal{E},\mathcal{D}]$):

$$
\begin{array}{ccc}
[\mathcal{C}^{\mathrm{op}}\times\mathcal{C}\times\mathcal{E}^{\mathrm{op}}\times\mathcal{E},\mathcal{D}] & [\mathcal{C}^{\mathrm{op}}\times\mathcal{C}\times\mathcal{E}^{\mathrm{op}}\times\mathcal{E},\mathcal{D}] & [\mathcal{C}^{\mathrm{op}}\times\mathcal{C}\times\mathcal{E}^{\mathrm{op}}\times\mathcal{E},\mathcal{D}]\\
\| & \| & \|\\
[\mathcal{C}^{\mathrm{op}}\times\mathcal{C},[\mathcal{E}^{\mathrm{op}}\times\mathcal{E},\mathcal{D}]] & [\mathcal{E}^{\mathrm{op}}\times\mathcal{E},[\mathcal{C}^{\mathrm{op}}\times\mathcal{C},\mathcal{D}]] & [(\mathcal{C}\times\mathcal{E})^{\mathrm{op}}\times(\mathcal{C}\times\mathcal{E}),\mathcal{D}]\\
{\scriptstyle [\mathcal{C}^{\mathrm{op}}\times\mathcal{C},\int^E]}\downarrow & \downarrow{\scriptstyle [\mathcal{E}^{\mathrm{op}}\times\mathcal{E},\int^C]} & \\
[\mathcal{C}^{\mathrm{op}}\times\mathcal{C},\mathcal{D}] & [\mathcal{E}^{\mathrm{op}}\times\mathcal{E},\mathcal{D}] & \downarrow{\scriptstyle \int^{(C,E)}}\\
{\scriptstyle\int^C}\downarrow & \downarrow{\scriptstyle\int^E} & \\
\mathcal{D} & \mathcal{D} & \mathcal{D}
\end{array} \tag{7.89}
$$

(of course, we can provide similar definitions for the iterated end functor).

Now that this has been clarified, we can deduce the isomorphisms (7.87) and (7.88) from the fact that the three functors $\int^{CE}, \int^{EC}, \int^{(C,E)}$ have right adjoints isomorphic to each other, and hence they must be isomorphic themselves.

This argument evidently mimics the one given in 1.3.1.

Proposition 7.3.7. The functor $R = \mathrm{Ran}_{\Sigma}(c(\_))$ acts as 'cotensoring with mapping space':

$$D \mapsto \Big((C,C') \mapsto \mathrm{Map}_{\mathcal{C}}(C,C') \pitchfork D\Big). \tag{7.90}$$

Dually, the functor $L = \mathrm{Lan}_{\Sigma}(c(\_))$ acts as 'tensoring with mapping space':

$$D \mapsto \Big((C,C') \mapsto \mathrm{Map}_{\mathcal{C}}(C,C') \otimes D\Big). \tag{7.91}$$

Proof We only prove the first statement about R; the other is dual.

It turns out that the statement relies on very well-known features of simplicial categories: we move to that setting, lacking an equally simple and conceptual quasicategorical proof.

Recall from 7.2.15 that by translating this result into the simplicially enriched setting, we get the *coherent end* of a simplicial functor $F : \mathcal{A} \to \underline{\mathcal{B}}$ of Kan-tensored and cotensored simplicial categories as the end

$$\oint^{A} F(A,A) := \int^{A',A''} \delta\mathcal{A}[A'|A''] \otimes F(A',A'') \tag{7.92}$$

defined in 7.2.15; the object $\delta\mathcal{A}[X|Y]$ is a cofibrant resolution of the hom functor, regarded as the identity hom : $\mathcal{A} \rightsquigarrow \mathcal{A}$. In view of this, and since the weighted colimit functor

$$\_ \otimes = : [\mathcal{X}^{\mathrm{op}}, \mathrm{Kan}] \times [\mathcal{X}, \mathcal{B}] \to \mathcal{B} : (G, F) \mapsto \mathrm{colim}^G F \tag{7.93}$$

is functorial in its F argument, it is also a left adjoint with right adjoint the cotensoring with the weight.

$$\mathcal{D}(W \otimes F, D) \cong [\mathcal{X}, \mathcal{D}](F, W \pitchfork D). \tag{7.94}$$

Since $\oint^A F(A, A) \cong \delta\mathcal{A} \otimes F$ is the weighted colimit with $\delta\mathcal{A}$ as a weight, it turns out that there is an adjunction

$$\oint^A \cong \delta\mathcal{A} \otimes \_ \quad \dashv \quad \lambda D.\lambda XY.\delta\mathcal{A}[X|Y] \pitchfork D \tag{7.95}$$

and by the uniqueness of adjoint functors (of course valid also in this setting) we obtain that $\mathrm{Ran}_\Sigma \circ c(D) \cong \delta\mathcal{A} \pitchfork D$. □

The Fubini rule now follows from uniqueness of adjoints: in the diagram

$$\begin{array}{c}
\lambda F. \int^C\!\!\int^E F \xrightarrow{\text{adjoint}}\!\!\dashv \lambda D.\lambda CC'.\lambda EE'.\mathrm{Map}(C, C') \pitchfork \Big(\mathrm{Map}(E, E') \pitchfork D\Big) \\
\wr\| \\
\lambda F. \int^E\!\!\int^C F \xrightarrow{\text{adjoint}}\!\!\dashv \lambda D.\lambda EE'.\lambda CC'.\mathrm{Map}(E, E') \pitchfork \Big(\mathrm{Map}(C, C') \pitchfork D\Big) \\
\wr\| \\
\lambda F. \int^{(C,E)} F \xrightarrow{\text{adjoint}}\!\!\dashv \lambda D.\lambda CEC'E'\Big(\mathrm{Map}(C, C') \times \mathrm{Map}(E, E')\Big) \pitchfork D \\
\wr\| \\
\mathrm{Map}\big((C, E), (C', E')\big) \pitchfork D
\end{array} \tag{7.96}$$

the vertical isomorphisms on the right are justified by (7.85). A completely analogous argument, using (7.84) and the left adjoints given by tensoring with the derived mapping space instead, gives the Fubini rule for (7.88).

7.4 (Co)ends in a Derivator

The theory of derivators provides a purely 2-categorical model for higher category theory, where all the coherence information is encoded in conditions that are imposed on suitable diagrams of 2-cells.

The theory was invented by A. Grothendieck in order to address the many shortcomings of triangulated categories, a categorical structure naturally arising in stable homotopy theory.

Here we only sketch some of the basic definitions needed to pave the way to 7.4.4 below.

Definition 7.4.1 (The 2-category of prederivators). We define a 2-category PDer such that

- an object of PDer, called a *prederivator*, is a strict 2-functor $\mathbb{D} : \mathsf{cat}^{\mathrm{op}} \to \mathsf{Cat}$;
- a *morphism of prederivators* is a pseudonatural transformation between pseudofunctors, $\eta : \mathbb{D} \Rightarrow \mathbb{D}'$;
- a *2-cell* between morphisms of prederivators is a modification (see 7.1.3) $\Theta : \eta \Rrightarrow \eta'$ between pseudonatural transformations.

Notation 7.4.2. Throughout this section we employ a local notation which is specific to the literature on derivator theory: the terminal category is often denoted e (perhaps French for 'ensemble avec un seul élément'); small categories are in Roman uppercase $I, J, K \dots$ and functors are in Roman lowecase $u : I \to J$. If $u : I \to J$, then its image along a prederivator $\mathbb{D}$ is denoted $u^* : \mathbb{D}(J) \to \mathbb{D}(I)$. A morphism of prederivators has components $F_I : \mathbb{D}(I) \to \mathbb{D}'(I)$ and it is specified by families of invertible 2-cells $\gamma_u : F_I \circ u^* \Rightarrow u^* \circ F_J$ subject to suitable coherence conditions.

The notion of derivator is a refinement of 7.4.1, originally motivated by the desire to provide a satisfactory axiomatisation for triangulated categories –and more generally, homotopy categories of model categories– that only appeals to 2-categorical language.

A *derivator* is then a prederivator that satisfies the following additional conditions (we adopt the labelling convention of [Gro13]).

DR1. The functor $\mathbb{D}(I \sqcup J) \to \mathbb{D}(I) \times \mathbb{D}(J)$ obtained from the canonical inclusions $i_I : I \to I \sqcup J \leftarrow J : i_J$ is an equivalence.

DR2. Each object $j : e \to J$ induces a family of functors $\mathbb{D}(J) \xrightarrow{j^*} \mathbb{D}(e)$; we ask that this family *jointly reflects isomorphisms*, i.e. a morphism $f \in \mathbb{D}(J)$ is invertible if and only if each $j^* f$ is invertible in $\mathbb{D}(e)$.

DR3. Each functor $u^* : \mathbb{D}(J) \to \mathbb{D}(I)$ induced by $u : I \to J$ admits both a left adjoint $u_!$ and a right adjoint u_*. These functors are called, respectively, the *homotopy left Kan extension* and *homotopy right Kan extension* along u.

DR4. Given a functor $u : J \to K$, and the two squares in the left column below, there exist two squares in Cat, in the right column below, induced by the colax pullbacks defining the slice and coslice categories:

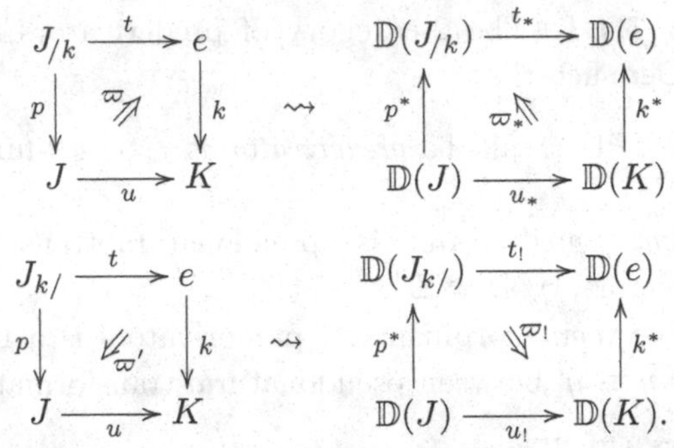

$$
\begin{CD}
J_{/k} @>{t}>> e \\
@V{p}VV @VV{k}V \\
J @>>{u}> K
\end{CD}
\quad \overset{\varpi}{\nearrow} \quad \rightsquigarrow \quad
\begin{CD}
\mathbb{D}(J_{/k}) @>{t_*}>> \mathbb{D}(e) \\
@A{p^*}AA @AA{k^*}A \\
\mathbb{D}(J) @>>{u_*}> \mathbb{D}(K)
\end{CD}
\quad \overset{\varpi_*}{\nwarrow}
$$

$$
\begin{CD}
J_{k/} @>{t}>> e \\
@V{p}VV @VV{k}V \\
J @>>{u}> K
\end{CD}
\quad \overset{\varpi'}{\swarrow} \quad \rightsquigarrow \quad
\begin{CD}
\mathbb{D}(J_{k/}) @>{t_!}>> \mathbb{D}(e) \\
@A{p^*}AA @AA{k^*}A \\
\mathbb{D}(J) @>>{u_!}> \mathbb{D}(K).
\end{CD}
\quad \overset{\varpi_!}{\searrow}
$$

We ask that these squares are filled by invertible 2-cells $\varpi'_! : t_! p^* \Rightarrow k^* u_!$, and $\varpi_* : k^* u_* \Rightarrow t_* p^*$.

Remark 7.4.3. Taken all together, the axioms of a derivator are meant to ensure that we can build a theory which is expressive enough for applications: the fundamental idea is that we can do category theory 'over cat' as a family of large categories $\mathbb{D}(J)$ contravariantly depending on functors $u : I \to J$.

Let us take a deeper look at the axioms: DR1 asks $\mathbb{D}$ to act independently on different (finite) connected components; DR2 asks that being an isomorphism in $\mathbb{D}(J)$ is a 'local' notion; the other two axioms are meant to express the fact that we can compute left and right Kan extensions for every functor $u : I \to J$ (this is DR3), and that these extensions are pointwise extensions (this is DR4). More precisely, Axiom DR4 (see [Gro13, 1.10]) states that these Kan extensions can always be computed with a *pointwise* formula. Since, according to our 4.1.7, Kan extensions can be identified with weighted (co)limits on representable weights, and more precisely because all the following concepts are equivalent:

- the left Kan extension of $F : \mathcal{A} \to \mathcal{C}$ along $G : \mathcal{A} \to \mathcal{B}$ computed in B;
- the weighted colimit of F with respect to the representable $\hom(G\_, B)$;
- the conical (co)limit of F over the category of elements of $\hom(G\_, B)$;
- the conical colimit of the diagram $(G/B) \to \mathcal{A} \xrightarrow{F} \mathcal{B}$ over the comma category of G and B;

we can try to express (co)end calculus using pointwise Kan extensions in a derivator. We exploit two basic facts: first, according to 1.2.3 (co)ends are colimits over twisted arrow categories (or suitable opposites thereof); second, (co)limits are Kan extensions along terminal arrows $\mathrm{p} : K \to *$.

With these remarks in mind, let $\text{TW}(K)$ be the twisted arrow category of K (see 1.2.2), or equivalently the category of elements (see A.5.9) of $\hom_K$, for a small category K; then there exists a functor $\Sigma_K = (t,s) : \text{TW}(K) \to K^{\text{op}} \times K$ (see A.5.13).

Definition 7.4.4 (Homotopy coend in a derivator). Let $\mathbb{D}$ be a derivator, and $K \in \mathsf{cat}$ a category. The *homotopy coend*

$$\oint^K : \mathbb{D}(J \times K^{\text{op}} \times K) \to \mathbb{D}(J) \tag{7.97}$$

is defined as the pseudonatural transformation with components obtained from the composition

$$\oint^{K,[J]} : \mathbb{D}(J \times K^{\text{op}} \times K) \xrightarrow{\Sigma_K^*} \mathbb{D}(J \times \text{TW}(K)) \xrightarrow{\mathrm{p}_!} \mathbb{D}(J). \tag{7.98}$$

Remark 7.4.5. If we rephrase the above definition in terms of the *shifted derivator* $\mathbb{D}(J|\_) : \mathsf{cat}^{\text{op}} \to \mathsf{Cat}$ of $\mathbb{D}$, i.e. the functor $I \overset{\times J}{\longmapsto} I \times J \overset{\mathbb{D}(\_)}{\mapsto} \mathbb{D}(I \times J)$, the homotopy coend $\oint^K$ defines a morphism between the shifted derivators $\mathbb{D}(K^{\text{op}} \times K|\_) \to \mathbb{D}$. (Pseudonaturality follows from a simple pasting rule between pseudonatural transformations: filling in the details is straightforward.)

Remark 7.4.6 (Homotopy ends as homotopy limits). We can state the dual notion of homotopy *end* in $\mathbb{D}$: it is enough to replace $\mathrm{p}_!$ with the right adjoint p_* computing limits instead of colimits in the definition above (the twisted arrow category is replaced by a suitable opposite thereof). In components,

$$\int_{K,[J]} : \mathbb{D}(J \times K^{\text{op}} \times K) \xrightarrow{J \times \Sigma_K^*} \mathbb{D}(J \times \text{TW}(K)) \xrightarrow{\mathrm{p}_*} \mathbb{D}(J). \tag{7.99}$$

Lemma 7.4.7. If $F : \mathbb{D} \to \mathbb{D}'$ is a morphism of derivators, there is a canonical 'comparison' morphism

$$\varsigma : \textstyle\int^K \circ F \to F \circ \int^K \tag{7.100}$$

obtained as the composition

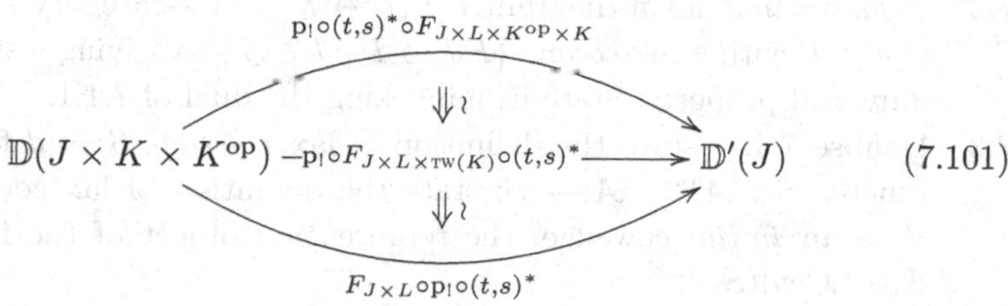

where the second morphism results from the pasting of 2-cells:

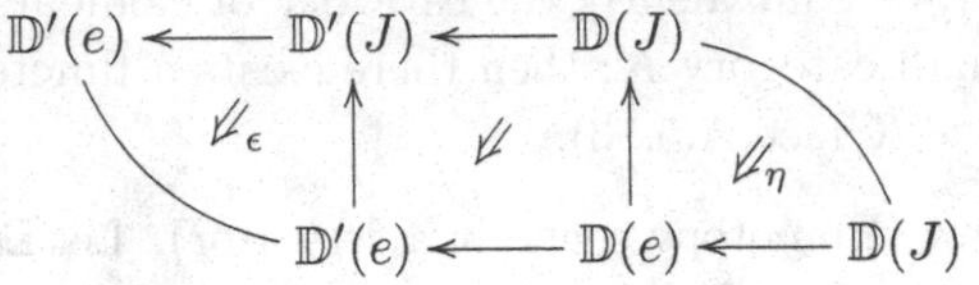

$$\begin{array}{ccccc} \mathbb{D}'(e) & \longleftarrow & \mathbb{D}'(J) & \longleftarrow & \mathbb{D}(J) \\ & \Swarrow_\epsilon & \uparrow & \Swarrow & \uparrow & \Swarrow_\eta \\ & & \mathbb{D}'(e) & \longleftarrow & \mathbb{D}(e) & \longleftarrow & \mathbb{D}(J) \end{array} \tag{7.102}$$

Proof The morphism $F : \mathbb{D} \to \mathbb{D}'$ has components $F_I : \mathbb{D}(I) \to \mathbb{D}'(I)$ and there is a 2-cell filling the central square in (7.102). This ensures (7.101) is a well-posed definition. The rest is an easy check of pseudonaturality conditions, following from conditions on the data defining ς. □

A derivator morphism F *preserves homotopy coends* (i.e. the above 2-cell is invertible) if (and only if? Is it still possible to describe a (co)limit as a certain (co)end?) it preserves colimits, or more generally left homotopy Kan extensions. Dually, F *preserves homotopy ends* if (and only if?) it preserves right homotopy Kan extensions.

Exercises

7.1 Study the category of 2-profunctors with the composition of 1-cells given by the lax coend

$$\mathfrak{p} \diamond \mathfrak{q}(A, C) = \oint^B \mathfrak{p}(A, B) \times \mathfrak{q}(B, C).$$

- The 3-category 2-Cat can be embedded into 2-Prof in various ways, using $F \mapsto \hom(F, 1)$, $F \mapsto \hom(F^\sharp, 1)$, etc.; which is the strictest of these embeddings? Are they fully faithful in a suitable sense?
- Is there an adjunction $\hom(F, 1) \dashv \hom(1, F)$? Or rather an adjunction $\hom(F^\sharp, 1) \dashv \hom(1, F^\flat)$?
- Does 2-Prof have a dualiser in the sense of 5.3.1? Does it have products, coproducts?

7.2 A *lax colimit* for a diagram $F : \mathcal{J} \to \mathcal{K}$ in a 2-category $\mathcal{K}$ is an object L with a *lax cocone* $\{FJ \to L \mid J \in \mathcal{J}\}$ satisfying a suitable universal property. State it, mimicking the dual of 7.1.1.

7.3 Dualise 7.1.1: state the definition of lax *cowedge* $S \Rightarrow d$ for a 2-functor $S : \mathcal{A}^{\mathrm{op}} \times \mathcal{A} \to \mathcal{B}$; state the definition of lax coend for S as an *initial* cowedge, the representing object of the functor $d \mapsto \mathrm{LCwd}(S, d)$.

7.4 Prove that the bar and cobar complexes define functors

$$B(\_,\mathcal{C},F) \quad B(G,\mathcal{C},\_) \quad C(\_,\mathcal{C},F) \quad C(G,\mathcal{C},\_).$$

Deduce that there exists a canonical morphism $B(G,\mathcal{C},F) \to N\mathcal{C}$ in the category of simplicial sets.

Prove that if $F, G : \mathcal{C} \to \text{Set}$ are parallel functors, there is an isomorphism

$$B(G,\mathcal{C},F)^{\text{op}}_{\bullet} \cong B(F,\mathcal{C}^{\text{op}},G)_{\bullet}.$$

Prove that the bar complex is isomorphic to the nerve of a certain category of elements (see A.5.9). Can you provide intuition for this identification? Dualise this statement.

7.5 Define (co)face and (co)degeneracy maps for the (co)simplicial objects $\boldsymbol{Y}(T)$ and $\boldsymbol{W}(T)$ of 7.2.17. (Hint: there is an isomorphism

$$\tau : T(X_0, X_n)^{\Pi\mathcal{A}[\vec{X}]} \cong \left(T(X_0, X_n)^{\mathcal{A}(X_0,X_1)}\right)^{\mathcal{A}[X_1|\vec{Y}|X_n]_{n-1}},$$

and you want to assemble a map $\boldsymbol{Y}(T)^{n-1} \to \boldsymbol{Y}(T)^n$ from its components $\Pi\mathcal{A}[\vec{X}] \pitchfork T(X_1, X_n) \to \boldsymbol{Y}(T)^n$; this defines d^0. The map d^n is defined via an isomorphism σ and a similar argument.)

7.6 ☉☉ Prove the Fubini theorem for simplicially coherent (co)ends: given a functor $T : \mathcal{A}^{\text{op}} \times \mathcal{A} \times \mathcal{B}^{\text{op}} \times \mathcal{B} \to \mathcal{C}$, then

$$\oint^{A}\left(\oint^{B} T(A,A,B,B)\right) \cong \oint^{(A,B)\in\mathcal{A}\times\mathcal{B}} T(A,B,A,B)$$
$$\cong \oint^{B}\left(\oint^{A} T(A,A,B,B)\right).$$

(Hint: it is a simple theorem about the relation between $\delta(\mathcal{A}\times\mathcal{B})$ and $\delta\mathcal{A}\times\delta\mathcal{B}$ and about 1.3.1.)

7.7 Prove that $[F,\overline{G}] \cong \text{Cat}_{\boldsymbol{\Delta}}(\!(F,G)\!) \cong [\underline{F},G]$, using 7.2.15 and a formal argument.

7.8 Prove that the isomorphisms (7.72) and (7.73) hold, and thus define coherent Kan extensions for simplicial functors.

7.9 Prove that the standard resolutions of 7.2.23 'absorb coherence' in coherent Kan extensions, showing that

$$\text{Ran}_G \overline{F} \cong \text{hoRan}_G F \qquad \text{Lan}_G \underline{F} \cong \text{hoLan}_G F.$$

(Show a preliminary lemma: that $\overline{F}(\_) \cong \int_A FA^{\delta\mathcal{A}(-,A)}$.)

7.10 Find an expression for $\mathrm{hoRan}_H F$, given a cospan of functors

$$\mathrm{Cat}_{\mathbf{\Delta}}(\mathcal{C}, \mathrm{sSet}) \xleftarrow{H} \mathcal{A} \xrightarrow{F} \mathrm{Cat}_{\mathbf{\Delta}}(\mathcal{B}, \mathrm{sSet}). \tag{7.103}$$

7.11 ◉◉ Prove that $\int^K : \mathbb{D}(K^{\mathrm{op}} \times K|-) \to \mathbb{D}$ defines a morphism of derivators (you can either prove that a functor $u : K \to L$ induces a morphism between the shifted derivators $\mathbb{D}(L|-) \to \mathbb{D}(J|-)$, or give an explicit argument if you prefer; both ways are considerably long).

7.12 Prove that 'coends in a derivator are pointwise', i.e. that given an arrow $j : e \to J$, there is a canonical isomorphism $j^*\big(\int^K X\big) \cong \int^K j^*X$ for each $X \in \mathbb{D}(J \times K^{\mathrm{op}} \times K)$.

7.13 State and prove the Fubini theorem for homotopy coends in $\mathbb{D}$: the diagram

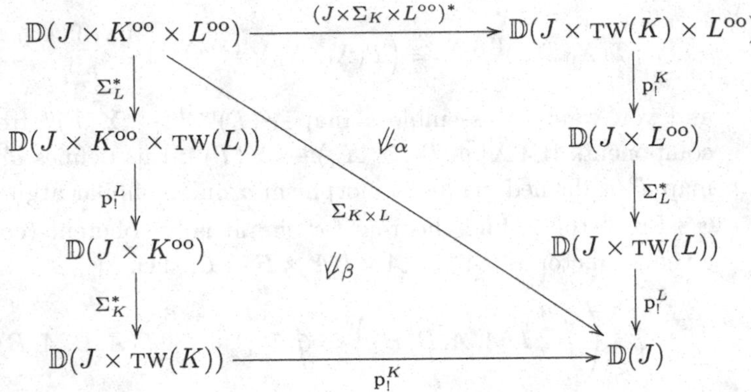

commutes for canonically determined 2-cells α and β.

7.14 We define a *bimorphism* between three derivators to be a family of functors

$$B_{IJ} : \mathbb{D}(I) \times \mathbb{E}(J) \to \mathbb{F}(I \times J)$$

in Cat endowed with 2-cells γ_{u_1,u_2} filling the diagrams

$$\begin{array}{ccc} \mathbb{D}(J_1) \times \mathbb{E}(J_2) & \longrightarrow & \mathbb{F}(J_1 \times J_2) \\ \downarrow & \Downarrow\gamma & \downarrow \\ \mathbb{D}(I_1) \times \mathbb{E}(I_2) & \longrightarrow & \mathbb{F}(I_1 \times I_2) \end{array}$$

These γ_{u_1,u_2} are subject to certain coherence conditions. For every pair (α_1, α_2) of natural transformations $\alpha_\epsilon : u_\epsilon \Rightarrow v_\epsilon : I_\epsilon \to J_\epsilon$ and

$\epsilon = 1, 2$, the two diagrams

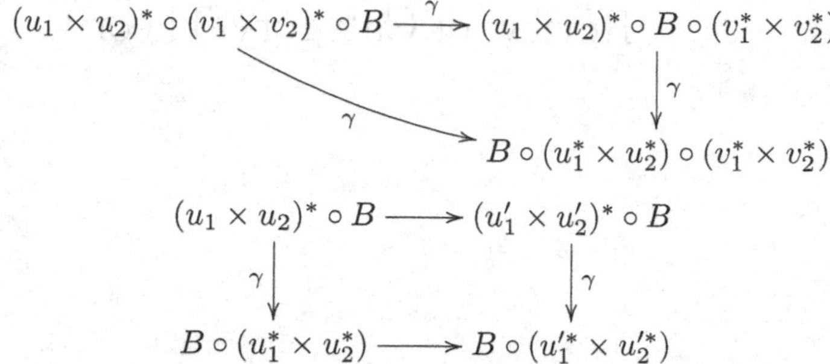

commute. Write suitable diagrams of 2-cells expressing these commutativities less briefly. Show that there is a category $\mathsf{Der}_b((\mathbb{D}, \mathbb{E}), \mathbb{F})$ of bimorphisms of derivators, and that there is an equivalence

$$\mathsf{Der}_b((\mathbb{D}, \mathbb{E}), \mathbb{F}) \cong \mathsf{Der}(\mathbb{D} \times \mathbb{E}, \mathbb{F}).$$

Is there a way to prove this result using the fact that the right side is a pseudo-end?

7.15 ◉◉ Let $\mathbb{D}^{\mathrm{Set}}$ be the *discrete* derivator sending a small category J into the functor category $\mathrm{Set}^J = \mathrm{Cat}(J, \mathrm{Set})$. State a definition and an existence theorem for *weighted colimits* in a derivator $\mathbb{D}$. Given a small category $I \in \mathsf{cat}$ and a bimorphism $\boxplus : (\mathbb{D}^{\mathrm{Set}}, \mathbb{D}) \to \mathbb{D}(-|I)$, we define the colimit of $X \in \mathbb{D}(J)$, weighted by $W \in \mathbb{D}^{\mathrm{Set}}(J^{\mathrm{op}})$, as the coend (in $\mathbb{D}$) $\int^J W \boxplus X$, i.e. as the image of the pair (W, X) under the composition

$$\mathbb{D}^{\mathrm{Set}}(J^{\mathrm{op}}) \times \mathbb{D}(J) \xrightarrow{\boxplus_{J^{\mathrm{op}},J}} \mathbb{D}(J^{\mathrm{op}} \times J|I) \xrightarrow{\int^{J,[I]}} \mathbb{D}(I).$$

Appendix A
Review of Category Theory

SUMMARY. The scope of the present appendix is to recall the bare minimum of category theory that we employ in the book. This is also meant to fix our notation beyond what is already done in the introduction. Even though we assume the typical reader of this book is already acquainted with basic category theory, we will still indulge in a certain desire for self-containment: as a rule of thumb, major results like A.5.3 or A.5.7 are proved in full detail; most of the other proofs are barely sketched, and some of the marginal results are not proved. It is in fact unrealistic to aim at such an ambitious target as a complete account of basic category theory in this appendix; the reader not feeling at ease while consulting the present section is warmly invited to parallel it with more classical references such as [Lei14, ML98, Rie17, Sim11].

> A major explanation for the cognitive advantages of diagrams is *computational offloading*: They shift some of the processing burden from the cognitive system to the perceptual system, which is faster and frees up scarce cognitive resources for other tasks.
>
> Daniel L. Moody [Moo09]

A.1 Categories and Functors

In simple terms, a category is a structure capable of abstracting a number of working assumptions of everyday mathematics:

- all objects of a given 'type' can be collected in a class;
- such objects form coherent conglomerates, allowing for relations between structures to form;
- far from being rare, these relational conglomerates are pretty common and arise at every corner of pure and applied mathematics.

Of course, it is not trivial at all to define what the 'type' of a structure is; nor is it easy to define properly the process leading to the formation of 'homomorphisms', understood as maps preserving the structure (why is a 'homomorphism of topological spaces' a continuous function and not an open map?). As always, only experience gives the right answer to such questions, but category theory is of great help in building and strengthening some sort of sixth sense for which mathematics among the many is 'the right one'.

Lacking sufficient authority to break the unspoken rule that a mathematician should talk about explicit theorems or concrete examples, instead of speculating, we shall refrain from these kinds of rambling on the philosophy of mathematics. The problem of what category theory is, and what it is for, and what there is to do in it, has however been addressed to some extent. Much better informed opinions than ours can be found in [Krö07, MZ10].[1] We warmly invite the interested reader to refer to these sources, but it is our humble opinion that too deep a study of the philosophical ground that made category theory possible quickly turns out to be counterproductive, when it is not backed up by at least an elementary knowledge of the techniques.

Definition A.1.1 (Category). A *(locally small) category* $\mathcal{C}$ consists of the following data.

- C1. A class $\mathcal{C}_o$ whose elements are termed *objects*, usually denoted with Latin letters like $A, B, \dots$.
- C2. A collection of sets $\mathcal{C}(A, B)$, indexed by the pairs $A, B \in \mathcal{C}_o$, whose elements are termed *morphisms* or *arrows* (see A.1.3 below) with *domain* A and *codomain* B.
- C3. An associative[2] *composition law*

$$\circ = \circ_{\mathcal{C},ABC} : \mathcal{C}(B,C) \times \mathcal{C}(A,B) \to \mathcal{C}(A,C) : (g,f) \mapsto g \circ f \quad \text{(A.1)}$$

[1] This shows how category theory fits into a solid track of prior philosophical and mathematical research; in a certain sense it is the pinnacle of such research, and the result of its declination in the field of pure mathematics. Category theory is what structuralism becomes when it is merged with mathematical craftmanship.

[2] If $\mathcal{C}$ is a class of sets, we say that a family of functions $\{f_{XYZ} : X \times Y \to Z\}$ indexed by the elements $X, Y, Z \in \mathcal{C}$ is *associative* if

$$f_{WZU}(w, f_{XYZ}(x, y)) = f_{ZYU}(f_{WXZ}(w, x), y)$$

for all tuples X, Y, Z, U, W and elements for which this is meaningful. When $f_{XYZ} = \circ$ is the composition map of a category this translates, of course, into the familiar associativity property $u \circ (v \circ w) = (u \circ v) \circ w$.

defined for any triple of objects A, B, C. The composition $\circ(g, f)$ is always denoted as an infix operator, $g \circ f := \circ(g, f)$.

C4. For every object $A \in \mathcal{C}_o$ there is an arrow $\mathrm{id}_A \in \mathcal{C}(A, A)$ such that for every $A, B \in \mathcal{C}_o$ and $f : A \to B$ we have $f \circ \mathrm{id}_A = f = \mathrm{id}_B \circ f$.

Remark A.1.2 (On composition). From time to time, the composition $g \circ f$ in a category may be denoted by similar 'monoid-like' infix operations as $g \cdot f$ or $g \bullet f$, or even mere juxtaposition gf. What will *never* happen is that we denote function application and morphism composition with an infix semicolon as in $f\ ;g$.

Remark A.1.3 (On arrows). The fact that for every $f \in \mathcal{C}(A, B)$ we call A the *domain* of f and B the *codomain* of f suggests how a morphism can be pictorially represented as an arrow $f : A \to B$ 'traveling', so to speak, from the domain to the codomain. Alternative notations for the set of arrows $A \to B$ are: $\hom_{\mathcal{C}}(A, B)$, $\hom(A, B)$ (when the category $\mathcal{C}$ is understood from the context), or more rarely $[A, B]$.

Remark A.1.4 (On morphism application). There are few illustrious exceptions to the tradition of accumulating function symbols to the left, when denoting the composition

$$f_n \circ f_{n-1} \circ \cdots \circ f_1 \tag{A.2}$$

of a tuple of morphisms $C_{i-1} \xrightarrow{f_i} C_i$. We stick to the most common notation that function application is on the left, without further mention; in this sense, a composition $A \xrightarrow{f} B \xrightarrow{g} C$ is written $g \circ f$.

The fact that composition is associative in a category $\mathcal{C}$ makes every such tuple of compositions well defined, and we will refer to the arrow $f_n \circ \cdots \circ f_1$, where the f_i are composed according to an arbitrary parenthesisation, as *the* composition of the tuple $(f_n, \ldots, f_1)$.

Definition A.1.5 (Functor). Let $\mathcal{C}$ and $\mathcal{D}$ be two categories; we define a *functor* $F : \mathcal{C} \to \mathcal{D}$ as a pair (F_0, F_1) consisting of the following data.

F1. F_0 is a function $\mathcal{C}_o \to \mathcal{D}_o$ sending an object $C \in \mathcal{C}_o$ to an object $FC \in \mathcal{D}_o$.

F2. F_1 is a family of functions $F_{AB} : \mathcal{C}(A, B) \to \mathcal{D}(FA, FB)$, one for each pair of objects $A, B \in \mathcal{C}_o$, sending each arrow $f : A \to B$ into an arrow $Ff : FA \to FB$, and such that:

- $F_{AA}(\mathrm{id}_A) = \mathrm{id}_{FA}$;
- $F_{AC}(g \circ_{\mathcal{C}} f) = F_{BC}(g) \circ_{\mathcal{D}} F_{AB}(f)$.

Remark A.1.6. Every family of arrows F_{AB} like in A.1.5.F1,F2 will be said to satisfy a *functoriality property.* From now on, we will always denote both the action on objects and on arrows of F with the same symbol F: so, F sends an object A into FA and an arrow $f : A \to B$ into an arrow $Ff : FA \to FB$.

Definition A.1.7 (Subcategory). Let $\mathcal{C}$ be a category. A *subcategory* $\mathcal{S}$ of $\mathcal{C}$ is a category defined by the following conditions.

SC1. The objects of $\mathcal{S}$ form a subclass of the class of objects of $\mathcal{C}$.

SC2. For every $A, B \in \mathcal{S}_o$ there is an injective function $\mathcal{S}(A,B) \subseteq \mathcal{C}(A,B)$.

If in the second condition above the inclusion is in fact an equality, or a bijection, the subcategory is called *full.*

Definition A.1.8 (Isomorphism). Let $\mathcal{C}$ be a category, and $f : A \to B$ one of its morphisms. We say that f is an *isomorphism* (or an *invertible morphism*) if there exists a morphism $g : B \to A$ going in the opposite direction, such that $f \circ g = \mathrm{id}_B$ and $g \circ f = \mathrm{id}_A$. When $A = B$ we call an arrow $f : A \to A$ an *endomorphism*, and an invertible arrow an *automorphism* of A.

Remark A.1.9. As happens for groups, if an inverse of $f : A \to B$ exists, then it is unique: if there are two such inverses,

$$g = g \circ \mathrm{id} = g \circ f \circ g' = g'$$

using associativity of the composition and the definition of inverses.[3] We can thus call an inverse of f *the* inverse f^{-1}.

Example A.1.10 (Examples of categories).

C1. Let us rule out all edge examples in a single item: the *empty* category, having no objects and morphisms, satisfies all axioms of A.1.1. So does the *singleton* category, having a single object $*$ and a single morphism, its identity id_*. Every set A can be regarded as a category A^δ, having as objects the elements of A, and where $A^\delta(x,y) = \varnothing$ if $x \neq y$, and where there is a unique arrow $x \to x$, which must be the identity id_x. This is called the *discrete* category on the set A. In a dual fashion, every set A can be regarded as a category A^χ, having as objects the elements of A, and where

[3] Note that this is actually showing something stronger: if f has a left inverse g, and a right inverse g', then f is invertible, and $g = g'$.

$A^{\chi}(x,y)$ has *exactly* one element for each pair $(x,y) \in A \times A$; this is called the *chaotic* category on the set A.

C2. The collection of sets, and functions between sets is a category Set. The set $\text{Set}(A,B)$ is the set of all functions $f : A \to B$, seen as the subset of the power-set of $A \times B$ of those relations that are functions (i.e. such that for every $a \in A$ such that $(a,b) \in f$ there is a unique such $b \in B$).

C3. The above category contains as (non-full) subcategories those of *structured sets*, having objects the sets endowed with operations like groups, rings, vector spaces, and morphisms the *homomorphisms* of these structures (homomorphisms of groups, rings, linear maps of vector spaces, ...). These categories are denoted with evocative terms like Grp (groups and their homomorphisms), Ab (*abelian* groups and their homomorphisms), $\text{Mod}(K)$ (vector spaces and linear maps), or more generally $\text{Mod}(R)$, Gph (graphs[4] and their homomorphisms).

C4. Moreover, there is a category of *topological spaces*, whose objects are pairs (X,τ) (τ is a topology on the set X) and the morphisms $f : X \to Y$ are continuous functions.

C5. Every partially ordered set (*po*set) P endowed with a relation '$\le$' can be seen as a category $\mathcal{P}$. Indeed, the objects of P are its elements, and there is an arrow $x \to y$ if and only if $x \le y$ in P (in all other cases, the set $P(x,y)$ is empty).

C6. Every monoid M can be regarded as a category $\mathcal{M}$ with a single object $*_M$, and where the set of arrows $m : *_M \to *_M$ is the set of elements of M; indeed, in this case, the monoid axioms and the axioms in A.1.1 translate perfectly into each other. The group of invertible elements of M identifies to the automorphism group of the unique object $*_M$. Given this, every group G can be regarded as a category with a single object, such that every morphism is invertible.

C7. Given a directed graph Γ, we can consider the *free category* generated by the graph. Its class of objects is the same collection of vertices of Γ; given two vertices U, V the set of morphisms $f : U \to V$ is the set of paths $U \to W_1 \to \cdots \to W_n \to V$ of

[4] A *graph* consists of a pair of sets E, V (*edges* and *vertices*) such that $E \subseteq V \times V$; a graph homomorphism is a pair of functions f_E, f_V between the sets of edges and vertices of two graphs. A graph $\underline{G} = (E, V)$ is *directed* if each element (e_0, e_1) in E is an ordered pair; in this case e_0 is the *source* or *domain* of the edge and e_1 is its *target* or *codomain*. Of course, a category can be regarded as a (possibly $\mho^+$-small) graph of a particular kind.

length n; composition of morphisms is concatenation of paths; the identity morphism is the (unique) empty path of length 0.

Definition A.1.11. Let $\mathcal{C}$ be a category; we say that an arrow $f : X \to Y$ is

ME1. a *monomorphism* if every two parallel arrows $u, v : A \rightrightarrows X$ such that $f \circ u = f \circ v$ are equal;

ME2. an *epimorphism* if every two parallel arrows $u, v : Y \rightrightarrows B$ such that $u \circ f = v \circ f$ are equal.

These conditions are in turn respectively equivalent to the following ones, stated in terms of the hom-sets of $\mathcal{C}$. The arrow $f : X \to Y$ is

EM1. a monomorphism if for every object A the function

$$\mathcal{C}(A, f) : \mathcal{C}(A, X) \to \mathcal{C}(A, Y) : u \mapsto f \circ u \tag{A.3}$$

is injective;

EM2. an epimorphism if for every object B the function

$$\mathcal{C}(f, B) : \mathcal{C}(Y, B) \to \mathcal{C}(X, B) : u \mapsto u \circ f \tag{A.4}$$

is injective.

Arrangements of objects and arrows in a category are called *diagrams*; to some extent, category theory is the art of making diagrams *commute*, i.e. the art of proving that two paths $X \to A_1 \to A_2 \to \cdots \to A_n \to Y$ and $X \to B_1 \to \cdots \to B_m \to Y$ result in the same arrow when they are fully composed.

We attempt the difficult task of providing a precise formalisation of what is a commutative diagram; it is with a certain surprise that we noticed how even reliable sources such as [AHS90, ML98] fail to provide more than a bland intuition for such a fundamental notion.

We begin by recording an easy and informal remark.

Remark A.1.12 (Diagrams and their commutation). Depicting morphisms as arrows allows regions of a given category $\mathcal{C}$ to be drawn as parts of a (possibly non-planar) graph; we call a *diagram* such a region in $\mathcal{C}$, the graph whose vertices are objects of $\mathcal{C}$ and whose edges are morphisms of suitable domains and codomains. For example, we can

consider the diagram

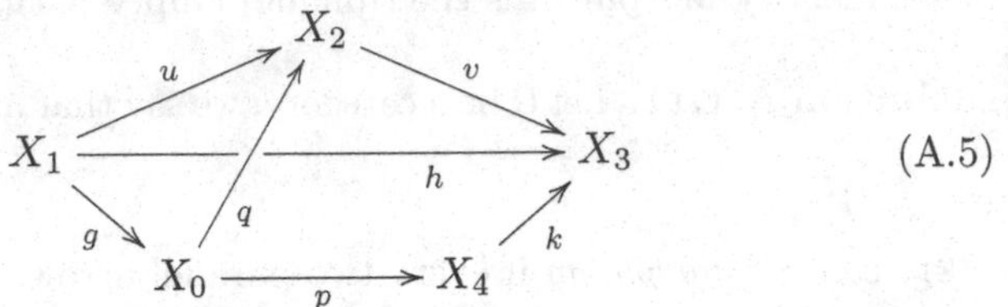

(A.5)

The presence of a composition rule in $\mathcal{C}$ entails that we can meaningfully compose *paths* $[u_0, \ldots, u_n]$ of morphisms of $\mathcal{C}$. In particular, we can consider diagrams having distinct paths between a fixed source and a fixed 'sink'. For example, in the diagram above, we can consider two different paths $\mathfrak{P} = [k, p, g]$ and $\mathfrak{Q} = [v, u]$; both paths go from X_1 to X_3, and we can ask the two compositions $\circ[k, p, g] = k \circ p \circ g$ and $v \circ u$ to be the same arrow $X_1 \to X_3$; we say that a diagram *commutes at* $\mathfrak{P}, \mathfrak{Q}$ if this is the case. We say that a diagram *commutes* (without mention of $\mathfrak{P}, \mathfrak{Q}$) if it commutes for every choice of paths for which this is meaningful.

Searching for a formalisation of this intuitive pictorial idea leads to the following.

Definition A.1.13. A *diagram* is a map of directed graphs ('digraphs') $D : J \to \mathcal{C}$ where J is a digraph and $\mathcal{C}$ is the digraph underlying a category.[5] Such a diagram D *commutes* if for every pair of parallel edges $f, g : i \rightrightarrows j$ in J one has $Df = Dg$.

Definition A.1.14 (Full, faithful, conservative functors). Let $F : \mathcal{C} \to \mathcal{D}$ be a functor between two categories.

FF1. F is called *full* if each $F_{XY} : \mathcal{C}(X, Y) \to \mathcal{D}(FX, FY)$ in A.1.5.F2 is surjective.

FF2. F is called *faithful* if each $F_{XY} : \mathcal{C}(X, Y) \to \mathcal{D}(FX, FY)$ in A.1.5.F2 is injective.

FF3. F is called *fully faithful* if it is full and faithful.

FF4. F is called *conservative* if whenever an arrow Fv is an isomorphism in $\mathcal{D}$, then the arrow v is already an isomorphism in $\mathcal{C}$.[6]

[5] Every small category has an underlying graph, obtained keeping objects and arrows and forgetting all compositions; there is of course a category of graphs, and regarding a category as a graph is another example of a forgetful functor. Of course, making this precise implies showing that the collection of categories and functors forms a category in its own right; we will come back to this.

[6] Note that the fact that F preserves invertibility of v is a consequence of the

Remark A.1.15. A subcategory $\mathcal{S} \subseteq \mathcal{C}$ is full if and only if the inclusion functor $\mathcal{S} \hookrightarrow \mathcal{C}$ is full in the sense of A.1.14.FF1 above.

A.2 Natural Transformations

Category theory was born in order to find the correct definition of a natural transformation.

The original motivation for this search lays in algebraic topology: there are two well-known ways to attach a homotopy invariant algebraic structure to a topological space X.

- The homotopy groups $\pi_n(X, x)$, obtained as homotopy classes of pointed maps $S^n \to X$; these are abelian groups if $n \geq 2$, and they are notoriously from-difficult-to-impossible to compute.
- The homology groups $H_n(X, \mathbb{Z})$ with integer coefficients, a family of abelian groups way easier to compute, and very well-behaved, but not as expressive and comprehensively descriptive as homotopy groups.

The two constructions are tightly linked: the homology group $H_n(X, \mathbb{Z})$ is obtained as the nth homology of the chain complex whose groups of n-simplices are free on the set of all continuous functions $s_n : D^n \to X$ (D^n is the n-dimensional ball, identified up to an obvious homeomorphism with the topological n-simplex of 3.1.2). Since $H_n(S^n, \mathbb{Z}) \cong \mathbb{Z}\langle u \rangle$, the homotopy class of a continuous function $f : S^n \to X$ determines a unique element $f_*(u) := H_n(f)(u) \in H_n(X)$, and this defines a function $\pi_n(X, x) \to H_n(X, \mathbb{Z})$ by sending f to $f_* u$.

Now, a capital observation is in order: this construction is compatible with the functoriality of π_n, H_n in the following sense. If $g : X \to Y$ is a continuous function of spaces, then the square

$$\begin{array}{ccc} \pi_n(X) & \xrightarrow{h} & H_n(X) \\ {\scriptstyle g_*}\downarrow & & \downarrow{\scriptstyle g_*} \\ \pi_n(Y) & \xrightarrow[h]{} & H_n(Y) \end{array} \tag{A.6}$$

is commutative, so that the maps $h_X : \pi_n(X, x) \to H(X, \mathbb{Z})$ 'coherently' or 'naturally' vary according to the action of π_n, H_n on morphisms (since

functoriality conditions in A.1.5; one also often says that a functor F is conservative if, besides preserving them, it *reflects* isomorphisms. A conservative functor is thus such that v is invertible if and only if Fv is invertible.

a category is just a bunch of objects linked by morphisms, it is 'natural' to ask functors to preserve composition, and families of morphisms $\alpha_X : FX \to GX$ between two functors to vary accordingly, in the same sense of (A.6)).

This leads directly to the notion of *natural transformation.*

Definition A.2.1 (Natural transformation). Let $F, G : \mathcal{C} \rightrightarrows \mathcal{D}$ be functors between two categories; a *natural transformation* $\tau : F \Rightarrow G$ consists of a family of morphisms $\tau_X : FX \to GX$, one for each object $X \in \mathcal{C}_o$, called the *components* of the transformation, such that for every morphism $f : X \to Y$ the diagram

$$\begin{array}{ccc} FX & \xrightarrow{F(f)} & FY \\ {\scriptstyle \tau_X}\downarrow & & \downarrow{\scriptstyle \tau_Y} \\ GX & \xrightarrow[G(f)]{} & GY \end{array} \tag{A.7}$$

commutes, i.e. we have the equation $\tau_Y \circ F(f) = G(f) \circ \tau_X$.

Definition A.2.2 (Natural equivalence). Let $F, G : \mathcal{C} \rightrightarrows \mathcal{D}$ be two parallel functors. A natural transformation $\tau : F \Rightarrow G$ such that every component is an isomorphism in $\mathcal{D}$ is called a *natural equivalence* or (less often) an *isomorphism of functors*, or a *functorial isomorphism.*

Note that if $\tau : F \Rightarrow G$ is a natural transformation, and each component $\tau_C : FC \to GC$ is an isomorphism in $\mathcal{D}$, then the family of maps τ_C^{-1} is also natural.

Definition A.2.3 (Whiskering). Let $\tau : F \Rightarrow G$ be a natural transformation between functors $F, G : \mathcal{C} \rightrightarrows \mathcal{D}$; given a third functor $H : \mathcal{H} \to \mathcal{C}$, we define the natural transformations

$$\tau * H : FH \Rightarrow GH : (\tau * H)_C = \tau_{HC} : FHC \to GHC. \tag{A.8}$$

Similarly, given a functor $K : \mathcal{D} \to \mathcal{K}$ we can define the natural transformation

$$K * \tau : KF \Rightarrow KG : (K * \tau)_C = K(\tau_C) : KFC \to KGC. \tag{A.9}$$

It is clear that the two equations

$$(\tau * H) * H' = \tau * (H \circ H') \tag{A.10}$$
$$K' * (K * \tau) = (K' \circ K) * \tau \tag{A.11}$$

hold true for every τ, H, H', K, K' that make them meaningful. It is equally obvious that $(K * \tau) * H = K * (\tau * H)$. This operation is called *whiskering*, since it acts 'drawing whiskers' on the left and on the right of τ:

$$\mathcal{H} \xrightarrow{H} \mathcal{C} \underset{G}{\overset{F}{\Downarrow \tau}} \mathcal{D} \xrightarrow{K} \mathcal{K}. \tag{A.12}$$

Finally, we make precise the idea that categories and functors form a category in their own right.

Remark A.2.4 (The category of functors). Let $\mathcal{C}, \mathcal{D}$ be two *small* categories; then the functors $F : \mathcal{C} \to \mathcal{D}$ form the object class of a category, whose morphisms $\alpha : F \Rightarrow G$ are the natural transformations. Two natural transformations can be joined component-wise, in such a way that if $F \overset{\alpha}{\Rightarrow} G \overset{\beta}{\Rightarrow} H$, we have that the components of the composite transformation $\beta \circ \alpha$ are

$$(\beta \circ \alpha)_C = \beta_C \circ \alpha_C. \tag{A.13}$$

The naturality square is of course

$$\begin{array}{ccccc} FC & \xrightarrow{\alpha_C} & GC & \xrightarrow{\beta_C} & HC \\ {\scriptstyle Ff}\downarrow & & {\scriptstyle Gf}\downarrow & & \downarrow{\scriptstyle Hf} \\ FC' & \xrightarrow[\alpha_{C'}]{} & GC' & \xrightarrow[\beta_{C'}]{} & HC' \end{array} \tag{A.14}$$

The identity in $\mathrm{Cat}(\mathcal{C}, \mathcal{D})(F, F)$ is the natural transformation having object-wise identity components. This is called the *vertical* composition of natural transformations, because α, β can be arranged in a diagram

$$\mathcal{C} \underset{H}{\overset{F}{\xrightarrow{\Downarrow\alpha \quad \Downarrow\beta}}} \mathcal{D} \tag{A.15}$$

and the natural transformation $\beta \circ \alpha$ is the result of stacking α, β one on top of the other.

Now, we shall show that there is another way to compose the natural transformations β and α called *horizontal composition*: let $\mathcal{A}, \mathcal{B}, \mathcal{C}$ be three categories, and F, G, H, K functors arranged as follows:

$$\mathcal{A} \underset{G}{\overset{F}{\rightrightarrows}} \Downarrow\alpha\ \mathcal{B} \underset{K}{\overset{H}{\rightrightarrows}} \Downarrow\beta\ \mathcal{C} \tag{A.16}$$

Definition A.2.5. Given categories $\mathcal{A}, \mathcal{B}, \mathcal{C}$, functors $F, G : \mathcal{A} \to \mathcal{B}$ and $H, K : \mathcal{B} \to \mathcal{C}$ we can define the *horizontal composition* of natural transformations $\alpha : F \Rightarrow G$ and $\beta : H \Rightarrow K$ to be the natural transformation $\beta \boxminus \alpha$, whose components are defined thanks to the fact that $(\beta * G) \circ (H * \alpha) = (K * \alpha) \circ (\beta * F)$ (as it is easy to check):

$$(\alpha \boxminus \beta)_X = \beta_{GX} \circ H(\alpha_X) = K(\alpha_X) \circ \beta_{FX}. \tag{A.17}$$

Applying the definition, we can show that the horizontal composition is (well defined and) associative, in the sense of A.1.1.C3.

Definition A.2.6 (Categorical equivalence)**.** Let $\mathcal{C}, \mathcal{D}$ be two categories. An *equivalence of categories* (F, G, ξ, η) between $\mathcal{C}$ and $\mathcal{D}$ consists of two functors $F : \mathcal{C} \to \mathcal{D}, G : \mathcal{D} \to \mathcal{C}$ endowed with two natural equivalences $\xi : F \circ G \to \mathrm{id}_{\mathcal{C}}$ and $\eta : G \circ F \Rightarrow \mathrm{id}_{\mathcal{D}}$ between the two compositions of F and G and the respective identity functors of $\mathcal{C}$ and $\mathcal{D}$ (see A.1.5.F1). If this is the case, the categories $\mathcal{C}$ and $\mathcal{D}$ are called *equivalent*; we write $\mathcal{C} \cong \mathcal{D}$.

Remark A.2.7. It is quite common to denote an equivalence with the only functor $F : \mathcal{C} \to \mathcal{D}$, and to say that it is an equivalence if there exists a functor G in the opposite direction, and natural transformations ξ, η such that the tuple (F, G, ξ, η) is an equivalence; this is customary and harmless since (as we will observe in A.4.2) such a G is unique up to natural isomorphism, provided it exists.

The richness of category theory is to some extent due to the fact that the correct way to assert 'sameness' for two categories is the above definition of equivalence, and not the stricter notion of isomorphism.

Definition A.2.8 (Isomorphism of categories)**.** We call two categories $\mathcal{C}, \mathcal{D}$ *isomorphic* if there is an invertible functor (called an *isomorphism* between the two categories) $F : \mathcal{C} \to \mathcal{D}$; this means that there exists $G : \mathcal{D} \to \mathcal{C}$ with the property that $F \circ G = \mathrm{id}_{\mathcal{D}}$ and $G \circ F = \mathrm{id}_{\mathcal{C}}$.

In a certain deep sense, equivalences and isomorphisms of categories stand in the same relation as *homotopy equivalences* and *homeomorphisms* of topological spaces (this path leads directly to higher category theory).

Another example is the following: observe that in A.2.7 we are saying that the 'space' of inverses to $F : \mathcal{C} \to \mathcal{D}$ is either empty or contractible.

Proposition A.2.9. Let $\mathcal{C}, \mathcal{D}$ be two categories and $F : \mathcal{C} \to \mathcal{D}$ a functor. Then, F is (part of) an equivalence of categories if and only if it is fully faithful and *essentially surjective*, i.e. if every object in $\mathcal{D}$ is isomorphic to some FC lying in the image of F.

A.2.1 Duality and Slices

As for every other algebraic structure, there are plenty of ways we can obtain new categories out of given ones.

- First of all, we observe that the shape of the axioms in A.1.1 entails that the structure obtained taking the same object of a category $\mathcal{C}$, but where 'all arrows have been reversed' remains a category $\mathcal{C}^{\text{op}}$. This is called the *dual*, or *opposite* category of $\mathcal{C}$.
- Then, we notice how given an object C of $\mathcal{C}$, the class of all arrows with fixed (co)domain C becomes a category, called the *(co)slice* of $\mathcal{C}$.
- Finally, as for every other algebraic structure, two categories can be arranged in a *cartesian product* and a *disjoint union*: these last two examples will appear as part of a more general theory of *(co)limits* in Section A.3.

The first procedure we introduce is *dualisation*: the *dual*, or *opposite* category of $\mathcal{C}$ is made by the same objects, and the same class of arrows, but we have interchanged domain and codomain. If an arrow is represented pictorially, as $f : A \to B$, then its namesake in $\mathcal{C}^{\text{op}}$ is $f^{\text{op}} : B \to A$.

To be more precise, there exists a functorial correspondence of Cat into itself called *duality involution* and denoted $(\_)^{\text{op}} : \text{Cat} \to \text{Cat}$ that has the following properties. It is the identity on objects, and it is the 'swapping' involution on arrows, as soon as morphisms $f : A \to B$ are identified with triples (A, B, f);

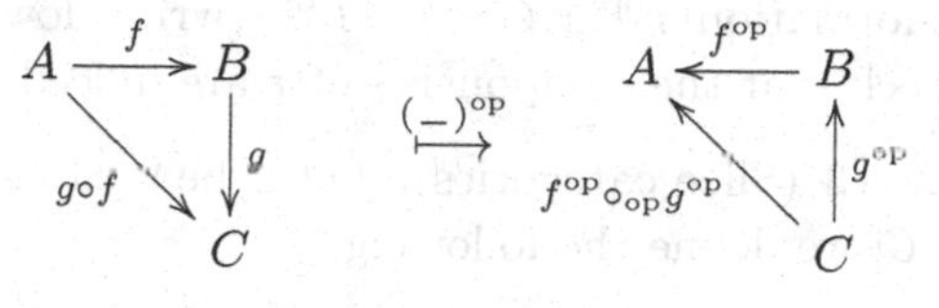

$$\begin{array}{ccc} A \xrightarrow{f} B & & A \xleftarrow{f^{\text{op}}} B \\ {\scriptstyle g\circ f}\searrow \quad \downarrow{\scriptstyle g} & \overset{(\_)^{\text{op}}}{\longmapsto} & {\scriptstyle f^{\text{op}}\circ_{\text{op}} g^{\text{op}}}\nwarrow \quad \uparrow{\scriptstyle g^{\text{op}}} \\ C & & C \end{array} \tag{A.18}$$

Remark A.2.10. A commutative triangle in $\mathcal{C}$ as in the left diagram gets modified as in the right diagram by the duality involution; the object $\mathcal{C}^{\text{op}}$ thus defined satisfies all the axioms of a category, as stated in A.1.1.

The reason why this construction is interesting is that every assertion made in $\mathcal{C}$ has a 'companion' in $\mathcal{C}^{\text{op}}$: in the words of [AHS90],

> The concept of category is well balanced, which allows an economical and useful duality. Thus in category theory the 'two for the price of one' principle holds: every concept is two concepts, and every result is two results.

In short, the situation goes as follows: in order to define a category, we use axioms regarding certain indefinite notions (object, morphism, domain, codomain...); now, every statement φ made in the language of category theory, involving solely relations between objects of $\mathcal{C}$ and the notions of object, arrow, domain, codomain, identity, composition, is valid in said language *if and only if* the statement φ^{op} obtained from φ substituting each occurrence of '$g \circ f$' with '$f^{\text{op}} \circ g^{\text{op}}$', and every occurrence of *domain* (respectively, *codomain*) with *codomain* (respectively, *domain*) is valid.

Of course, this does not mean that a statement φ about a commutative diagram is true in $\mathcal{C}$ if and only if φ^{op} is true in $\mathcal{C}$! Exercise A.4 will show that (for example) the opposite category of Set is not equivalent to Set, exhibiting a property that Set^{op} does enjoy, while Set does not.

Definition A.2.11 (Contravariant functor)**.** Let $\mathcal{C}, \mathcal{D}$ be two categories. A *contravariant functor* is a functor $F : \mathcal{C}^{\text{op}} \to \mathcal{D}$ (in the sense of A.1.5); more explicitly, a contravariant functor amounts to the same data of A.1.5, where the second condition in F2 is replaced by $F(g \circ f) = Ff \circ Fg$, for every pair of composable morphisms f, g.

Functors in the sense of A.1.5 are called *covariant.*

Remark A.2.12. If we call $(-)^{\text{op}} : \mathcal{C} \mapsto \mathcal{C}^{\text{op}}$ the duality involution, its functoriality amounts to saying that every functor $F : \mathcal{C} \to \mathcal{D}$ induces a functor $F^{\text{op}} : \mathcal{C}^{\text{op}} \to \mathcal{D}^{\text{op}}$ (so a contravariant functor $F : \mathcal{C} \to \mathcal{D}^{\text{op}}$):

$$F^{\text{op}}(g \circ f) = F((g \circ f)^{\text{op}}) = F(g^{\text{op}} \circ f^{\text{op}}) = F^{\text{op}} f \circ F^{\text{op}} g. \qquad \text{(A.19)}$$

Similarly, to every natural transformation $\tau : F \Rightarrow G$ is associated a natural transformation $\tau^{\text{op}} : G^{\text{op}} \Rightarrow F^{\text{op}}$ (write down the relevant diagram and check that the components of τ are indeed reversed).

Definition A.2.13 (Slice categories)**.** Let $\mathcal{C}$ be a category and $C \in \mathcal{C}$ be an object of $\mathcal{C}$; we define the following.

- The 'slice' category $\mathcal{C}/C$ of *arrows over* C having class of objects the arrows with codomain C, and morphisms between $f : C' \to C$ and $g : C'' \to C$ the arrows $h : C' \to C''$ such that $g \circ h = f$.

- The 'coslice' category $C/\mathcal{C}$ of *arrows under* C having class of objects the arrows with domain C, and morphisms between $f : C \to C'$ and $g : C \to C''$ the arrows $h : C' \to C''$ such that $h \circ f = g$.

Note that $C/\mathcal{C} \cong (\mathcal{C}^{\mathrm{op}}/C)^{\mathrm{op}}$ and similarly $\mathcal{C}/C = (C/\mathcal{C}^{\mathrm{op}})^{\mathrm{op}}$.

Definition A.2.14 (Comma categories)**.** The *comma category* of a diagram $\mathcal{S} \xrightarrow{F} \mathcal{C} \xleftarrow{G} \mathcal{T}$ of functors is the category having objects the tuples $(S \in \mathcal{S}, T \in \mathcal{T}, \varphi \in \mathcal{C}(FS, GT))$, and morphisms the pairs $(u : S \to S', v : T \to T')$ such that the square

$$\begin{array}{ccc} FS & \xrightarrow{Fu} & FS' \\ {\scriptstyle\varphi}\downarrow & & \downarrow{\scriptstyle\varphi'} \\ GT & \xrightarrow[Gv]{} & GT' \end{array} \tag{A.20}$$

commutes. This category is denoted $(F \downarrow G)$ or (F/G).

The reader is invited to study a few properties of the comma category (F/G) in Exercise A.5; we now turn our attention to the *product* and *coproduct* of categories.

Definition A.2.15 (Product of categories)**.** Let $\mathcal{C}, \mathcal{D}$ be two categories. We define the *product category* $\mathcal{C} \times \mathcal{D}$ to be the following structure:

- it has as objects the product of the classes of objects $\mathcal{C}_o \times \mathcal{D}_o$;
- it has as sets of morphisms $(C, D) \to (C', D')$ the cartesian product $\mathcal{C}(C, C') \times \mathcal{D}(D, D')$.

All the axioms of a category are satisfied, as compositions and identities are defined factor-wise and thus act independently of each other. Of course, the definition inductively extends to the product of any finite number of categories $\mathcal{C}_1 \times \cdots \times \mathcal{C}_n$.

Dually, we can define the *coproduct* of categories.

Definition A.2.16 (Coproduct of categories)**.** Let $\mathcal{C}, \mathcal{D}$ be two categories. We define the *coproduct category* $\mathcal{C} \coprod \mathcal{D}$ to be the following structure:

- it has as objects the set-theoretical disjoint union of the classes of objects $\mathcal{C}_o \times \mathcal{D}_o$;
- it has as sets of morphisms $X \to Y$ the set $\mathcal{C}(X, Y)$ if $X, Y \in \mathcal{C}$, $\mathcal{D}(X, Y)$ if $X, Y \in \mathcal{D}$, and the empty set otherwise.

All the axioms of a category are satisfied, as compositions and identities are defined summand-wise and thus act independently one from the other. Of course, the definition inductively extends to the coproduct of any finite number of categories $\mathcal{C}_1 \amalg \cdots \amalg \mathcal{C}_n$.

Observe that the following property holds true for the product $\mathcal{C} \times \mathcal{D}$:

> Let $\mathcal{E}$ be a category and $\mathcal{C} \xleftarrow{F} \mathcal{E} \xrightarrow{G} \mathcal{D}$ be two functors; then there exists a unique functor $\mathcal{E} \to \mathcal{C} \times \mathcal{D}$ such that the compositions $\mathcal{E} \to \mathcal{C} \times \mathcal{D} \xrightarrow{P} \mathcal{C}$ and $\mathcal{E} \to \mathcal{C} \times \mathcal{D} \xrightarrow{Q} \mathcal{D}$ with the projections on the factors of the product correspond respectively to F and G.

Observe that $\mathcal{C} \coprod \mathcal{D}$ enjoys the same property in the dual category $\mathrm{Cat}^{\mathrm{op}}$, or rather, the dual property in Cat:

> Let $\mathcal{E}$ be a category and $\mathcal{C} \xrightarrow{F} \mathcal{E} \xleftarrow{G} \mathcal{D}$ be two functors; then there exists a unique functor $\mathcal{C} \coprod \mathcal{D} \to \mathcal{E}$ such that the compositions $\mathcal{C} \to \mathcal{C} \coprod \mathcal{D} \to \mathcal{E}$ and $\mathcal{D} \to \mathcal{C} \coprod \mathcal{D} \to \mathcal{E}$ with the embeddings into the coproduct correspond respectively to F and G.

The scope of the following section is to formalise the intuition that these two constructions are 'universal' and 'dual' to each other.

A.3 Limits and Colimits

Right after having introduced the definition of an algebraic structure, it is often shown that given one or more such structures one can build others from it; one can for example assemble the *product* of two groups, sets or topological spaces.

The objects that are built in this way are often characterised by some sort of uniqueness; for example, there is only one way to assemble two vector spaces V, W into a third space $V \oplus W$ that contains 'no more than V, W'; this vector space is made by pairs of vectors (v, w), and the vector sum $(v, w) + (v', w')$, as well as scalar multiplication $\alpha(v, w)$, done component-wise. Moreover, there is a diagram

$$V \xrightarrow{i_V} V \oplus W \xleftarrow{i_W} W \tag{A.21}$$

such that *every other* similar diagram $V \xrightarrow{f} U \xleftarrow{g} W$ factors through the first. There is a unique $[f /\!/ g] : V \oplus W \to U$ that coincides with f, g when restricted to the summands, i.e. such that $[f /\!/ g] \circ i_V = f$ and $[f /\!/ g] \circ i_W = g$.

Category theory provides a very neat way to organise these data and explain what the essential features of this phenomenon are, through the theory of (co)limits.

We assume that our reader is already familiar with elementary algebra, and in particular that they are familiar with the simple idea that algebraic structures of the same kind (e.g., groups or vector spaces) can be assembled together.

Definition A.3.1 (Initial and terminal objects)**.** Let $\mathcal{C}$ be a category.

- An object $\varnothing \in \mathcal{C}_o$ is called *initial* if for every other object $C \in \mathcal{C}_o$ there is a single morphism $i_C : \varnothing \to C$.
- Dually, an object 1 is called *final* or *terminal* if for every object $C \in \mathcal{C}_o$ there is a unique morphism $t_C : C \to 1$.

(Note the substantial difference between 'there is at most one morphism' and 'there is exactly one morphism'!)

Remark A.3.2. As a consequence of the definition, if $\varnothing$ is an initial object in $\mathcal{C}$, then there is a single arrow $i_\varnothing : \varnothing \to \varnothing$, the identity of $\varnothing$. Similarly, if 1 is terminal, there is a unique arrow $t_1 : 1 \to 1$, the identity of 1. If $\mathcal{C}$ has both an initial and a terminal object, then there is a unique arrow $z : \varnothing \to 1$; we say that $\mathcal{C}$ has a *zero* object if z is an isomorphism.

The simple proof of the following statement will illuminate the nature of the notion of universal property. An initial object $\varnothing \in \mathcal{C}$ enjoys what is called a *universal property.*

Remark A.3.3. Let $\mathcal{C}$ be a category with an initial object $\varnothing$. If $\varnothing'$ is another initial object, then there is a unique isomorphism $\varnothing \cong \varnothing'$.

Proof Assume that there are two initial objects $\varnothing, \varnothing'$; then, by the respective universal properties of $\varnothing$ and $\varnothing'$, there is a unique arrow $u : \varnothing \to \varnothing'$, and similarly a unique arrow $v : \varnothing' \to \varnothing$. The compositions $v \circ u$ and $u \circ v$ must be the identities of $\varnothing$ and $\varnothing'$ respectively, thus showing that u, v are mutually inverse isomorphisms □

The notion of *universal property* arises to generalise this phenomenon; if $\mathcal{J}$ is a category, a functor $D : \mathcal{J} \to \mathcal{D}$ can be thought as a diagram ('of shape $\mathcal{J}$') representing the category $\mathcal{J}$ in its codomain. For example if $\mathcal{J}$ is the category having three objects $0, 1, 2$ and whose non-identity

morphisms are

(A.22)

then a diagram of shape $\mathcal{J}$ is simply a triple of objects D_0, D_1, D_2 linked by similar morphisms $D_0 \to D_1, D_0 \to D_2$. The key observation of the theory of (co)limits is that to each diagram D one can associate a category of *cones* (and dually, a category of *cocones*), obtained as suitable extensions of D to a 'cone category' $\tilde{\mathcal{J}} \supset \mathcal{J}$, adding to $\mathcal{J}$ a new initial (or terminal) object.

Extensions of this kind can be organised in categories $\mathcal{J}^{\triangleright}$ if we add a terminal object, and $\mathcal{J}^{\triangleleft}$ if we add an initial object, of (co)cones over $\mathcal{J}$.[7]

This brief procedure directly leads to the following.

Definition A.3.4. The *limit* of a diagram D is (whenever it exists) the *terminal* object in the category of its *cones*, and dually the colimit of D is the *initial* object in its category of *cocones*.

Of course, we will not leave the reader alone deciphering this mysterious axiomatics; let us start to make this construction precise by defining the following.

Definition A.3.5 (Cone completions of $\mathcal{J}$)**.** Let $\mathcal{J}$ be a small category; we denote $\mathcal{J}^{\triangleright}$ the category obtained by adding to $\mathcal{J}$ a single terminal object ∞; in more detail, $\mathcal{J}^{\triangleright}$ has objects $\mathcal{J}_o \cup \{\infty\}$, where $\infty \notin \mathcal{J}$, and it is defined by

$$\mathcal{J}^{\triangleright}(J, J') = \mathcal{J}(J, J')$$
$$\mathcal{J}^{\triangleright}(J, \infty) = \{*\}$$

and it is empty otherwise. This category is called the *right cone* of $\mathcal{J}$.

Dually, we define a category $\mathcal{J}^{\triangleleft}$, the *left cone* of $\mathcal{J}$, as the category obtained by adding to J a single *initial* object $-\infty$; this means that $\mathcal{J}^{\triangleleft}(J, J') = \mathcal{J}(J, J')$, $\mathcal{J}^{\triangleleft}(-\infty, J) = \{*\}$, and it is empty otherwise.

Definition A.3.6 (Cone of a diagram)**.** Let $\mathcal{J}$ be a small category, $\mathcal{C}$ a category, and $D : \mathcal{J} \to \mathcal{C}$ a functor. Throughout this section, an

[7] The reader will surely appreciate the origin of the name: if X is a topological space, the *cone* of X is the space obtained by adding a distinguished point ∞, disconnected from X, and then by adding a path from ∞ to each point of X.

idiosyncratic way to refer to D will be as a *diagram of shape* $\mathcal{J}$. We call a *cone for* D any extension of the diagram D to the left cone category of $\mathcal{J}$ defined in A.3.5, so that the diagram

$$\begin{array}{ccc} \mathcal{J} & \xrightarrow{D} & \mathcal{C} \\ {\scriptstyle i_{\triangleleft}}\big\downarrow & \nearrow_{\bar{D}} & \\ \mathcal{J}^{\triangleleft} & & \end{array} \tag{A.23}$$

commutes.

Every such extension is thus forced to coincide with D on all objects in $\mathcal{J} \subseteq \mathcal{J}^{\triangleleft}$; the value of $\bar{D}$ on $-\infty$ is called the base of the cone. Dually, the value of an extension of D to $\mathcal{J}^{\triangleright}$ coincides with D on $\mathcal{J} \subseteq \mathcal{J}^{\triangleright}$, and $\bar{D}(\infty)$ is called the *tip* of the cone.

There is of course a similar definition of a *cocone* for D: it is an extension of the diagram D to the right cone category of $\mathcal{J}$ so that the diagram

$$\begin{array}{ccc} \mathcal{J} & \xrightarrow{D} & \mathcal{C} \\ {\scriptstyle i_{\triangleright}}\big\downarrow & \nearrow_{\bar{D}} & \\ \mathcal{J}^{\triangleright} & & \end{array} \tag{A.24}$$

commutes.

Remark A.3.7.

- The class of cones for D forms a category $\mathrm{Cn}(D)$, whose morphisms are the natural transformations $\alpha : D' \Rightarrow D'' : \mathcal{J} \to \mathcal{C}$ such that the right whiskering of α with $i : \mathcal{J} \to \mathcal{J}^{\triangleleft}$ coincides with the identity natural transformation of D; this means that a morphism α of this sort is a natural transformation such that

$$\begin{array}{ccccc} & & \mathcal{J} & & \\ & {\scriptstyle i}\swarrow & & \searrow{\scriptstyle D} & \\ \mathcal{J}^{\triangleleft} & & \overset{D'}{\underset{D''}{\Downarrow \alpha}} & & \mathcal{C} \end{array} \quad = \mathrm{id}_D \tag{A.25}$$

  as a 2-cell $D \Rightarrow D$.
- Dually, the class of cocones for D forms a category $\mathrm{Ccn}(D)$, whose morphisms are the natural transformations $\alpha : D' \Rightarrow D'' : \mathcal{J} \to \mathcal{C}$ such that the right whiskering of α with $i : \mathcal{J} \to \mathcal{J}^{\triangleright}$ coincides with the

identity natural transformation of D; this means that a morphism α of this sort is a natural transformation such that

$$\begin{array}{c} \mathcal{J} \\ {}^{i}\swarrow \quad \searrow^{D} \\ \mathcal{J}^{\triangleright} \overset{D'}{\underset{D''}{\Downarrow\alpha}} \mathcal{C} \end{array} = \mathrm{id}_D \tag{A.26}$$

as a 2-cell $D \Rightarrow D$.

Definition A.3.8 (Colimit, limit). The *limit* of a diagram $D : \mathcal{J} \to \mathcal{C}$ is the terminal object denoted '$\lim_{\mathcal{J}} D$' in the category of cones for D; dually, the *colimit* of D is the initial object denoted '$\mathrm{colim}_{\mathcal{J}} D$' in the category of cocones for D.

Proposition A.3.9. Let $\mathcal{C}$ be a category; the following conditions are equivalent:

- L1. $\mathcal{C}$ has all limits of shape $\mathcal{J}$;
- L2. $\mathcal{C}$ has products indexed by every set $\mathcal{J}$, and equalisers;
- L3. $\mathcal{C}$ has a terminal object, and pullback of every co-span $X \to Z \leftarrow Y$.

Dually, the following conditions are equivalent:

- C1. $\mathcal{C}$ has all colimits of shape $\mathcal{J}$;
- C2. $\mathcal{C}$ has coproducts indexed by every set $\mathcal{J}$, and coequalisers;
- C3. $\mathcal{C}$ has an initial object, and pushout of every span $X \leftarrow Z \to Y$.

Proof See [Bor94a, 2.8.2] (and a suitable dual statement). We only record that the limit of a diagram D has the same universal property of the equaliser of the pair

$$u, v : \prod_A DA \rightrightarrows \prod_{f:A\to B} DB \tag{A.27}$$

where the arrows (u, v) are defined as

$$\pi_{(f:A\to B)} \circ u = \pi_B$$
$$\pi_{(f:A\to B)} \circ v = (Df)\pi_A$$

if $\pi_X : \prod_A DA \to DX$ denotes the projection at X coordinate. The terminal cone exhibiting $\lim D$ is of course the composition $\lim D \to \prod_A DA \to DA$.

Dually, the coequaliser of

$$(u', v') : \coprod_{f:A\to B} DA \rightrightarrows \coprod_{A} DA \tag{A.28}$$

for a similar pair of arrows (u', v') has the same universal property of colim D. □

Definition A.3.10 (Preservation of (co)limits). Let $\mathcal{J}, \mathcal{C}, \mathcal{D}$ be categories, $\mathcal{J}$ small, $D : \mathcal{J} \to \mathcal{C}$ be a diagram, and $F : \mathcal{C} \to \mathcal{D}$ a functor. Assume that the limit $\lim_{\mathcal{J}} D$ exists in $\mathcal{C}$; then, applying the functor F to the terminal cone $\lambda : \mathcal{J}^{\triangleleft} \to \mathcal{C}$ of D, we get a cone $F * \lambda$ for the composed diagram $\mathcal{J} \xrightarrow{D} \mathcal{C} \xrightarrow{F} \mathcal{D}$. We say that F *preserves* the limit of D if $F * \lambda$ is the limit of $F \circ D$, i.e. if

$$\lim_{\mathcal{J}}(F \circ D) \cong F(\lim_{\mathcal{J}} D). \tag{A.29}$$

Definition A.3.11 ((Co)complete category). A category $\mathcal{C}$ *has all limits of shape* $\mathcal{J}$ if every diagram $D : \mathcal{J} \to \mathcal{C}$ has a limit; dually, we define a category having all colimits of shape $\mathcal{J}$. A category is said to *have all* κ*-(co)limits* if it has (co)limits of shape $\mathcal{J}$ for every category $\mathcal{J}$ with less than κ objects. A category is said to have *all small (co)limits*, or simply *all (co)limits* (but the implicit smallness requirement on $\mathcal{J}$ is needed!) if every diagram $D : J \to \mathcal{C}$ with small domain has a (co)limit.[8]

In view of A.3.9 above, a category $\mathcal{C}$ has κ-(co)limits if and only if it has (co)equalisers and (co)products of every family of less than κ objects.

Proposition A.3.12. The category Cat of small categories and functors has all small limits and colimits.

Definition A.3.13. Let $\mathcal{C}, \mathcal{D}$ be two categories; we define the *join* of $\mathcal{C}$ and $\mathcal{D}$ to be the category having objects the disjoint union $\mathcal{C}_o \coprod \mathcal{D}_o$ of $\mathcal{C}_o$ and $\mathcal{D}_o$, and where morphisms from X to Y are defined as follows:

$$\begin{cases} \mathcal{C}(X, Y) & \text{if } X, Y \in \mathcal{C} \\ \mathcal{D}(X, Y) & \text{if } X, Y \in \mathcal{D} \\ \{*\} & \text{if } X \in \mathcal{C}, Y \in \mathcal{D} \end{cases} \tag{A.30}$$

where $\{*\}$ is a singleton set. In every other case, there are no morphisms $X \to Y$.

[8] Every category admitting 'too large' (co)limits must be a (large) poset; this is a theorem by P. Freyd, and constitutes the main reason why the smallness assumption on $\mathcal{J}$ cannot be dropped.

Remark A.3.14. The join operation gives the category Cat a monoidal structure, having the empty category as monoidal unit; we shall observe that the associator is easily determined, but the structure is highly non-symmetric.

As already observed, in the terminology of Exercise 5.4, the join of $\mathcal{C}$ and $\mathcal{D}$ coincides with the *collage* of $\mathcal{C}$ and $\mathcal{D}$ along the terminal profunctor $* : \mathcal{C} \rightsquigarrow \mathcal{D}$, i.e. with the category of elements of the functor $* : \mathcal{C}^{\text{op}} \times \mathcal{D} \to \text{Set}$ constant in the singleton set. The join operation can thus be regarded as a particular case of an operation in the bicategory of profunctors.

Definition A.3.15 (Creation of (co)limits [Rie17, 3.3]). Let $F : \mathcal{C} \to \mathcal{D}$ be a functor; we say that F *creates* the (co)limit of a diagram $D : \mathcal{J} \to \mathcal{C}$ if for every (co)limit (co)cone for FD in $\mathcal{D}$ there exists a (co)limit (co)cone for D in $\mathcal{C}$, and every (co)cone for D that is sent to a (co)limit (co)cone by F is itself a (co)limit (co)cone.

The above definition admits a simple restriction in case D is the empty diagram: a functor $F : \mathcal{C} \to \mathcal{D}$

- preserves the initial object if $F(\varnothing_{\mathcal{C}})$ is an initial object in $\mathcal{D}$,
- reflects the initial object if the fact that $F(A)$ is initial in $\mathcal{D}$ entails that A is initial in $\mathcal{C}$,
- creates the initial object if the existence of an initial object in $\mathcal{D}$ of the form FA entails the existence of an initial object in $\mathcal{C}$, and every object such that $FA = \varnothing$ is itself initial.

Of course, the word 'initial' can be replaced with 'terminal', obtaining preservation, reflection and creation of terminal objects.

The notion of preserving, reflecting and creating an initial object is sufficient to capture A.3.15 by virtue of the same argument in A.4.6: F creates the colimit of D if and only if the functor $F_* : \text{Ccn}(D) \to \text{Ccn}(FD)$ creates the initial object. A simple dualisation yields that F creates the limit of D if and only if the functor $F_* : \text{Cn}(D) \to \text{Cn}(FD)$ creates the terminal object.

A.4 Adjunctions

The notion of adjunction lies at the very core of category theory.

Definition A.4.1 (Adjunction). Let $\mathcal{C}, \mathcal{D}$ be two categories. We define an *adjunction* between $\mathcal{C}, \mathcal{D}$ to be a pair of functors $F : \mathcal{C} \to \mathcal{D}$, $G : \mathcal{D} \to \mathcal{C}$ endowed with a collection of bijections φ_{CD},

$$\varphi_{CD} : \mathcal{D}(FC, D) \cong \mathcal{C}(C, GD) \tag{A.31}$$

one for each $C \in \mathcal{C}_o$, $D \in \mathcal{D}_o$, natural in both arguments C, D.

We denote the presence of an adjunction (F, G) between $\mathcal{C}$ and $\mathcal{D}$ writing

$$F : \mathcal{C} \leftrightarrows \mathcal{D} : G \tag{A.32}$$

and we say that F is a *left adjoint* to G, or that G is a *right adjoint* to F, or that (F, G) is an adjoint pair, or that F, G are mutually adjoint, etc.

Given an adjunction as in A.4.1, we write $F \dashv G$ in brief to denote this (highly non-symmetric) relation among F and G. Once we have introduced the Yoneda lemma in A.5.3, we will be able to instantly deduce the following uniqueness result.

Proposition A.4.2. If a functor $F : \mathcal{C} \to \mathcal{D}$ has a left adjoint $G : \mathcal{D} \to \mathcal{C}$, such an adjoint is unique up to a unique natural isomorphism; in other words, if we have adjunctions $F \dashv G$ and $F \dashv G'$ then there exists a *unique* natural isomorphism $\tau : G \cong G'$.

Of course, a similar result holds for the uniqueness of a left adjoint F.

Example A.4.3 (Examples of adjunctions).

AD1. Let $f : M \to N$ be a morphism of monoids; according to A.1.10.C6 we can regard it as a functor between one-object categories; it is easy to see that such a functor has a left adjoint $g : N \to M$ if and only if it has a right adjoint, if and only if it is an equivalence of categories, if and only if it is an isomorphism of monoids.

AD2. Let P be a poset regarded as a category. An adjunction $f : P \leftrightarrows Q : g$ consists of a pair of monotone functions f, g such that $fp \le q$ if and only if $p \le gq$. These pairs of monotone mappings are called *Galois connections* (the name is motivated by Exercise A.9).

AD3. More generally, let $i : \mathcal{S} \hookrightarrow \mathcal{C}$ be the embedding of a full subcategory. We say that $\mathcal{S}$ is *reflective* into $\mathcal{C}$, or a *reflective subcategory* of $\mathcal{C}$, if i admits a left adjoint $L : \mathcal{C} \to \mathcal{S}$. Almost all familiar subcategories of algebraic structures (magmas, semigroups, monoids, groups, abelian groups, R-modules, ...) fit into an adjunction with the category of sets, but almost none of these categories is reflective; this is because the forgetful functor $U : \mathcal{C} \to \text{Set}$ is rarely full. (Difficult exercise: generalise the construction of unit and counit maps to an abstract algebraic structure.)

AD4. Given three sets A, B, C there exists a bijection between the set $\text{Set}(A \times B, C)$ of functions $f : A \times B \to C$ and the set $\text{Set}(A, C^B)$ of functions from A to $\text{Set}(B, C)$; given $f : A \times B \to C$, we can define a *transposed* (or *curried*) function $\hat{f} : A \to C^B$ sending $a \in A$ into the function $f_a = f(a, \_) : B \to C$. Currying a function f into $\hat{f}$ is a bijection, and the inverse sends a function $g : A \to C^B$ into $\tilde{g} : A \times B \to C$ via an *uncurrying* operation.

If for every $B \in \text{Set}$ we define $(\_) \times B : \text{Set} \to \text{Set} : A \mapsto A \times B$ and $(\_)^B : \text{Set} \to \text{Set} : C \mapsto C^B$ we get two functors (defined accordingly on arrows) that form an adjunction $\_ \times B \dashv (\_)^B$.

Every category $\mathcal{C}$ having finite products and exhibiting the same adjunction

$$\mathcal{C}(A \times B, C) \cong \mathcal{C}(A, C^B) \cong \mathcal{C}(B, C^A) \qquad \text{(A.33)}$$

is called *cartesian closed.* The category of sets, and the category of categories are both cartesian closed; no category with a zero object can be cartesian closed without being the terminal category $*$; the category of all topological spaces and continuous maps is notoriously not cartesian closed, but many subcategories of 'nice' spaces are.

AD5. Let $\mathcal{C}$ be a category. Then $\mathcal{C}$ has a terminal object (see A.3.1) if and only if the unique functor $t_{\mathcal{C}} : \mathcal{C} \to *$ has a right adjoint; dually, $\mathcal{C}$ has an initial object if and only if $t_{\mathcal{C}}$ has a left adjoint.

A.4.1 Unit and Counit, Triangle Identities

From every adjoint pair (F, G) we can obtain a pair of natural transformations, the *unit* and the *counit* of the adjunction: if in (A.31) we put $D = FC$ we have

$$\varphi_{CFC} : \mathcal{D}(FC, FC) \to \mathcal{C}(C, GFC). \qquad \text{(A.34)}$$

Now, the domain $\mathcal{D}(FC, FC)$ is non-empty (by A.1.1 it must contain at least the identity of FC) thus φ is not the empty function, and it is well defined as the image $\eta_C = \varphi_{CFC}(\text{id}_{FC})$ of the identity under φ; η_C is the component at C of a natural transformation $\eta : \text{id}_{\mathcal{C}} \to G \circ F$ called the *unit* of the adjunction. Indeed, given $h : C' \to C$ we have that

$$GF(h) \circ \eta_{C'} = GF(h) \circ \varphi_{C'FC'}(\text{id}_{FC'})$$

$$
\begin{aligned}
&= \varphi \circ F(h) \circ \mathrm{id}_{FC'} \\
&= \varphi_{C'FC} \circ \mathrm{id}_{FC} \circ F(h) \\
&= \varphi_{CFC}(\mathrm{id}_{FC}) \circ h \\
&= \eta_C \circ h. \qquad (A.35)
\end{aligned}
$$

This is equivalent to the joint commutativity of both parts in the diagram

$$
\begin{array}{ccccc}
\mathcal{D}(FC', FC) & \xrightarrow{\mathcal{D}(FC', F(h))} & \mathcal{D}(FC', FC) & \xleftarrow{\mathcal{D}(F(h), FC)} & \mathcal{D}(FC, FC) \\
\downarrow{\scriptstyle \varphi_{C'FC'}} & & \downarrow{\scriptstyle \varphi_{C'FC}} & & \downarrow{\scriptstyle \varphi_{CFC}} \\
\mathcal{C}(C', GFC') & \xrightarrow[\mathcal{C}(C', G(F(h)))]{} & \mathcal{C}(C', G(FC)) & \xleftarrow[\mathcal{C}(h, GFC)]{} & \mathcal{C}(C, GFC)
\end{array}
\qquad (A.36)
$$

Dually, if in A.31 we put $C = GD$ we have

$$
\varphi_{GDD} : \mathcal{D}(FGD, D) \to \mathcal{C}(GD, GD) \qquad (A.37)
$$

(or equivalently $\varphi^{-1}_{GDD} : \mathcal{C}(GD, GD) \to \mathcal{D}(F(GD), D)$). Again, the set $\mathcal{C}(GD, GD)$ is non-empty, thus we can define $\epsilon_D = \varphi^{-1}_{GDD}(\mathrm{id}_{GD})$; this arrow is the component at D of a natural transformation $\epsilon : F \circ G \to \mathrm{id}_{\mathcal{D}}$ called the *counit* of the adjunction, in such a way that for every $k : D \to D'$ the diagram

$$
\begin{array}{ccc}
F(GD) & \xrightarrow{\epsilon_D} & D \\
\downarrow{\scriptstyle F(G(k))} & & \downarrow{\scriptstyle k} \\
F(G(D')) & \xrightarrow[\epsilon_{D'}]{} & D'
\end{array}
\qquad (A.38)
$$

commutes.

Notation A.4.4. A compact way to denote all the information of an adjunction is

$$
F : \mathcal{C} \underset{\eta}{\overset{\epsilon}{\dashv}} \mathcal{D} : G \qquad (A.39)
$$

(to be parsed as 'there is an adjunction between the functors F and G, $F : \mathcal{C} \to \mathcal{D}$ and $G : \mathcal{D} \to \mathcal{C}$, with counit ϵ and unit η'): one of the first important results about adjunctions is that we can characterise them solely through their unit and counit instead of using the adjunction isomorphisms in A.4.1.

Proposition A.4.5 (Zig-zag identities). Let $F : \mathcal{C} \leftrightarrows \mathcal{D} : G$ be an adjunction, having η and ϵ as unit and counit; using the whiskering operation of A.2.3 we have

$$(G * \epsilon) \circ (\eta * G) = \mathrm{id}_G$$
$$(\epsilon * F) \circ (F * \eta) = \mathrm{id}_F \tag{A.40}$$

(id_F denotes, here, the identity natural transformation of F into itself, and similarly for id_G).

Proposition A.4.6 (Adjoint preserve (co)limits). Let $F \dashv G$ be an adjunction, and let $F : \mathcal{C} \to \mathcal{D}$ be the left adjoint. Then

- F preserves all colimits that exist in $\mathcal{C}$, in the sense of A.3.10,
- dually, the right adjoint G preserves all limits that exist in $\mathcal{D}$, in the sense of (the dual) of A.3.10.

Theorem A.4.7. Let $\mathcal{C}$ be a category, $F : \mathcal{C} \dashv \mathcal{D} : G$ a pair of adjoint functors. Then, the following conditions are equivalent.

EQ1. F, G are both fully faithful (see A.1.14.FF3).
EQ2. Te unit η and the counit ϵ of the adjunction are both natural equivalences (see A.2.2).
EQ3. F is an equivalence of categories between $\mathcal{C}, \mathcal{D}$, whose inverse is G (see A.2.6).

If one of these conditions holds true, then there is also an adjunction

$$G : \mathcal{D} \underset{\epsilon^{-1}}{\overset{\eta^{-1}}{\dashv}} \mathcal{C} : F. \tag{A.41}$$

A.5 The Yoneda Lemma

A.5.1 Presheaves

Definition A.5.1 (Category of presheaves). A *presheaf* is a functor $F : \mathcal{C}^{\mathrm{op}} \to \mathrm{Set}$; the category of presheaves has morphisms the natural transformations between functors, as defined in A.2.1, i.e. the families of maps $\{\alpha_A : FA \to GA\}$ such that the square of functions

$$\begin{array}{ccc} FA & \xrightarrow{\alpha_A} & GA \\ {\scriptstyle Ff}\uparrow & & \uparrow{\scriptstyle Gf} \\ FB & \xrightarrow[\alpha_B]{} & GB \end{array} \tag{A.42}$$

commutes for every arrow $f : A \to B$ in $\mathcal{C}$.

Definition A.5.2 (Representable presheaf). Every object $X \in \mathcal{C}$ defines a presheaf obtained by sending $A \in \mathcal{C}$ into the set of morphisms $u : A \to X$ in $\mathcal{C}$. This defines a functor, called the presheaf *associated* to X, thanks to the associative property of composition in $\mathcal{C}$. The action of $\mathcal{C}(\_, X)$ on morphisms is given by the function that sends $f : X \to Y$ in the natural transformation $\mathcal{C}(\_, X) \Rightarrow \mathcal{C}(\_, Y)$ having components

$$f_{*,A} : \mathcal{C}(A, X) \to \mathcal{C}(A, Y) : u \mapsto f \circ u. \tag{A.43}$$

A completely analogous definition can be given for a functor $X \mapsto \mathcal{C}(X, \_)$, only in this case the functor is contravariant. We call a presheaf F *representable* if it is isomorphic to $\mathcal{C}(\_, X)$ or $\mathcal{C}(X, \_)$ (depending on its variance), for some object $X \in \mathcal{C}$.

A.5.2 The Yoneda Lemma

The Yoneda lemma is one of the tautologies on which our understanding of reality is built.

Lemma A.5.3 (Yoneda lemma). Let よ $=$ よ$_{\mathcal{C}}$ $: \mathcal{C} \to \mathrm{Cat}(\mathcal{C}^{\mathrm{op}}, \mathrm{Set})$ be the functor sending $X \in \mathcal{C}$ to the presheaf associated to X; then, for every $F \in \mathrm{Cat}(\mathcal{C}^{\mathrm{op}}, \mathrm{Set})$, there exists a bijection between the set of natural transformations よ$X \Rightarrow F$ and the set FX. This bijection is moreover natural in the object X.

Proof The proof is elementary, in the algebraic sense of the word: we do the only possible thing with the data we have, and it works.

We define a function of sets

$$Y : [\mathcal{C}^{\mathrm{op}}, \mathrm{Set}](\text{よ}X, F) \to FX \tag{A.44}$$

and we show that it is a bijection. Given a natural transformation α : よ$X \Rightarrow F$ we can consider its X-component $\alpha_X : \mathcal{C}(X, X) \to FX$; as such, it is a function of sets, and it is not equal to the empty function (because $\mathcal{C}(X, X)$ contains at least one element called id_X, thanks to Axiom C4 of A.1.1).

Thus we can define $Y(\alpha) := \alpha_X(\mathrm{id}_X) \in FX$. We shall show

- Y is injective. Assume that $\alpha_X(\mathrm{id}_X) = \beta_X(\mathrm{id}_X)$ for a pair of natural transformations α, β : よ$X \Rightarrow F$. The naturality requirement then entails that for every $u : A \to X$ one has

$$\begin{aligned} \alpha_A(u) &= \alpha_A(\mathrm{id}_X \circ u) \\ &= Fu \circ \alpha_X(\mathrm{id}_X) \end{aligned}$$

$$
\begin{aligned}
&= Fu \circ \beta_X(\mathrm{id}_X) \\
&= \beta_A(u)
\end{aligned}
$$

(if there are no maps $u : A \to X$, then α_A, β_A coincide vacuously, as they both are the unique empty function $\varnothing \to FX$).

- Y is surjective. Given an element $x \in FX$ we can define, for every $A \in \mathcal{C}$ and every $u : A \to X$ the function $\alpha^x_A : \mathcal{C}(A, X) \to FA : u \mapsto Fu(x)$; this is well defined, because Fu is a function $FX \to FA$, given that F is contravariant. Now, we shall show that this is the A-component of a natural transformation $\text{よ}X \Rightarrow F$ (once this has been proved, the fact that $Y(\alpha^x) = x$ is obvious in view of the definition of α^x).

  The naturality of this correspondence follows from the fact that

$$
Ff \circ \alpha^x_B(v) = \alpha^x_A(v \circ f) \tag{A.45}
$$

  for every $f : A \to B$, i.e. from the fact that $Ff(Fv(x)) = F(vf)(x)$, true because F is a contravariant functor.

This concludes the proof. □

There are many ways to read this result.

- A natural transformation $\text{よ}X \to F$ is uniquely determined by the value that its X-component $\alpha_X : \mathcal{C}(X, X) \to FX$ takes on id_X.
- The identity arrow id_X can be thought as the *universal element* witnessing the representability of a functor; thanks to the Yoneda lemma every element $t \in FX$ induces a unique natural transformation $\hat{t}_X : \text{よ}X \Rightarrow F$, and we call t a *universal element* if $\hat{t}_X$ is invertible. The Yoneda lemma can thus be thought as the statement that $\mathrm{id}_X \in \mathcal{C}(X, X)$ is a universal element.

This leads to an immediate corollary.

Corollary A.5.4. The Yoneda map is a fully faithful functor $\text{よ} : \mathcal{C} \to \mathrm{Cat}(\mathcal{C}^{\mathrm{op}}, \mathrm{Set})$.

As a consequence, we call よ the *Yoneda embedding* of $\mathcal{C}$ in $\mathrm{Cat}(\mathcal{C}^{\mathrm{op}}, \mathrm{Set})$.

Proof We can apply A.5.3 to the special case of a representable F: if $F = \hom(\_, A)$ for an object $A \in \mathcal{C}$, we have

$$
\mathrm{Cat}(\mathcal{C}^{\mathrm{op}}, \mathrm{Set})(\text{よ}X, \text{よ}A) \cong \text{よ}AX = \mathcal{C}(X, A). \tag{A.46}
$$

According to the way in which the bijection Y is defined above, in this case it coincides with the action of the functor よ on arrows, and this allows the conclusion. The proof that Y_X defines the X-component of a natural transformation $\text{よ} \Rightarrow F$ is an easy exercise for the reader stated in Exercise A.12. □

Remark A.5.5. There is of course an analogue of the Yoneda lemma for *covariant* functors $G : \mathcal{C} \to \text{Set}$; given such G, there is a natural bijection

$$\text{Cat}(\mathcal{C}, \text{Set})(\check{\text{よ}}_{\mathcal{C}} A, G) \cong GA \tag{A.47}$$

given by evaluating at the universal element id_A for the representable $\check{\text{よ}}_{\mathcal{C}} A = \mathcal{C}(A, \_)$; a similar argument to that in A.5.4 yields that $\check{\text{よ}}_{\mathcal{C}} : \mathcal{C}^{\text{op}} \to \text{Cat}(\mathcal{C}, \text{Set})$ is fully faithful. The same happens for the next result, which can be seen as another fundamental piece of basic category theory: the essential image of $\text{よ}_{\mathcal{C}}$ generates all the category $[\mathcal{C}^{\text{op}}, \text{Set}]$ under colimits.

Lemma A.5.6. The natural transformation $\hat{t}_X$ is the X-component of a cocone for the diagram

$$F \textstyle\int \mathcal{C} \xrightarrow{\Sigma} \mathcal{C} \xrightarrow{\text{よ}} \text{Cat}(\mathcal{C}^{\text{op}}, \text{Set}). \tag{A.48}$$

This means that for every $f \in \mathcal{C}(X, Y)$ which is a morphism in $F \int \mathcal{C}$ between objects $(X, x \in FX)$ and $(Y, y \in FY)$, i.e. such that $Ff(y) = x$, the diagram

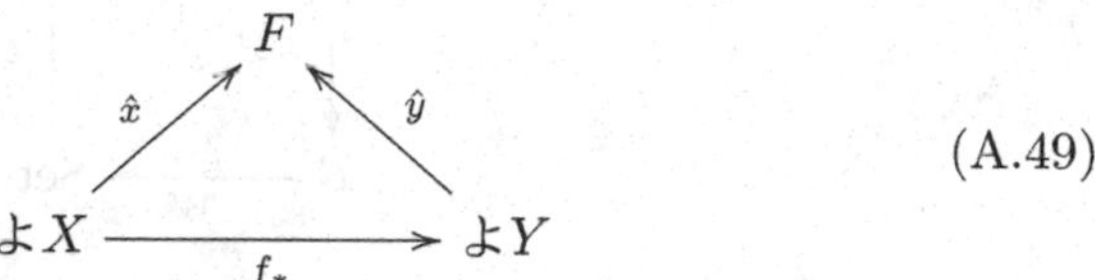

(A.49)

commutes.

Theorem A.5.7 (Density of the Yoneda embedding)**.** Let $F : \mathcal{C}^{\text{op}} \to \text{Set}$ be a presheaf; then F is canonically isomorphic to the colimit of the diagram

$$\mathcal{C} \textstyle\int F \xrightarrow{\Sigma_F} \mathcal{C} \xrightarrow{\text{よ}} \text{Cat}(\mathcal{C}^{\text{op}}, \text{Set}) \tag{A.50}$$

or in other words, the presheaf F is isomorphic to $\text{colim}_{(X,x) \in \mathcal{C} \int F} \text{よ} X$.

Remark A.5.8. The density theorem can be rephrased using coend calculus, as we know from Chapter 4: every presheaf is isomorphic to the weighted colimit of the Yoneda embedding, and it is its own weight.

Alternatively, the ninja Yoneda lemma of Chapter 2 constitutes an elegant rewriting of the density theorem: the isomorphism given by the coend

$$\int^{A} FA \times \mathcal{C}(\_, A) \cong F(\_) \tag{A.51}$$

describes explicitly the way in which the presheaf F is a colimit of all representable presheaves $\mathcal{C}(\_, A)$.

A.5.3 An Alternative View on the Yoneda Lemma

Definition A.5.9. Let $\mathcal{C}$ be an ordinary category, and let $W : \mathcal{C} \to \text{Set}$ be a functor; the *category of elements* $\mathcal{C} \int W$ of W is the category having objects the pairs $(C \in \mathcal{C}, u \in WC)$, and morphisms $(C, u) \to (C', v)$ those $f \in \mathcal{C}(C, C')$ such that $W(f)(u) = v$.

Notation A.5.10. The notation $\mathcal{C} \int W$ for the category of elements of W is borrowed from [Gra80]. Other references call it $\int W$ (it is obvious why we cannot stick to this more compact notation) or $\text{El}(W)$.

Proposition A.5.11. The category $\mathcal{C} \int W$ defined in A.5.9 can be equivalently characterised as each of the following objects.

EL1. The category which results from the pullback

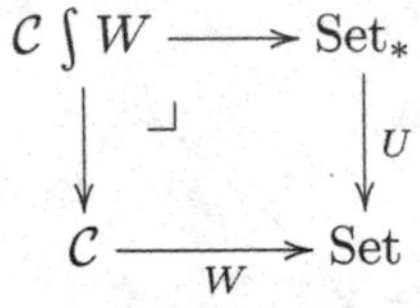

$$\begin{array}{ccc} \mathcal{C} \int W & \longrightarrow & \text{Set}_* \\ \downarrow & \lrcorner & \downarrow U \\ \mathcal{C} & \xrightarrow[W]{} & \text{Set} \end{array} \tag{A.52}$$

where $U : \text{Set}_* \to \text{Set}$ is the forgetful functor which sends a pointed set to its underlying set.

EL2. The comma category $(* \downarrow W)$ of the cospan $\{*\} \to \text{Set} \xleftarrow{W} \mathcal{C}$, where $\{*\} \to \text{Set}$ chooses the terminal object of Set.

EL3. The opposite of the comma category $(\check{よ}_{\mathcal{C}} \downarrow \ulcorner W \urcorner)$, where $\ulcorner W \urcorner : \{*\} \to [\mathcal{C}, \text{Set}]$ is the *name* of the functor W, i.e. the unique functor choosing the presheaf $W \in [\mathcal{C}, \text{Set}]$:

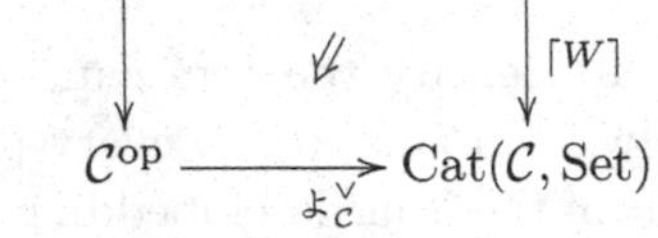

$$\begin{array}{ccc} (\mathcal{C} \int W)^{\text{op}} & \longrightarrow & * \\ \downarrow & \swarrow & \downarrow \ulcorner W \urcorner \\ \mathcal{C}^{\text{op}} & \xrightarrow[\check{よ}_{\mathcal{C}}]{} & \text{Cat}(\mathcal{C}, \text{Set}) \end{array} \tag{A.53}$$

Proof The proof that these categories are all canonically isomorphic to $\mathcal{C} \int W$ is an exercise relying on the Yoneda lemma and universal properties, that we leave to the reader. □

Definition A.5.12 (Discrete opfibration). A *discrete opfibration* of categories is a functor $G : \mathcal{E} \to \mathcal{C}$ with the property that for every object $E \in \mathcal{E}$ and every arrow $p : C \to GE$ in $\mathcal{C}$ there is a unique $q : E' \to E$ 'over p', i.e. such that $Gq = p$.

With a straightforward definition of morphism between discrete opfibrations, we can define the category $\mathrm{DFib}(\mathcal{C})$ of discrete opfibrations over $\mathcal{C}$. The nomenclature here comes from the fact that there exists a dual notion of discrete *fibration*: it is a functor $G : \mathcal{E} \to \mathcal{C}$ such that for every $E \in \mathcal{E}$ and $p : GE \to C$, there exists a unique $q : E \to E'$ such that $Gq = p$. Of course, there is a category of discrete opfibrations over $\mathcal{C}$. As we will see later on, discrete *op*fibrations determine contravariant functors, while fibrations determine covariant ones out of $\mathcal{C}$.

Proposition A.5.13. The category of elements $\mathcal{C} \int W$ of a functor $W : \mathcal{C} \to \mathrm{Set}$ comes equipped with a canonical discrete fibration to the domain of W, which we denote $\Sigma : \mathcal{C} \int W \to \mathcal{C}$, defined forgetting the distinguished element $u \in Wc$.

Now, let S be any set. It is well-known (see for example [MLM92]) that the category of functors $S \to \mathrm{Set}$ (viewing the set S as a discrete category) is equivalent to the category Set/S of *functions over S*. This seemingly innocuous result is a particular instance of what is called *Grothendieck construction.* If possible, the Grothendieck construction is equally as important as the Yoneda lemma, because it clarifies the way in which categories are intrinsically *geometric* entities. A presheaf $F : \mathcal{C}^{\mathrm{op}} \to \mathrm{Set}$ is equivalent to some sort of generalised space 'lying over' the category $\mathcal{C}$, similar to the way in which the total space of a fibre bundle lies over its base. Now, we can consider the *category of elements* A.5.9 of a presheaf $F : \mathcal{C}^{\mathrm{op}} \to \mathrm{Set}$; this sets up a functor from $\mathrm{Cat}(\mathcal{C}^{\mathrm{op}}, \mathrm{Set})$ to the category of discrete opfibrations over $\mathcal{C}$. The Grothendieck construction asserts that this is an equivalence of categories, as defined in A.2.6.

Theorem A.5.14. There is an equivalcnce of categories

$$\mathrm{Cat}(\mathcal{C}^{\mathrm{op}}, \mathrm{Set}) \to \mathrm{DFib}(\mathcal{C}) \tag{A.54}$$

defined by the correspondence sending $F \in \mathrm{Cat}(\mathcal{C}^{\mathrm{op}}, \mathrm{Set})$ to its *fibration of elements* $\Sigma_F : \mathcal{C} \int F \to \mathcal{C}$.

Proof First of all, let us show that Σ_F is a discrete opfibration; then, we shall show that there is a correspondence in the opposite direction assigning to a discrete opfibration $G : \mathcal{E} \to \mathcal{C}$ a presheaf of sets $p^G : \mathcal{C}^{\mathrm{op}} \to \mathrm{Set}$. To this end, given a functor G having codomain $\mathcal{C}$, we define a correspondence that sends an object $X \in \mathcal{C}$ into the category $G^{\leftarrow}X$, where the notion of fibre is understood as follows: we take the objects $E \in \mathcal{E}$ such that $GE = X$, and morphisms $E \to E'$ such that $Gf = \mathrm{id}_X$.

It is easy to notice that the fibre of G over X is a discrete category if G is a discrete opfibration; this means that each fibre of G over X can be regarded as a set (incidentally, this motivates the *discrete* in the name). It is equally easy to see that a morphism $f : X \to X'$ induces a function of sets between the fibres $G^{\leftarrow}X'$ and $G^{\leftarrow}X$ (see Figure A.1). This allows us to show the correspondence with domain $\mathrm{DFib}(\mathcal{C})$ that sends G into $p^G : X \mapsto G^{\leftarrow}X$; the above argument entails that p^G is a presheaf of sets on $\mathcal{C}$ (the discrete opfibration condition entails that it is a contravariant functor).

It is a straightforward check that the composition of the two correspondences is the identity in both directions.

- Given a discrete opfibration $G : \mathcal{E} \to \mathcal{C}$, the category $\mathcal{C} \int p^G$ receives a natural morphism of fibrations from G, in a commutative triangle

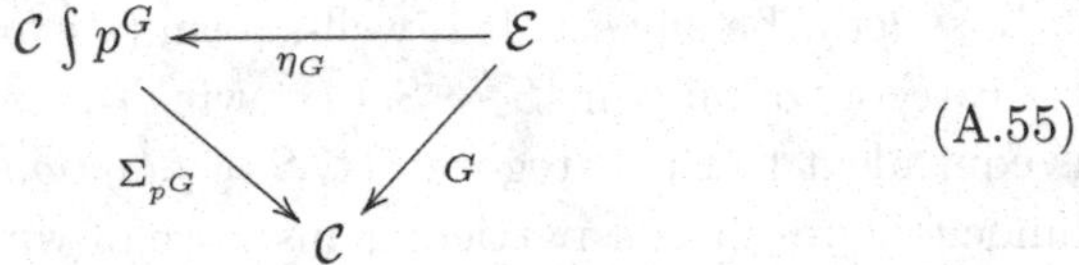

(A.55)

  This is the unit of an adjunction $(\_)^G \dashv \mathcal{C} \int \_$, and it is quite easy to show that η_G is an isomorphism over $\mathcal{C}$ (i.e. an isomorphism of categories over $\mathcal{C}$).
- Given a presheaf F, and given the definition of $p^{\mathcal{C} \int F}$, the identity maps work as components of a natural transformation $p^{\mathcal{C} \int F} \Rightarrow F$.

□

With A.5.14 in our hand, we shall now attempt to answer the very natural question: what does the Yoneda lemma become, if a presheaf $F : \mathcal{C}^{\mathrm{op}} \to \mathrm{Set}$ is regarded as a discrete opfibration over $\mathcal{C}$? First of all, the proof of the following result is an immediate consequence of the definition of $\mathcal{C} \int \mathcal{C}(\_, C)$:

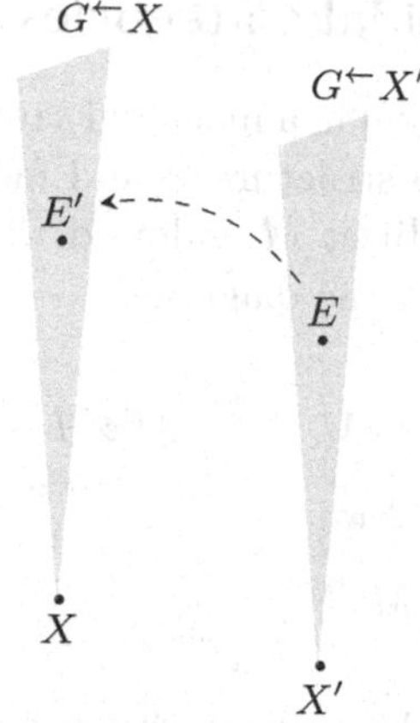

A discrete opfibration $G : \mathcal{E} \to \mathcal{C}$ induces a presheaf, sending $X \in \mathcal{C}$ to the fibre $G^{\leftarrow}X$. A morphism $f : X \to X'$ induces a function of sets $p^G X' \to p^G X$ sending $E \in p^G X'$ to the unique E' such that $GE' = X$.

Lemma A.5.15. The fibration of elements $\Sigma_{\mathcal{C}(\_,C)} = \Sigma_C$ of a representable presheaf coincides with the slice category $\mathcal{C}/C$ of arrows over C, defined in A.2.13; the functor Σ_C coincides with the 'src' functor sending $u : A \to C$ into A.

We are now ready to investigate the form of A.5.3 in terms of fibrations.

Proposition A.5.16 (Yoneda lemma, the geometric way)**.** Let $G : \mathcal{E} \to \mathcal{C}$ be a discrete opfibration over a small category $\mathcal{C}$; let $X \in \mathcal{C}$ be an object; then, there is a bijection between the set of functors $H : \mathcal{C}/X \to \mathcal{E}$ such that $G \circ H = \text{src}$, i.e. of discrete opfibration maps from the element opfibration of $\text{よ}X$ and G, and the set $G^{\leftarrow}X$.

Proof The opfibration of elements of a representable functor $\text{よ}X$ is the category $\mathcal{C}/X$ of arrows having codomain X, the functor $\text{src} : \mathcal{C}/X \to \mathcal{C}$ being the opfibration Σ_C in the notation of A.5.15. A morphism of discrete opfibrations is a functor $H : \mathcal{C}/X \to \mathcal{C}$ such that $GH(\text{id}_X) = X$, so $H \mapsto H(\text{id}_X)$ defines a function $\text{DFib}(\mathcal{C}) \to G^{\leftarrow}X$. This assignment determines the desired bijection, as is easy to see using a similar argument to that of A.5.3. □

A.6 Monoidal Categories and Monads

Definition A.6.1 (Monoid in a monoidal category)**.** Let $\mathcal{C}$ be a monoidal category with monoidal structure $\otimes$ and monoidal unit I; we define a *monoid* in $\mathcal{C}$ to be an object M endowed with maps $m : M \otimes M \to M$ and $u : I \to M$ such that the diagrams

$$\begin{array}{ccc} M\otimes M\otimes M & \xrightarrow{M\otimes m} & M\otimes M \\ {\scriptstyle m\otimes M}\downarrow & & \downarrow{\scriptstyle m} \\ M\otimes M & \xrightarrow[m]{} & M \end{array} \qquad \begin{array}{ccccc} M\otimes I & \xrightarrow{M\otimes u} & M\otimes M & \xleftarrow{u\otimes M} & I\otimes M \\ & \searrow{=} & \downarrow{\scriptstyle m} & {=}\swarrow & \\ & & M & & \end{array} \tag{A.56}$$

commute; these testify the associativity and unitality property of (m, u).

Remark A.6.2. The category $[\mathcal{C}, \mathcal{C}]$ of endofunctors of a category $\mathcal{C}$ has a natural choice of monoidal structure given by composition, whose monoidal unit is the identity functor $\mathrm{id}_{\mathcal{C}}$. It is a very useful exercise to verify the monoidal category axioms one by one; all coherence morphisms are in fact identity, so this is an example of a *strict* monoidal structure (the composition of functors is strictly associative, and $\mathrm{id}_{\mathcal{C}}$ is a strict unit). Of course, this monoidal structure is highly non-symmetric.

Definition A.6.3 (Monad)**.** Let $\mathcal{C}$ be a category; a *monad* on $\mathcal{C}$ consists of an endofunctor $T : \mathcal{C} \to \mathcal{C}$ endowed with two natural transformations

- $\mu : T \circ T \Rightarrow T$, the *multiplication* of the monad, and
- $\eta : \mathrm{id}_{\mathcal{C}} \Rightarrow T$, the *unit* of the monad,

such that the following axioms are satisfied:

- the multiplication is associative, i.e. the diagram

$$\begin{array}{ccc} T\circ T\circ T & \xrightarrow{T*\mu} & T\circ T \\ {\scriptstyle \mu*T}\downarrow & & \downarrow{\scriptstyle \mu} \\ T\circ T & \xrightarrow[\mu]{} & T \end{array} \tag{A.57}$$

  is commutative, i.e. the equality of natural transformations $\mu \circ (\mu * T) = \mu \circ (T * \mu)$ holds;

- the multiplication has the transformation η as unit, i.e. the diagram

$$\begin{array}{ccccc} T & \xrightarrow{\eta * T} & T \circ T & \xleftarrow{T * \eta} & T \\ & \searrow\!\!\!\searrow & \downarrow{\mu} & \swarrow\!\!\!\swarrow & \\ & & T & & \end{array} \tag{A.58}$$

is commutative, i.e. the equality of natural transformations $\mu \circ (\eta * T) = \mu \circ (T * \eta) = \mathrm{id}_T$ holds.

Proposition A.6.4. Let $F \underset{\eta}{\overset{\epsilon}{\dashv}} G$ be an adjunction between two categories, say $F : \mathcal{C} \to \mathcal{D}$; then the composition GF is the endofunctor of a monad on $\mathcal{C}$. The monad multiplication is given by the whiskering $G * \epsilon * F : GFGF \Rightarrow GF$, and the monad unit coincides with the unit of the adjunction, $\eta : \mathrm{id}_{\mathcal{C}} \Rightarrow GF$.

The correspondence sending an adjunction $F \underset{\eta}{\overset{\epsilon}{\dashv}} G$ into a monad $(T, G * \epsilon * F, \eta)$ is a functor towards a suitable category of monads; but it is predictably highly non-bijective, and in fact there is an entire *category* $\mathsf{Spli}(T)$ of adjunctions inducing the same monad. The category $\mathsf{Spli}(T)$ is in general quite difficult to describe; we know, however, that it always has a terminal and an initial object.

Definition A.6.5. Let $T : \mathcal{C} \to \mathcal{C}$ be a monad; we define a T-*algebra* as a pair (A, a), where $A \in \mathcal{C}$ and $a : TA \to A$ is a morphism (the *algebra map*) in $\mathcal{C}$, such that the following properties hold:

- compatibility with the multiplication, namely the equation $a \circ Ta = a \circ \mu_A$;
- compatibility with the unit, namely the equation $a \circ \eta_A = \mathrm{id}_A$.

A *morphism of* T-*algebras* is a morphism $f : A \to B$ that commutes with the algebra maps a, b of the objects A, B, namely such that the square

$$\begin{array}{ccc} TA & \xrightarrow{Tf} & TB \\ {\scriptstyle a}\downarrow & & \downarrow{\scriptstyle b} \\ A & \xrightarrow[f]{} & B \end{array} \tag{A.59}$$

commutes.

Proposition A.6.6. The category of T-algebras, so defined, has the universal property of a terminal object in $\mathsf{Spli}(T)$.

Definition A.6.7. Every pair (TX, μ_X) is, trivially, a T-algebra; these objects define a (full) subcategory of *free T-algebras.*

Proposition A.6.8. The category of free T-algebras has the universal property of an initial object in $\mathsf{Spli}(T)$.

The universal properties A.6.6 and A.6.8 of $\mathrm{Kl}(T)$ and $\mathrm{Alg}(T)$ yields a unique functor $\mathrm{Kl}(T) \to \mathrm{Alg}(T)$, corresponding to the inclusion of free algebras into all algebras. Exercise A.13 makes this characterisation precise.

More in general, the universal property of $\mathrm{Alg}(T)$ as a terminal object yields a unique functor for every other adjunction $F \dashv G$ splitting a given monad:

Definition A.6.9. Given a pair of adjoint functors $(F \dashv G)$ in $\mathsf{Spli}(T)$ we call the functor K, associated with the adjunction and fitting in the diagram below, the *comparison functor.*

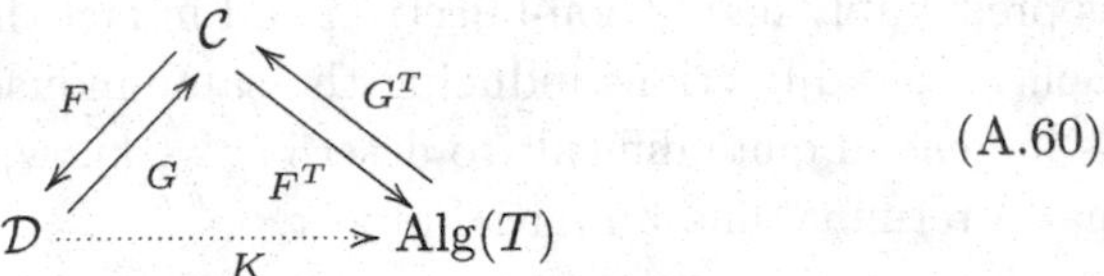

Definition A.6.10. An adjunction $(F \dashv G)$ is *monadic* if the associated comparison functor K is an equivalence.

Beck's theorem gives a necessary and sufficient condition to recognise if a functor is the right adjoint of a monadic adjunction.

Definition A.6.11. Let $U : \mathcal{C} \to \mathcal{D}$ be a functor; we say that a diagram

$$X \underset{g}{\overset{f}{\rightrightarrows}} Y \xrightarrow{h} Z \tag{A.61}$$

is a *split coequaliser* if there exists a fourth and a fifth arrows $s : Z \to Y$ and $t : Y \to X$ such that $h \circ s = \mathrm{id}_Z$, $g \circ t = s \circ h$ and $f \circ t = 1_Y$.

Note that if a diagram like the one above is a split coequaliser, then h is forced to be the coequaliser of the pair f, g, so that h is uniquely determined up to isomorphism by the pair f, g; note also that every functor $F : \mathcal{D} \to \mathcal{E}$ sends a split coequaliser in $\mathcal{D}$ into a split coequaliser in $\mathcal{E}$.

We say that a pair of arrows f, g in $\mathcal{C}$ extends to a split coequaliser if there is an h such that the diagram $X \underset{g}{\overset{f}{\rightrightarrows}} Y \xrightarrow{h} Z$ is a split coequaliser; given the above remark, this is a well defined notion.

Definition A.6.12. Let $U : \mathcal{C} \to \mathcal{D}$ be a functor; then U is monadic if and only if it admits a left adjoint and

MF1. U is conservative (i.e. Uf is an isomorphism if and only if f is),

MF2. $\mathcal{D}$ has, and U preserves, the coequalisers of U-*split pairs*, i.e. those parallel pairs of morphisms in $\mathcal{C}$,

$$X \underset{g}{\overset{f}{\rightrightarrows}} Y \tag{A.62}$$

such that the pair Uf, Ug has a split coequaliser in $\mathcal{D}$.

Definition A.6.13. An *idempotent monad* is a monad (T, μ, η) such that the multiplication $\mu : TT \Rightarrow T$ is an isomorphism. If T is an idempotent monad, each of these properties is true (and equivalent to the requirement of idempotency).

IM1. The natural transformations $T * \eta$ and $\eta * T$ are (invertible and) equal.

IM2. If A is a T-algebra, then there is a unique algebra map $TA \to A$, and it is invertible.

IM3. The category of T-algebras embeds into the category $\mathcal{C}$ via its forgetful functor $U : \mathrm{Alg}(T) \to \mathcal{C}$, which is thus full and faithful.

IM4. Every adjunction $(F, G) \in \mathsf{Spli}(T)$ has invertible counit.

Being endofunctors of $\mathcal{C}$, any two monads T, S on $\mathcal{C}$ can be composed, but the composition ST is often not a monad. In order for such a functor to have a multiplication $\mu : STST \Rightarrow ST$ we must provide an 'intertwining' operator $\lambda : TS \Rightarrow ST$.

Definition A.6.14. A *distributive law* between two monads consists of a 2-cell $\lambda : TS \Rightarrow ST$ such that the following commutativities hold:

DL1. $\lambda \circ (\eta^{(T)} * S) = S * \eta^{(T)}$;

DL2. $(S * \mu^{(T)}) \circ (\lambda * T) \circ (T * \lambda) = \lambda \circ (\mu^{(T)} * S)$;

DR1. $\lambda \circ (T * \eta^{(S)}) = \eta^{(S)} * T$;

DR2. $\lambda \circ (T * \mu^{(S)}) = (\mu^{(S)} * T) \circ (S * \lambda) \circ (\lambda * S)$.

These conditions are expressed by the commutativity of diagrams

$$\begin{array}{ccc} TS & \xrightarrow{\lambda} & ST \\ & {\scriptstyle \eta^{(T)}S}\nwarrow \quad \nearrow{\scriptstyle S\eta^{(T)}} & \\ & S & \end{array} \qquad \begin{array}{ccccc} TTS & \xrightarrow{T\lambda} & TST & \xrightarrow{\lambda T} & STT \\ {\scriptstyle \mu^{(T)}S}\downarrow & & & & \downarrow{\scriptstyle S\mu^{(T)}} \\ TS & & \xrightarrow{\lambda} & & ST \end{array} \tag{A.63}$$

$$\begin{array}{ccc} TS & \xrightarrow{\lambda} & ST \\ & {\scriptstyle T\eta^{(S)}}\nwarrow \quad \nearrow{\scriptstyle \eta^{(S)}T} & \\ & T & \end{array} \qquad \begin{array}{ccccc} TSS & \xrightarrow{\lambda S} & STS & \xrightarrow{S\lambda} & SST \\ {\scriptstyle T\mu^{(S)}}\downarrow & & & & \downarrow{\scriptstyle \mu^{(S)}T} \\ TS & & \xrightarrow{\lambda} & & ST \end{array} \tag{A.64}$$

Remark A.6.15. The notion of distributive law can be seen as just a particular instance of a *monad morphism*: in particular, as a certain kind of endomorphism (S, λ) of the monad T. We now provide the reader with the precise definition of such a notion. Let S, T be two monads, respectively $S : \mathcal{C} \to \mathcal{C}$ and $T : \mathcal{D} \to \mathcal{D}$; a *monad morphism* $(X, \lambda) : S \to T$ consists of a pair $X : \mathcal{C} \to \mathcal{D}$ and $\lambda : TX \Rightarrow XS$ such that the following diagrams are commutative:

$$\begin{array}{ccc} & X & \\ & {\scriptstyle \eta^{(S)}X}\swarrow \quad \searrow{\scriptstyle X\eta^{(T)}} & \\ SX & \xrightarrow{\lambda} & XT \end{array} \qquad \begin{array}{ccccc} SSX & \xrightarrow{S\lambda} & SXT & \xrightarrow{\lambda T} & XTT \\ {\scriptstyle \mu^{(S)}X}\downarrow & & & & \downarrow{\scriptstyle X\mu^{(T)}} \\ SX & & \xrightarrow{\lambda} & & XT \end{array} \tag{A.65}$$

Moreover, a 2-cell between two parallel monad morphisms $(X, \lambda), (Y, \sigma) : (S, \mathcal{C}) \to (T, \mathcal{D})$ consists of a natural transformation $\nu : X \Rightarrow Y$ such that the square

$$\begin{array}{ccc} SX & \xrightarrow{S\nu} & SY \\ {\scriptstyle \lambda}\downarrow & & \downarrow{\scriptstyle \sigma} \\ XT & \xrightarrow[\nu T]{} & YT \end{array} \tag{A.66}$$

is commutative.

Now, let $\mathcal{C}$ be a 2-category. A distributive law between two monads can be characterised as a monad in the 2-category of monads on $\mathcal{C}$, monad morphisms and monad 2-cells. The reader is invited to make this statement precise as an exercise: how do monad morphisms compose? What is the object over which a distributive law is a monad?

Remark A.6.16. Some authors prefer to restrict to a more rigid definition of monad morphism, where the endofunctor X of A.6.15 above is the identity: we call such special monad morphisms *restrained.*

Remark A.6.17. There is a dual theory of *comonads*, these are comonoids in the monoidal category of endofunctors $[\mathcal{C}, \mathcal{C}]$; outlining the basic definitions is left as Exercise A.15.

A.7 2-Categories

Definition A.7.1. A *category enriched over the monoidal base* $\mathcal{V}$, or briefly a $\mathcal{V}$*-category* $\underline{\mathcal{A}}$, consists of the following data.

MC1. a class of objects $\underline{\mathcal{A}}_o$;
MC2. an object $\underline{\mathcal{A}}[A, B] \in \mathcal{V}$ for each pair of objects $A, B \in \mathcal{A}_0$;
MC3. a family of *composition maps* $c_{ABC} : \underline{\mathcal{A}}[A, B] \otimes \underline{\mathcal{A}}[B, C] \to \underline{\mathcal{A}}[A, C]$, one for each triple of objects $A, B, C \in \underline{\mathcal{A}}_o$;
MC4. a family of *identity arrows* $i_A : I \to \underline{\mathcal{A}}[A, A]$, one for each object $A \in \underline{\mathcal{A}}_o$.

These data satisfy the following axioms.

A1. Composition is associative, where associativity is defined via the associator of $\mathcal{V}$: the diagram

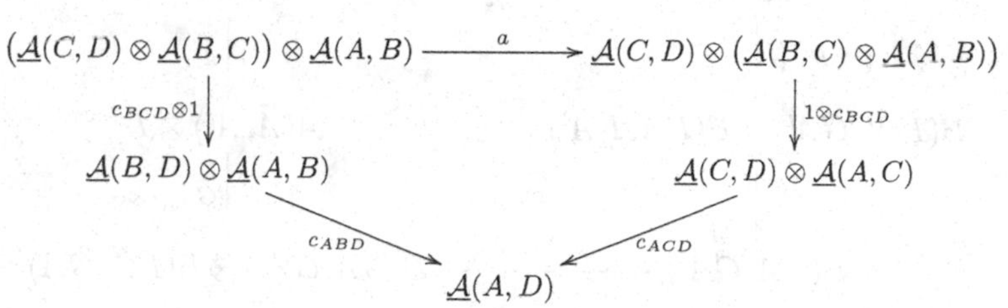

$$\begin{array}{ccc} (\underline{\mathcal{A}}(C,D) \otimes \underline{\mathcal{A}}(B,C)) \otimes \underline{\mathcal{A}}(A,B) & \xrightarrow{a} & \underline{\mathcal{A}}(C,D) \otimes (\underline{\mathcal{A}}(B,C) \otimes \underline{\mathcal{A}}(A,B)) \\ \downarrow{\scriptstyle c_{BCD}\otimes 1} & & \downarrow{\scriptstyle 1\otimes c_{BCD}} \\ \underline{\mathcal{A}}(B,D) \otimes \underline{\mathcal{A}}(A,B) & & \underline{\mathcal{A}}(C,D) \otimes \underline{\mathcal{A}}(A,C) \\ & {\scriptstyle c_{ABD}} \searrow \quad \swarrow {\scriptstyle c_{ACD}} & \\ & \underline{\mathcal{A}}(A,D) & \end{array}$$

is commutative.

A2. Composition has the identities i_A as neutral elements, i.e. the two diagrams

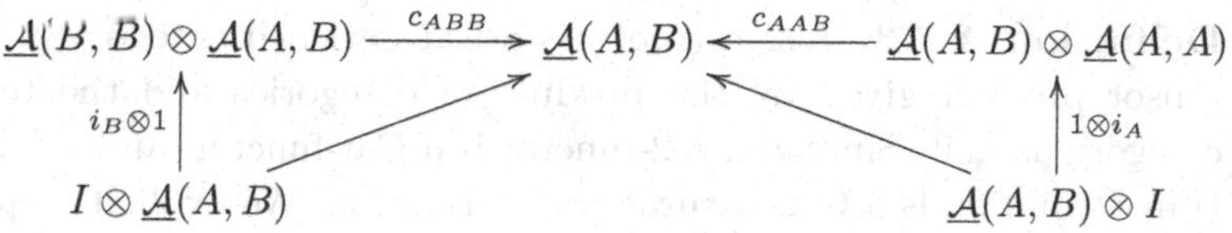

$$\begin{array}{ccccc} \underline{\mathcal{A}}(B,B) \otimes \underline{\mathcal{A}}(A,B) & \xrightarrow{c_{ABB}} & \underline{\mathcal{A}}(A,B) & \xleftarrow{c_{AAB}} & \underline{\mathcal{A}}(A,B) \otimes \underline{\mathcal{A}}(A,A) \\ \uparrow{\scriptstyle i_B\otimes 1} & \nearrow & & \nwarrow & \uparrow{\scriptstyle 1\otimes i_A} \\ I \otimes \underline{\mathcal{A}}(A,B) & & & & \underline{\mathcal{A}}(A,B) \otimes I \end{array}$$

commute.

Definition A.7.2 ($\mathcal{V}$-functor and $\mathcal{V}$-natural transformation). If $\underline{\mathcal{A}}$ and $\underline{\mathcal{B}}$ are two $\mathcal{V}$-categories, a *$\mathcal{V}$-functor* $F : \underline{\mathcal{A}} \to \underline{\mathcal{B}}$ consists of

- a function $F_o : \mathcal{A}_o \to \mathcal{B}_o$,
- a family of $\mathcal{V}$-morphisms $F_{AA'} : \mathcal{A}(A, A') \to \mathcal{B}(FA, FA')$, one for each pair $A, A' \in \mathcal{A}_o$.

These data are such that the following diagrams commute:

$$\begin{array}{ccc} \mathcal{A}(A,A') \otimes \mathcal{A}(A',A'') & \xrightarrow{F_{AA'} \otimes F_{A'A''}} & \mathcal{B}(FA,FA') \otimes \mathcal{B}(FA',FA'') \\ \downarrow c & & \downarrow c \\ \mathcal{A}(A,A'') & \xrightarrow[F_{AA''}]{} & \mathcal{B}(FA,FA'') \end{array}$$

$$\begin{array}{ccc} \mathcal{B}(FA,FA) & \longleftarrow & I \\ & \nwarrow & \downarrow \\ & & \mathcal{A}(A,A) \end{array} \tag{A.67}$$

A $\mathcal{V}$-natural transformation α between two $\mathcal{V}$-functors $F, G : \underline{\mathcal{A}} \to \underline{\mathcal{B}}$ consists of a family of maps $\alpha_A : I \to \mathcal{B}(FA, GA)$ such that the following diagrams commute:

$$\begin{array}{ccc} I \otimes \mathcal{A}(A,A') & \xleftarrow[\sim]{\lambda} & \mathcal{A}(A,A') \\ \downarrow \alpha_{A'} \otimes F_{AA'} & & \wr \downarrow \rho \\ \mathcal{B}(FA',GA') \otimes \mathcal{B}(FA,FA') & & \mathcal{A}(A,A') \otimes I \\ \downarrow c & & \downarrow G_{AA'} \otimes \alpha_A \\ \mathcal{B}(FA,GA') & \xleftarrow[c]{} & \mathcal{B}(GA,GA') \otimes \mathcal{B}(FA,GA) \end{array} \tag{A.68}$$

These diagrams express the fact that α is 'natural' in the sense that $\alpha_A \circ Ff = Gf \circ \alpha_{A'}$ for every $f : A \to A'$; of course, there is no such thing as $f : A \to A'$ here, because $\mathcal{A}(A, A')$ does not have 'elements' *strictu senso*.

Definition A.7.3. A *2-category* is a Cat-enriched category, with the tensor product given by the product of categories and the terminal category as unit. Similarly, a 2-functor is a Cat-functor, and a 2-natural transformation is a Cat-natural transformation. We explicitly spell out the definition of a 2-category, leaving the definition of 2-functor to the reader, once they have understood how to (easily) argue by analogy.

Spelling out the definition above, a 2-category $\mathcal{C}$ consists of

MA1. a class of objects $\mathcal{C}_o$,
MA2. for any pair $X, Y \in C$ a small category $\mathcal{C}(X,Y)$,
MA3. for any triple $X, Y, Z \in C$ a functor $\mu\colon \mathcal{C}(X,Y) \times \mathcal{C}(Y,Z) \to \mathcal{C}(X,Z)$ called composition law,
MA4. a unit functor $\mathcal{I} \to \mathcal{C}(X,X)$, that is to say an object $\mathrm{id}_X \in \mathcal{C}(X,X)$ for every object $X \in \mathcal{C}$.

Furthermore, these data are subject to the following conditions.

MA1. for every $X \in \mathcal{C}$, the functors

$$\mu(\_,\mathrm{id}_Y)\colon \mathcal{C}(X,Y) \to \mathcal{C}(X,Y)$$
$$\mu(\mathrm{id}_X,\_)\colon \mathcal{C}(X,Y) \to \mathcal{C}(X,Y)$$

are the identity functors,
MA2. for every $X, Y, Z, W \in \mathcal{C}$, the diagram

$$\begin{array}{ccc} \mathcal{C}(X,Y) \times \mathcal{C}(Y,Z) \times \mathcal{C}(Z,W) & \xrightarrow{\mu\times\mathrm{id}} & \mathcal{C}(X,Z) \times \mathcal{C}(Z,W) \\ {\scriptstyle \mathrm{id}\times\mu}\downarrow & & \downarrow{\scriptstyle \mu} \\ \mathcal{C}(X,Y) \times \mathcal{C}(Y,W) & \xrightarrow[\mu]{} & \mathcal{C}(X,W) \end{array} \tag{A.69}$$

commutes.

The objects of $\mathcal{C}$ are called 0-cells, the objects of $\mathcal{C}(X,Y)$ are called 1-cells and the morphisms of $\mathcal{C}(X,Y)$ are called 2-cells. The notations are the same as for categories, functors and natural transformations in Cat, which is the prototypical example of 2-category.

Remark A.7.4. The vertical composition of 2-cells is defined by means of the composition law of the hom-categories, while the functor μ recovers the composition of 1-cells and the horizontal composition of 2-cells. Moreover, the functoriality of μ can be used to prove the interchange law for 2-cells.

Definition A.7.5. A 1-cell $f\colon X \to Y$ inside a 2-category $\mathcal{C}$ is said to be an *equivalence* if there exists another 1-cell $g\colon Y \to X$ together with two invertible 2-cells $1_X \Rightarrow gf$ and $fg \Rightarrow 1_Y$.

Naturally, the next step is to describe how the notion of enriched functor specialises to this case.

Remark A.7.6. The previous example defines *strict* 2-functors, so called to distinguish them from other weaker versions of 2-functors between 2-categories. Indeed, a 2-functor sends identities to identities and respects compositions, just as an ordinary functor does, with an extra action on 2-cells. Nevertheless, it makes sense to ask for a weak version of the coherences, whose diagrams commute only up to a 2-cell. If these 2-cells are invertible we get a *pseudofunctor*, otherwise we have a *lax* or *colax functor* (depending on the direction of the 2-cell). Further details can be found in [Bor94a, §7.5], and in A.7.12 below. We say that a (co)lax functor is *normal* if it preserves identities strictly.

Example A.7.7. The simplest example of 2-functors is the 2-dimensional analogue of a hom functor. For instance, the correspondence $\mathcal{C}(X, \_)\colon \mathcal{C} \to \mathrm{Cat}$, for a fixed $X \in \mathcal{C}_o$ defines a 2-functor. The action of this functor is really simple:

2F1. it sends every 0-cell Y to the small category $\mathcal{C}(X,Y)$,

2F2. it maps every 1-cell $f\colon Y \to Z$ to the functor

$$\begin{aligned} f \circ \_ \colon \mathcal{C}(X,Y) &\to \mathcal{C}(X,Z) \\ g &\mapsto f \circ g \\ (\gamma\colon g \Rightarrow g') &\mapsto 1_f * \gamma \end{aligned}$$

where $*$ is the horizontal composition of 2-cells,

2F3. and finally it sends every 2-cell $\alpha\colon f \Rightarrow g$ to the horizontal post-composition $\alpha * \_$.

The contravariant case $\mathcal{C}(\_, Y)\colon \mathcal{C}^{\mathrm{op}} \to \mathrm{Cat}$ is completely analogous.

Definition A.7.8. Let $F, G\colon \mathcal{C} \to \mathcal{D}$ be 2-functors between 2-categories. A 2-natural transformation $\alpha\colon F \Rightarrow G$ is the datum of a 1-cell

$$\alpha_C\colon FC \to GC \tag{A.70}$$

for every $C \in \mathcal{C}$, in such a way that the following diagram

$$\begin{array}{ccc} \mathcal{C}(C,C') & \xrightarrow{F_{CC'}} & \mathcal{D}(FC,FC') \\ {\scriptstyle G_{CC'}}\downarrow & & \downarrow{\scriptstyle \alpha_{C'}\circ\_} \\ \mathcal{D}(GC,GC') & \xrightarrow[\_\circ\alpha_C]{} & \mathcal{D}(FC,GC') \end{array} \tag{A.71}$$

commutes.

In Remark A.7.6 we said that 2-functors have weaker counterparts, namely pseudofunctors and lax functors. So it happens for 2-natural transformations, which in turn can be weakened into *pseudonatural transformations* and *lax natural transformations*. For the sake of simplicity, we prefer to give the definitions in the strict case. An explicit definition can be found, again, in [Bor94a, §7.5].

Definition A.7.9. Let $F, G\colon \mathcal{C} \to \mathcal{D}$ be 2-functors and $\alpha, \beta\colon F \Rightarrow G$ 2-natural transformations. A *modification* $\Xi\colon \alpha \Rrightarrow \beta$ is a family of 2-cells

$$\Xi_C\colon \alpha_C \Rightarrow \beta_C \tag{A.72}$$

such that for any two 1-cells $f, g\colon C \rightrightarrows C'$ and any 2-cell $\gamma\colon f \Rightarrow g$, we have that

$$\Xi_{C'} * F\gamma = G\gamma * \Xi_C. \tag{A.73}$$

The definition above applies, basically unchanged, also to pseudonatural and lax natural transformations. By its very definition, a modification is a kind of 'morphism in dimension 3'.

Theorem A.7.10. Let $\mathcal{C}$ be a 2-category, $F\colon \mathcal{C}^{\mathrm{op}} \to \mathrm{Cat}$ a 2-functor and C a 0-cell in $\mathcal{C}$, then there exists an isomorphism of categories

$$[\mathcal{C}^{\mathrm{op}}, \mathrm{Cat}](\mathcal{C}(C, \_), F) \cong FC \tag{A.74}$$

where $[\mathcal{C}^{\mathrm{op}}, \mathrm{Cat}](\mathcal{C}(C, \_), F)$ is the category whose objects are the 2-natural transformations $\mathcal{C}(C, \_) \Rightarrow F$ and with the modifications between those 2-natural transformations as morphisms.

Definition A.7.11 (Bicategory). A *(locally small) bicategory* $\mathcal{B}$ consists of the following data.

- BC1. A class $\mathcal{B}_o$ of *objects*, denoted with Latin letters like $A, B, \ldots$.
- BC2. A collection of (small) categories $\mathcal{B}(A, B)$, one for each $A, B \in \mathcal{B}_o$, whose objects are called *1-cells* or *arrows* with *domain* A and *codomain* B, and whose morphisms $\alpha : f \Rightarrow g$ are called *2-cells* or *transformations* with domain f and codomain g; the composition law $\circ$ in $\mathcal{B}(A, B)$ is called *vertical composition* of 2-cells.
- BC3. A *horizontal composition* of 2-cells

  $$\boxminus_{\mathcal{B},ABC} : \mathcal{B}(B, C) \times \mathcal{B}(A, B) \to \mathcal{B}(A, C) : (g, f) \mapsto g \boxminus f \tag{A.75}$$

  defined for any triple of objects A, B, C. This is a family of functors between hom-categories.

BC4. For every object $A \in \mathcal{B}_o$ there is an arrow $\mathrm{id}_A \in \mathcal{B}(A, A)$ such that for every $A, B \in \mathcal{C}_o$ and $f : A \to B$ we have $f \boxminus \mathrm{id}_A = f = \mathrm{id}_B \boxminus f$.

To this basic structure we add

BS1. a family of invertible maps $\alpha_{fgh} : (f \boxminus g) \boxminus h \cong f \boxminus (g \boxminus h)$ natural in all its arguments f, g, h, which taken together form the *associator* isomorphisms;

BS2. a family of invertible maps $\lambda_f : \mathrm{id}_B \boxminus f \cong f$ and $\varrho_f : f \boxminus \mathrm{id}_A \cong f$ natural in its component $f : A \to B$, which taken together form the *left unitor* and *right unitor* isomorphisms.

Finally, these data are subject to the following axioms.

BA1. For every quadruple of 1-cells f, g, h, k we have that the diagram

$$\begin{array}{ccccc} ((f \boxminus g) \boxminus h) \boxminus k & \xrightarrow{\alpha_{fg,h,k}} & (f \boxminus g) \boxminus (h \boxminus k) & \xrightarrow{\alpha_{f,g,hk}} & f \boxminus (g \boxminus (h \boxminus k)) \\ \downarrow{\scriptstyle \alpha_{f,g,h} \boxminus k} & & & & \uparrow{\scriptstyle f \boxminus \alpha_{g,h,k}} \\ (f \boxminus (g \boxminus h)) \boxminus k & & \xrightarrow{\alpha_{f,gh,k}} & & f \boxminus ((g \boxminus h) \boxminus k) \end{array} \tag{A.76}$$

commutes.

BA2. For every pair of composable 1-cells f, g,

$$\begin{array}{ccc} (f \boxminus \mathrm{id}_A) \boxminus g & \xrightarrow{a_{A,\mathrm{id}_A,g}} & f \boxminus (\mathrm{id}_A \boxminus g) \\ & {\scriptstyle \varrho_f \boxminus g} \searrow \quad \swarrow {\scriptstyle f \boxminus \lambda_g} & \\ & f \boxminus g & \end{array} \tag{A.77}$$

commutes.

Definition A.7.12 (Pseudofunctor, (co)lax functor). Let $\mathcal{B}, \mathcal{C}$ be two bicategories; a *pseudofunctor* consists of

PF1. a function $F_o : \mathcal{B}_o \to \mathcal{C}_o$,

PF2. a family of functors $F_{AB} : \mathcal{B}(A, B) \to \mathcal{C}(FA, FB)$,

PF3. an invertible 2-cell $\mu_{fg} : Ff \circ Fg \Rightarrow F(fg)$ for each $A \xrightarrow{g} B \xrightarrow{f} C$, natural in f (with respect to vertical composition) and an invertible 2-cell $\eta : \eta_f : \mathrm{id}_{FA} \Rightarrow F(\mathrm{id}_A)$, also natural in f.

These data are subject to the following commutativity conditions for every 1-cell $A \to B$:

$$\begin{array}{ccc} Ff \circ \mathrm{id}_A & \xrightarrow{\varrho_{Ff}} & Ff \\ {\scriptstyle Ff*\eta}\downarrow & & \downarrow{\scriptstyle F(\varrho_f)} \\ Ff \circ F(\mathrm{id}_A) & \xrightarrow[\mu_{f,\mathrm{id}_A}]{} & F(f \circ \mathrm{id}_A) \end{array} \qquad \begin{array}{ccc} \mathrm{id}_B \circ Ff & \xrightarrow{\lambda_{Ff}} & Ff \\ {\scriptstyle \eta*Ff}\downarrow & & \downarrow{\scriptstyle F(\lambda_f)} \\ F(\mathrm{id}_B) \circ Ff & \xrightarrow[\mu_{\mathrm{id}_B,f}]{} & F(\mathrm{id}_B \circ f) \end{array}$$

$$\begin{array}{ccc} (Ff \circ Fg) \circ Fh & \xrightarrow{\alpha_{Ff,Fg,Fh}} & Ff \circ (Fg \circ Fh) \\ {\scriptstyle \mu_{fg}*Fh}\downarrow & & \downarrow{\scriptstyle Ff*\mu_{gh}} \\ F(fg) \circ Fh & & Ff \circ F(gh) \\ {\scriptstyle \mu_{fg}*Fh}\downarrow & & \downarrow{\scriptstyle \mu_{f,gh}} \\ F((fg)h) & \xrightarrow[F\alpha_{fgh}]{} & F(f(gh)) \end{array}$$

(we denote invariably α, λ, ϱ the associator and unitor of $\mathcal{B}, \mathcal{C}$).

A *lax* functor is defined by the same data, but both the 2-cells $\mu : Ff \circ Fg \Rightarrow F(fg)$ and $\eta : \mathrm{id}_{FA} \Rightarrow F(\mathrm{id}_A)$ can be non-invertible; the same coherence diagrams A.7.12 hold. A *colax* functor reverses the direction of the cells μ, η, and the commutativity of the diagrams in A.7.12 changes accordingly.

A.8 Higher Categories

Definition A.8.1 (The simplex category). The *simplex category*, denoted $\mathbf{\Delta}$, is defined as the category having

- objects the non-empty finite sets that are totally ordered. The typical object of $\mathbf{\Delta}$ is denoted $[n] = \{0 < \cdots < n\}$, in this way, $[0]$ is the terminal object of $\mathbf{\Delta}$;
- morphisms $[n] \to [m]$ the order preserving functions.

Remark A.8.2. If in $\mathbf{\Delta}$ we consider

- the $n+1$ injective functions $\delta_{n,k} \colon [n] \to [n+1]$ whose image misses k (called *cofaces*),
- the $n+1$ surjective functions $\sigma_{n,k} \colon [n] \to [n-1]$ assuming the value k twice (called *codegeneracies*),

we obtain that every $f \colon [m] \to [n]$ can be written as a composition

$$f = \delta_{n_1,k_1} \circ \cdots \circ \delta_{n_r,k_r} \circ \sigma_{m_1,h_1} \circ \cdots \circ \sigma_{m_s,h_s} \tag{A.78}$$

for some indices n_i, m_j and $0 \leq k_i \leq n_i, 0 \leq h_j \leq m_j$.

Remark A.8.3. The functions $\delta_{n,k}$ and $\sigma_{n,k}$ satisfy the *cosimplicial identities*, namely they fit into commutative diagrams

$$
\begin{CD}
[n-1] @>{\delta_i}>> [n] \\
@V{\delta_{j-1}}VV @VV{\delta_j}V \\
[n] @>>{\delta_i}> [n+1]
\end{CD}
\ (i<j)
\qquad
\begin{CD}
[n-1] @>{\delta_i}>> [n] \\
@V{\sigma_j}VV @VV{\sigma^j}V \\
[n-2] @>>{\delta_i}> [n-1]
\end{CD}
\ (i<j)
\qquad
\begin{CD}
[n-1] @>{\delta_j}>> [n] \\
@V{\delta_{j+1}}VV @VV{\sigma_j}V \\
[n] @>>{\sigma_j}> [n-1]
\end{CD}
$$

$$
\begin{CD}
[n-1] @>{\delta_i}>> [n] \\
@V{\sigma_j}VV @VV{\sigma_j}V \\
[n-2] @>>{\delta_{i-1}}> [n-1]
\end{CD}
\ (i>j+1)
\qquad
\begin{CD}
[n+1] @>{\sigma_i}>> [n] \\
@V{\sigma_i}VV @VV{\sigma_j}V \\
[n] @>>{\sigma_{j+1}}> [n-1]
\end{CD}
\ (i\le j)
\tag{A.79}
$$

Throughout the discussion, we consider $\mathbf{\Delta}$ as a full subcategory of Cat, via the identification

$$\{0 < \cdots < n\} = \{0 \to 1 \to \cdots \to n\}. \tag{A.80}$$

This secretly defines an identity-on-objects functor $\iota : \mathbf{\Delta} \to \mathrm{Cat}$.

Definition A.8.4 (The category of simplicial sets)**.** The category of *simplicial sets* is defined as the category $[\mathbf{\Delta}^{\mathrm{op}}, \mathrm{Set}]$ of presheaves on $\mathbf{\Delta}$.

A simplicial set can equivalently be specified by a collection of sets (a *graded set*) $\{X_n \mid n \geq 0\}$ with maps

$$d_{n,i} \colon X_n \to X_{n-1} \qquad s_{n,j} \colon X_n \to X_{n+1}, \qquad 0 \leq i,j \leq n \tag{A.81}$$

(respectively called the *faces* and *degeneracies* of X) satisfying the dual identities of (A.8.3), called the *simplicial identities*

$$
\begin{cases}
d_i d_j = d_{j-1} d_i & i < j \\
d_i s_j = s_{j-1} d_i & i < j \\
d_j s_j = \mathrm{id} = d_{j+1} s_j & \\
d_i s_j = s_j d_{i-1} & i > j+1 \\
s_i s_j = s_{j+1} s_i & i \leq j.
\end{cases}
\tag{A.82}
$$

In the following discussion, we will freely employ, and without further mention, this identification. The elements of X_n are called *n-simplices* of X.

The Yoneda embedding $よ_{\mathbf{\Delta}} \colon \mathbf{\Delta} \to \mathrm{sSet}$ sends every object $\mathbf{\Delta}$ in the *representable* simplicial set $よ[n] = \Delta(\_, [n])$, acting on objects and morphisms of $\mathbf{\Delta}$ in the expected way. The usual notation for the representable よ$[n]$ on $[n]$ is $\Delta[n]$. In order to familiarise the reader with the fundamentals of simplicial combinatorics, we propose two simple exercises:

- Show that the m-simplices of $\Delta[n]$ are in bijection with the tuples $(a_0, \dots, a_m)$ of elements of $\{0, \dots, n\}$, such that $a_0 \leq a_1 \leq \dots \leq a_m$.
- Given a simplicial set X and one of its n-simplices, say $x \in X_n$, x is called *non degenerate* if it cannot be expressed as degeneracy of some lower dimensional simplex $y \in X_k$. Show that the non degenerate m-simplices of $\Delta[n]$ are in bijection with the subsets of $\{0, \dots, n\}$ having cardinality $m + 1$.

Definition A.8.5 (Boundaries and horns). We define the *boundary* of $\Delta[n]$ as the union

$$\partial\Delta[n] = \bigcup_{i=0}^{n} d_{n,i}(\Delta[n-1]), \tag{A.83}$$

and we define the *kth n-dimensional horn* as the union

$$\Lambda^k[n] = \bigcup_{i \neq k} d_{n,i}(\Delta[n-1]). \tag{A.84}$$

A.9 Miscellaneous Definitions

The already hard endeavour to write a self-contained introduction to category theory is made harder by the vastness of the subject. As a consequence, we must necessarily leave out important fragments of theory that often constitute research areas in their own right; these subtheories are here only touched in one or two lines of the book, in so tiny a space that it is impossible to do them justice.

While referring the interested reader to the customary sources, we employ this last section of the book to give a rapid glance at the category theory we have left out of it.

Definition A.9.1 (Ordinals, cardinals). For us, an *ordinal number* will be any well-ordered set, and a *cardinal number* is any ordinal which is not in bijection with a smaller ordinal. Every set X has a unique *cardinality*, i.e. a cardinal κ with a bijection $\kappa \cong X$ such that there are no bijections from a smaller ordinal. We freely employ results that depend on the axiom of choice when needed. A cardinal κ is *regular* if no set of cardinality κ is the union of fewer than κ sets of cardinality less than κ; all infinite cardinals are assumed regular without further mention.

Definition A.9.2 (Filtered category). Let κ be a cardinal; we say that a category $\mathcal{A}$ is *κ-filtered* if for every category $\mathcal{J} \in \mathrm{Cat}_{<\kappa}$ with less than κ objects, $\mathcal{A}$ is injective with respect to the cone completion $\mathcal{J} \to \mathcal{J}^{\triangleright}$; this means that every diagram

$$\begin{array}{ccc} \mathcal{J} & \xrightarrow{D} & \mathcal{A} \\ \downarrow & \nearrow_{\bar{D}} & \\ \mathcal{J}^{\triangleright} & & \end{array} \tag{A.85}$$

has a dotted filler $\bar{D} : \mathcal{J}^{\triangleright} \to \mathcal{A}$.

We say that a category $\mathcal{C}$ admits filtered colimits if for every filtered category $\mathcal{A}$ and every diagram $D : \mathcal{A} \to \mathcal{C}$, the colimit $\operatorname{colim} D$ exists as an object of $\mathcal{C}$. Of course, whenever an ordinal α is regarded as a category, it is a filtered category, so a category that admits all κ-filtered colimits admits all colimits of chains

$$C_0 \to C_1 \to \cdots \to C_\alpha \to \cdots \tag{A.86}$$

with less than κ terms. A useful, completely elementary result is that the existence of colimits over all ordinals less than κ implies the existence of κ-filtered colimits; this relies on the fact that every filtered category $\mathcal{A}$ admits a cofinal functor from an ordinal $\alpha_{\mathcal{A}}$.

Definition A.9.3. Let $\mathcal{C}$ be a category.

- We say that $\mathcal{C}$ is *κ-accessible* if it admits κ-filtered colimits, and if it has a *small* subcategory $\mathcal{S} \subset \mathcal{A}$ of κ-presentable objects such that every $A \in \mathcal{A}$ is a κ-filtered colimit of objects in $\mathcal{S}$.
- We say that $\mathcal{C}$ is *(locally) κ-presentable* if it is accessible and cocomplete.

There is a 2-category whose objects are κ-accessible categories, whose morphisms are functors that preserve κ-filtered colimits (also called κ-ary functors, or functors of rank κ), and all natural transformations between these. In particular we call *finitary* the functors of rank ω.

The theory of presentable and accessible categories is a cornerstone of *categorical logic*, i.e. of the translation of model theory into the language of category theory.

Accessible and presentable categories admit so-called *representation theorems*.

- A category $\mathcal{C}$ is accessible if and only if it is equivalent to the ind-completion $\text{Ind}_\kappa(\mathcal{S})$ of a small category, i.e. to the completion of a small category $\mathcal{S}$ under κ-filtered colimits.
- A category $\mathcal{C}$ is presentable if and only if it is a full reflective subcategory of a category of presheaves $i : \mathcal{C} \to \text{Cat}(\mathcal{S}^{\text{op}}, \text{Set})$, such that the embedding functor i commutes with κ-filtered colimits.

All categories of usual algebraic structures are (finitely) accessible, and they are locally (finitely) presentable as soon as they are cocomplete; an example of a category which is $\aleph_1$-presentable but not $\aleph_0$-presentable is the category $\mathcal{M}]\sqcup$ of *metric spaces* and short maps.

We now glance at *topos theory*:

Definition A.9.4. An *elementary topos* is a category $\mathcal{E}$

- which is *cartesian closed*, i.e. each functor $\_ \times A$ has a right adjoint $[A, \_]$,
- having a *subobject classifier*, i.e. an object $\Omega \in \mathcal{E}$ such that the functor $\text{Sub} : \mathcal{E}^{\text{op}} \to \text{Set}$ sending A into the set of isomorphism classes of monomorphisms $\left[\begin{smallmatrix} U \\ \downarrow \\ A \end{smallmatrix}\right]$ is representable by the object Ω.

The natural bijection $\mathcal{E}(A, \Omega) \cong \text{Sub}(A)$ is obtained by pulling back the monomorphism $U \subseteq A$ along a *universal arrow* $t : 1 \to \Omega$, as in the diagram

$$\begin{array}{ccc} U & \longrightarrow & 1 \\ \downarrow & \lrcorner & \downarrow t \\ A & \xrightarrow[\chi_U]{} & \Omega \end{array} \tag{A.87}$$

so, the bijection is induced by the map $\left[\begin{smallmatrix} U \\ \downarrow \\ A \end{smallmatrix}\right] \mapsto \chi_U$.

Definition A.9.5. A *Grothendieck topos* is an elementary topos that, in addition, is locally finitely presentable.

Whenever we spoke about sheaves on a topological space or a Grothendieck site, we were secretly talking about topos theory; the notion of Grothendieck topos is intimately connected with (co)end calculus, as we have seen throughout Chapter 3, and especially in 3.2.16.

In fact, Giraud theorem gives a proof for the difficult implication of the following *recognition principle* for Grothendieck toposes.

Theorem A.9.6. Let $\mathcal{E}$ be a category; then $\mathcal{E}$ is a Grothendieck topos if and only if it is a left exact reflection of a category $\text{Cat}(\mathcal{A}^{\text{op}}, \text{Set})$ of presheaves on a small category $\mathcal{A}$.

Recall that a *left exact reflection* of $\mathcal{C}$ is a reflective subcategory $\mathcal{R} \hookrightarrow \mathcal{C}$ such that the reflector $r : \mathcal{C} \to \mathcal{R}$ preserves finite limits. It is a reasonably easy exercise to prove that a left exact reflection of a Grothendieck topos is again a Grothendieck topos; Giraud proved that all Grothendieck toposes arise this way.

Next, we mention the existence of *abelian categories.*

Albeit tangential to the (co)end calculus exposed in this book, the notion of abelian category is historically relevant: the definition of (co)integration was given by Yoneda [Yon60] in the setting of module categories, and these constitute the main motivating example for the abstract definition of abelian category.

Definition A.9.7. A category $\mathcal{A}$ is called *abelian* if it satisfies the following list of axioms.

- A1. It is enriched (see A.7.1) over abelian groups, i.e. every $\mathcal{A}(A, A')$ is an abelian group and the composition operation is $\mathbb{Z}$-bilinear, thus can be represented as an abelian group homomorphism $\mathcal{A}(A, A') \otimes \mathcal{A}(A', A'') \to \mathcal{A}(A, A'')$.
- A2. $\mathcal{A}$ has all finite limits and all finite colimits.
- A3. In $\mathcal{A}$, every monomorphism is a kernel, and every epimorphism is a cokernel (this means that if $m : A \to B$ is a monomorphism, then it appears in an equaliser diagram $A \xrightarrow{m} B \underset{0}{\overset{f}{\rightrightarrows}} C$, and dually, if $e : B \to C$ is an epimorphism, then it appears in a coequaliser diagram $A \underset{0}{\overset{g}{\rightrightarrows}} B \xrightarrow{e} C$).

Remark A.9.8. The notion of abelian category admits many equivalent definitions; so, A.9.7 above is not the only possible way to define it. An alternative presentation of the axioms disassembles A2 into

- A2.1. $\mathcal{A}$ has finite products and finite coproducts,
- A2.2. every morphism $f : A \to B$ in $\mathcal{A}$ has a kernel and a cokernel.

A category satisfying only A1 is called *preadditive* (note that this entails that $\mathcal{A}$ has a zero object if and only if it has an initial object, if and only if it has a terminal object); a category satisfying A1 and A2.1 is called *additive*; a category satisfying A1, A2.1 and A2.2 is called *preabelian.*

A merit of A.9.7 above is that all axioms are visibly auto-dual, i.e. $\mathcal{A}$ satisfies A1–A3 if and only if the opposite category $\mathcal{A}^{\text{op}}$ satisfies A1–A3.

Throughout this book, there are few explicit mentions of abelian categories: see for example 3.2.10, or the discussion in 7.2.2; derived categories appearing in 7.2.2 are seldom abelian, because they lack finite (co)limits.

Exercises

A.1 Show that for every fixed cardinal number κ, there exists an abelian group of cardinality κ. If κ is infinite, is it true that there is a field of cardinality κ (of course, it is false if κ is finite and not a power p^n of a prime number p)?

A.2 Let $\mathcal{C}$ be a category admitting an initial and a terminal object; prove that the initial object is the colimit of the unique empty diagram $\varnothing \to \mathcal{C}$ from the empty category, using A.3.8, and dually that a terminal object is the limit of the same empty diagram.

A.3 Prove that a category $\mathcal{C}$ has a terminal object if and only if the unique functor $\mathcal{C} \to *$ to the terminal category has a right adjoint; dually, a category $\mathcal{C}$ has an initial object if and only if the unique functor $\mathcal{C} \to *$ to the terminal category has a left adjoint.

A.4 Prove that the category Set^{op}, i.e. the opposite category of sets and functions, is not equivalent to Set.

A.5 Consider the comma category of A.2.14.

- Let $\text{id}_\mathcal{C}$ be the identity functor: characterise the category $(\text{id}_\mathcal{C} / G)$. Let C be the functor $\mathcal{J} \to \mathcal{C}$ assuming the value C constantly: characterise the category $(C/\,\text{id}_\mathcal{C})$.
- What is the relation between (F/G) and (G/F)?
- The *iso-comma* category of the functors F and G is given by a similar definition to A.2.14; the only difference is that we only consider arrows $Fs \to Gt$ that are invertible; we denote it as $[F/G]$. What is the relation, now, between $[F/G]$ and $[G/F]$? Are they isomorphic?
- Show that there are two functors $\mathcal{S} \xleftarrow{p_\mathcal{S}} (F/G) \xrightarrow{p_\mathcal{T}} \mathcal{T}$ such that the square

$$\begin{array}{ccc} (F/G) & \xrightarrow{p_\mathcal{S}} & \mathcal{S} \\ {\scriptstyle p_\mathcal{T}}\downarrow & & \downarrow{\scriptstyle F} \\ \mathcal{T} & \xrightarrow[G]{} & \mathcal{C} \end{array}$$

commutes.

- Consider the functor $J_1 : (F/G) \xrightarrow{p_S} \mathcal{S} \xrightarrow{F} \mathcal{C}$. Determine the category (J_1/G). Does the composition of the projections above in the triangle $\star$ equal the projection $(J_1/G) \to \mathcal{T}$ (in other words, does the triangle commute)?

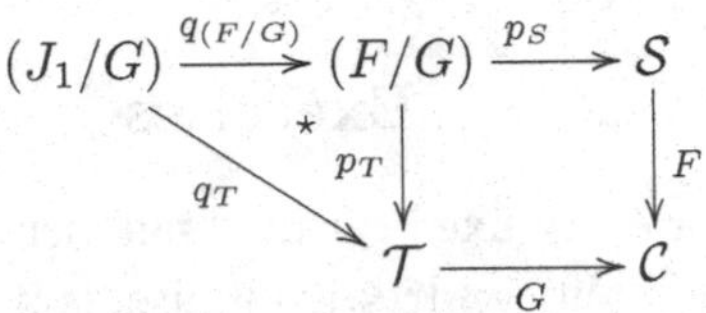

- Does the category (F/G) have a universal property?

A.6 Let Ś be the Sierpiński space, where a two-point set $J = \{a, b\}$ has topology $\{\varnothing, \{a\}, J\}$. Let C be the codiscrete space, where $\{a, b\}$ has trivial topology $\{\varnothing, J\}$, and D the discrete space where J has discrete topology $\{\varnothing, \{a\}, \{b\}, J\}$. Show that

- the functor $O : \text{Spc}^{\text{op}} \to \text{Set}$ that sends a topological space into its set of open subsets is representable, and that Ś is its representing object;
- the functor $D : \text{Spc}^{\text{op}} \to \text{Set}$ that sends a topological space X into its set of *disconnections*, i.e. the set of pairs (U, V) such that $U \cup V = X$ and $U \cap V = \varnothing$, is representable by the discrete space D;
- the functor $S : \text{Spc}^{\text{op}} \to \text{Set}$ that sends a topological space into its set of subspaces is representable by the codiscrete space C.

Show that the set of natural transformations $[\text{Spc}^{\text{op}}, \text{Set}](O, S)$ has exactly four elements.

A.7 Show that a cocone $\bar{D} : \mathcal{J}^{\triangleright} \to \mathcal{C}$ for a diagram $D : \mathcal{J} \to \mathcal{C}$ is exactly a cone $\bar{D}^{\text{op}} : \mathcal{J}^{\triangleleft} \to \mathcal{C}^{\text{op}}$ for the opposite functor $D^{\text{op}} : \mathcal{J}^{\text{op}} \to \mathcal{C}^{\text{op}}$. Prove this directly, but then notice that A.3.5 and in particular (A.23) give a slick argument to conclude.

A.8 Let $f : [n] \to [m]$ be a map in $\boldsymbol{\Delta}$ (see A.8.1). We regard $\boldsymbol{\Delta}$ as a subcategory of Cat in the obvious way. Show that

- a morphism $f\colon [n] \to [m]$ in $\boldsymbol{\Delta}$ has a left adjoint if and only if $f(n) = m$; in such a situation the adjoint f_L sends i into $f_L(i) = \min\{j \in [n] \mid f(j) \geq i\}$;
- dually, a morphism $f\colon [n] \to [m]$ in $\boldsymbol{\Delta}$ has a right adjoint if and only if $f(0) = 0$; in such a situation, the adjoint f_R sends i into $f_R(i) = \max\{j \in [n] \mid f(j) \leq i\}$.

Deduce that for every $j \in [n]$ the function $\sigma_{n,j}$ has both a left and a right adjoint, and in particular

$$\delta_{n,j+1} \dashv \sigma_{n-1,j} \dashv \delta_{n,j}\colon\ [n] \xrightarrow{\sigma_{n-1,j}} [n-1].$$

A.9 Let $F \hookrightarrow E$ be a homomorphism of rings between fields; it is thus a monomorphism. We define

$$\mathsf{Fix}(E|F) = \{\sigma : E \xrightarrow{\sim} E \mid \sigma|_F = \mathrm{id}_F\}$$

as the set of automorphisms of E that become the identity map when restricted to F, and

$$\mathsf{Ext}(E|F) = \{K \mid F \le K \le E\}$$

the set of intermediate extensions between F and E. We define two functions $\mathrm{i} : \mathsf{Fix}(E|F) \to \mathsf{Ext}(E|F)$ and $\mathrm{j} : \mathsf{Ext}(E|F) \to \mathsf{Fix}(E|F)$ sending respectively $\sigma \in \mathsf{Fix}(E|F)$ in the intermediate field $\{a \in E \mid \sigma(a) = a\}$ and the intermediate field $F \le K \le E$ in the group of those E-automorphisms that become the identity when restricted to K. Show that i, j set up an adjunction (thus the name *Galois connections* for these adjunctions) when the posets $\mathsf{Fix}(E|F)$ and $\mathsf{Ext}(E|F)$ are regarded as categories; what is the right adjoint, and what is the left adjoint? What is the unit, and what is the counit?

A.10 Show that the limit of a constant diagram $\Delta_{\mathcal{J}}A : \mathcal{J} \to \mathcal{A}$ is a product of copies of A indexed by the set $\pi_0\mathcal{J}$ of connected components of the domain $\mathcal{J}$: prove that there is a bijection

$$\mathrm{Cn}(X, A) \cong \mathrm{Set}(\pi_0\mathcal{J}, \mathcal{A}(X, A))$$

and conclude, using (2.16).

A.11 Show that the following conditions are equivalent in a category $\mathcal{C}$ with a terminal object.

- The terminal arrow $\begin{bmatrix} X \\ \downarrow \\ 1 \end{bmatrix}$ is right orthogonal to $f : A \to B$.
- The functor $\mathcal{C}(\_, X)$ sends f to an isomorphism.
- Every arrow $A \to X$ in $\mathcal{C}$ has a unique extension to $B \to X$ along f.

A.12 Show that in the same notation of (A.44), Y_X is the X-component of a natural transformation $\mathrm{Cat}(\mathcal{C}^{\mathrm{op}}, \mathrm{Set})(よ, F) \Rightarrow F$, that is natural in both its arguments; in other words, show that for every arrow

$X \to X'$ the square

$$\begin{array}{ccc} \mathrm{Cat}(\mathcal{C}^{\mathrm{op}}, \mathrm{Set})(よX, F) & \xrightarrow{Y} & FX \\ {\scriptstyle [\mathcal{C}^{\mathrm{op}},\mathrm{Set}](よ(f),F)} \uparrow & & \uparrow {\scriptstyle Ff} \\ \mathrm{Cat}(\mathcal{C}^{\mathrm{op}}, \mathrm{Set})(よ(X'), F) & \xrightarrow[Y]{} & FX' \end{array}$$

is commutative.

Similarly, show that for every natural transformation $\tau : F \Rightarrow G$ the square

$$\begin{array}{ccc} \mathrm{Cat}(\mathcal{C}^{\mathrm{op}}, \mathrm{Set})(よX, F) & \xrightarrow{Y} & FX \\ {\scriptstyle \mathrm{Cat}(\mathcal{C}^{\mathrm{op}},\mathrm{Set})(よX,\tau)} \downarrow & & \downarrow {\scriptstyle \tau_X} \\ \mathrm{Cat}(\mathcal{C}^{\mathrm{op}}, \mathrm{Set})(よX, G) & \xrightarrow[Y]{} & GX \end{array}$$

is commutative.

A.13 Denote $\mathcal{C}_T$ the category defined in A.6.7. Define the category $\mathrm{Kl}(T)$ as follows: it has the same objects of $\mathcal{C}$, and

$$\mathrm{Kl}(T)(X, Y) := \mathcal{C}(X, TY). \tag{A.88}$$

Composition of $X \xrightarrow{f} TY$ with $Y \xrightarrow{g} TZ$ is defined by the rule

$$g \bullet_T f := \left(X \xrightarrow{f} TY \xrightarrow{Tg} TTZ \xrightarrow{\mu_Z} TZ\right).$$

- Prove that this really defines a category, if the identity map of an object A is the unit $\eta_A : A \to TA$ (prove that the composition is associative and that $\eta_B \bullet_T f = f = f \bullet_T \eta_A$ for every $f : A \to TB$).
- Defines a functor $W : \mathrm{Kl}(T) \to \mathcal{C}_T$ that acts as T on objects, and such that $Wf = \mu_B \circ Tf$. Prove that W, so defined, is fully faithful, and that its essential image is made by free T-algebras.

A.14 Show that each square

$$\begin{array}{ccc} [n+1] & \xrightarrow{\sigma_i} & [n] \\ {\scriptstyle \sigma_i} \downarrow & {\scriptstyle i \le j} & \downarrow {\scriptstyle \sigma_j} \\ [n] & \xrightarrow[\sigma_{j+1}]{} & [n-1] \end{array}$$

is an absolute pushout, and that every square

$$\begin{array}{ccc} [n-1] & \xrightarrow{\delta_i} & [n] \\ {\scriptstyle \delta_{j-1}}\downarrow & {\scriptstyle i<j} & \downarrow{\scriptstyle \delta_j} \\ [n] & \xrightarrow[\delta_i]{} & [n+1] \end{array}$$

is an absolute pullback.

A.15 A comonoid in a monoidal category is an object $M \in \mathcal{C}$ endowed with maps $c : M \to M \otimes M$ and $e : M \to I$ such that the following diagrams commute

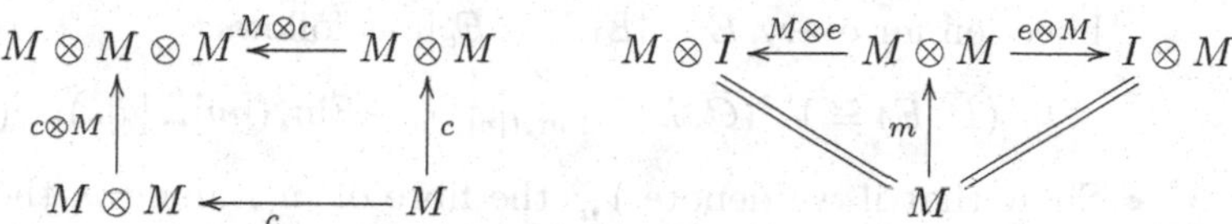

witnessing *coassociativity* and *counitality* properties for (c, e). A *comonad* on a category $\mathcal{C}$ is now a functor $S : \mathcal{C} \to \mathcal{C}$ which is a comonoid with respect to the monoidal structure $([\mathcal{C}, \mathcal{C}], \circ)$.

- Write down the axioms of *coassociativity* and counitality for S.
- Show that if $F \frac{\epsilon}{\eta}\!\!\mid G$ is an adjunction then $S = FG$ is a comonad with comultiplication $F * \eta * G$ and counit ϵ.
- Define a category of *coalgebras* for S, and of *free coalgebras* for S; show that they enjoy suitable universal properties.

A.16 Let

$$\mathcal{C} \underset{\xleftarrow[L]{}}{\overset{\xleftarrow{R}}{\xrightarrow{F}}} \mathcal{D}$$

be a triple of adjoint functors, meaning that $L \dashv F$ and $F \dashv R$; show that L is a fully faithful functor if and only if R is a fully faithful functor.

A.17 [Seg74] Given a monoidal category $(\mathcal{V}, \otimes)$, show that there exists a category $\mathcal{V}^{\otimes}$ defined as follows.

- The objects of $\mathcal{V}^{\otimes}$ are n-tuples of objects in $\mathcal{V}$, denoted as the tuple $[C_1, \ldots, C_n]$ (this follows the convention that if $n = 0$, the tuple is empty).

- Morphisms $[C_1, \dots, C_n] \to [D_1, \dots, D_m]$ are defined as pairs $(\alpha, \{f_j\})$ where α is a partial function $[n] \to [m]$ having domain S_α and
$$\{f_j : \bigotimes_{\{i|\alpha(i)=j\}} C_i \to D_j \mid 1 \le j \le m\}$$
is a family of morphisms in $\mathcal{V}$.
- Define the composition law of two morphisms $(\alpha, f), (\beta, g)$.
- Let Fin_* be the category $*/\mathrm{Set}_{<\omega}$ of pointed finite sets. Show that there is a functor $p : \mathcal{V}^\otimes \to \mathrm{Fin}_*$ sending $[C_1, \dots, C_n]$ into $[n]_*$. Show that p is an *opfibration*, i.e. for every object $\overline{C} = [C_1, \dots, C_n] \in \mathcal{V}^\otimes$ and every arrow $f : p(\overline{C}) \to [m]_*$ in Fin_* there is an arrow $(\theta_f, \bar{f}) : \overline{C} \to \overline{D} = [D_1, \dots, D_m]$ such that $p(\theta_f, \bar{f}) = f$, and such that the composition with $\bar{f}$ induces a bijection for every $\overline{E} = [E_1, \dots, E_d]$, as follows:
$$\mathcal{V}^\otimes(\overline{D}, \overline{E}) \cong \mathcal{V}^\otimes(\overline{C}, \overline{E}) \times_{\mathrm{Fin}_*([n]_*, [d]_*)} \mathrm{Fin}_*([m]_*, [d]_*). \quad \text{(A.89)}$$
- Show that if we denote $\mathcal{V}_n^\otimes$ the fibre of $[n]_*$ along p, then the functor p induces a functor $\mathcal{V}_m^\otimes \to \mathcal{V}_n^\otimes$ among the fibres, for every $f : [m]_* \to [n]_*$ in Fin_*.
- Show that $\mathcal{V}_0^\otimes \cong \{0\}$, $\mathcal{V}_1^\otimes \cong \mathcal{V}$, and that more in general $\mathcal{V}_n^\otimes \cong \mathcal{V} \times \dots \times \mathcal{V}$ (n times).
- Show that the correspondence sending $\mathcal{V}$ into $\mathcal{V}^\otimes$ is functorial in $\mathcal{V}$; are strong monoidal functors enough?
- Show that $\mathrm{Fin}_* \cong \{0\}^\otimes$ with respect to the unique monoidal structure that exists on the terminal category $\{0\}$; show that p is the functor induced by the unique functor $\mathcal{V} \to \{0\}$.

Appendix B
Table of Notable Integrals

Fubini rule	$\int^C \int^D T(C,D,C,D) \cong \int^D \int^C T(C,D,C,D) \cong \int^{(C,D)} T(C,C,D,D)$ $\forall n \geq 2, \sigma \in \mathrm{Sym}(n), \quad \int^{C_{\sigma 1}} \dots \int^{C_{\sigma n}} T \cong \int^{(C_1,\dots,C_n)} T$
Natural transformations	$\mathrm{Cat}(\mathcal{C},\mathcal{D})(F,G) \cong \int_C \mathcal{D}(FC,GC)$
Tensor product of functors	$G \boxtimes F = \int^C GC \otimes FC$
Yoneda lemma	$FX \cong \int^A [X,A] \times FA \qquad FX \cong \int_A \mathrm{Set}([A,X],FA)$ $GX \cong \int^A [A,X] \times GA \qquad GX \cong \int_A \mathrm{Set}([X,A],GA)$

Kan extensions: $\mathcal{C} \xleftarrow{G} \mathcal{A} \xrightarrow{F} \mathcal{B}$	$\mathrm{Lan}_G F(C) \cong \int^A \mathcal{C}(GA, C) \otimes FA$ $\mathrm{Ran}_G F(A) \cong \int_A \mathcal{C}(C, GA) \pitchfork FA$
Density comonad:	$T_F \cong \int_A \mathcal{C}(\_, FA) \pitchfork FA$
Codensity monad:	$S^F \cong \int^A \mathcal{C}(FA, \_) \otimes FA$
Weighted (co)limits	$\mathrm{colim}^W F \cong \int^A WA \otimes FA$ $\lim^W F \cong \int_A WA \pitchfork FA$
Profunctor theory	$\mathfrak{p} \bullet \mathfrak{q} = \int^X \mathfrak{p}(\_, X) \otimes \mathfrak{q}(X, =)$ $\mathrm{Ran}_{\mathfrak{p}} \mathfrak{q} = \int_A \hom(\mathfrak{p}_A, \mathfrak{q}_A) \qquad \mathrm{Rift}_{\mathfrak{p}} \mathfrak{q} = \int_A \hom(\mathfrak{r}_A, \mathfrak{p}_A)$

$$F * G := \int^{XY} \mathcal{C}(X \oplus Y, \_) \otimes FX \otimes GY$$

Operads

$$F \odot G := \int^{m} Fm \otimes G^{*m} \qquad \underset{m \text{ times}}{G^{*m} = G * \cdots * G}$$

$$\frac{F \diamond G \to H}{F \to \{G, H\}} \qquad \{G, H\}m = \int_k [G^{*m}k, Hk]$$

$$\oint_A T(A, A) := \int_{A', A''} \delta\mathcal{A}(A', A'') \pitchfork T(A', A'')$$

Simplicial coends

$$\oint^A T(A, A) := \int^{A', A''} \delta\mathcal{A}(A', A'') \otimes T(A', A'')$$

$$\delta\mathcal{A}(A', A'') = \int^{n \in \boldsymbol{\Delta}} \coprod_{X_0, \dots, X_n \in \mathcal{A}} \Delta[n] \times \mathcal{A}(A, X_0) \times \cdots \times \mathcal{A}(X_n, B)$$

$$[F *_{\mathfrak{P}} G]C = \int^{AB} P(A, B; C) \times FA \times GB$$

Promonoidal structures

$$\int^{Z} P_D^{AY} P_Y^{BC} \cong \int^{Z} P_X^{AB} P_D^{XC}$$

$$\int^{YZ} J_Z H_Y^A P_B^{YZ} \cong \hom(A, B)$$

Tambara module: $\quad \varphi_P(X, Y) = \int^{C,U,V} \mathcal{C}(X, C \otimes U) \times \mathcal{C}(C \otimes V, Y) \times P(U, V)$

References

[ACU10] T. Altenkirch, J. Chapman, and T. Uustalu, *Monads need not be endofunctors*, International Conference on Foundations of Software Science and Computational Structures, Lecture Notes in Comput. Sci., vol. 6014, Springer, Berlin, 2010, pp. 297–311.

[Ada78] J.F. Adams, *Infinite loop spaces*, Annals of Mathematics Studies, vol. 90, Princeton University Press, Princeton, N.J., 1978.

[AHS90] J. Adámek, H. Herrlich, and G.E. Strecker, *Abstract and concrete categories: the joy of cats*, Pure and Applied Mathematics, John Wiley & Sons, Inc., New York, 1990.

[AR94] J. Adámek and J. Rosický, *Locally presentable and accessible categories*, London Mathematical Society Lecture Note Series, vol. 189, Cambridge University Press, Cambridge, 1994.

[AR20] ———, *How nice are free completions of categories?*, Topology and its Applications **273** (2020), 106972.

[AV08] J. Adámek and J. Velebil, *Analytic functors and weak pullbacks*, Theory Appl. Categ. **21** (2008), No. 11, 191–209.

[Bak07] T.K. Bakke, *Hopf algebras and monoidal categories*, Master's thesis, Universitetet i Tromsø, 2007.

[Ber07] J. Bergner, *A model category structure on the category of simplicial categories*, Transactions of the American Mathematical Society **359** (2007), no. 5, 2043–2058.

[BK72] A.K. Bousfield and D.M. Kan, *Homotopy limits, completions and localizations*, Lecture Notes in Mathematics, Vol. 304, Springer-Verlag, Berlin-New York, 1972.

[BM13] C. Berger and I. Moerdijk, *On the homotopy theory of enriched categories*, Q.J. Math. **64** (2013), no. 3, 805–846.

[Bor94a] F. Borceux, *Handbook of categorical algebra. 1*, Encyclopedia of Mathematics and its Applications, vol. 50, Cambridge University Press, Cambridge, 1994.

[Bor94b] ———, *Handbook of categorical algebra. 2*, Encyclopedia of Mathematics and its Applications, vol. 51, Cambridge University Press, Cambridge, 1994.

[Boz76] S. Bozapalides, *Théorie formelle des bicatégories*, Esquisses Mathématiques 25, Ehresmann, Bastiani, 1976.

[Boz80] ———, *Some remarks on lax-presheaves*, Illinois Journal of Mathematics **24** (1980), no. 4, 676–680.

[Bre97] G. E. Bredon, *Sheaf theory*, second ed., Graduate Texts in Mathematics, vol. 170, Springer-Verlag, New York, 1997.

[BS00] J. Bénabou and T. Streicher, *Distributors at work*, Lecture notes written by Thomas Streicher, 2000.

[BSP11] C. Barwick and C. Schommer-Pries, *On the unicity of the homotopy theory of higher categories*, preprint arXiv:1112.0040, 2011.

[Cam] T. Campion, *Is Stokes' theorem natural in the sense of category theory?*, Mathematics Stack Exchange, `https://math.stackexchange.com/q/1229249` (version: 2015-04-10).

[CEG+] B. Clarke, D. Elkins, J. Gibbons, F. Loregian, B. Milewski, E. Pillmore, and M. Román, *Profunctor optics: a categorical update*, `http://events.cs.bham.ac.uk/syco/strings3-syco5/slides/roman.pdf`.

[CKS03] J.R.B. Cockett, J. Koslowski, and R.A.G. Seely, *Morphisms and modules for poly-bicategories*, Theory and Applications of Categories **11** (2003), no. 2, 15–74.

[CP89] J.-M. Cordier and T. Porter, *Shape theory: Categorical methods of approximation*, Dover books on mathematics, Ellis Horwood (Chichester, West Sussex, England and New York), 1989.

[CP97] ———, *Homotopy coherent category theory*, Transactions of the American Mathematical Society **349** (1997), no. 1, 1–54.

[Cra95] S.E. Crans, *Quillen closed model structures for sheaves*, Journal of Pure and Applied Algebra **101** (1995), no. 1, 35–57.

[CS10] G.S.H. Cruttwell and M. Shulman, *A unified framework for generalized multicategories*, Theory and Applications of Categories **24** (2010), no. 21, 580–655, arXiv:0907.2460.

[Cur12] P.L. Curien, *Operads, clones, and distributive laws*, Operads and universal algebra, World Scientific, Singapore, 2012, pp. 25–49.

[CV06] C. Centazzo and E.M. Vitale, *A classification of geometric morphisms and localizations for presheaf categories and algebraic categories*, Journal of Algebra **303** (2006), 77–96.

[CW96] G.L. Cattani and G. Winskel, *Presheaf models for concurrency*, International Workshop on Computer Science Logic, Springer, 1996, pp. 58–75.

[CW01] M. Cáccamo and G. Winskel, *A higher-order calculus for categories*, International Conference on Theorem Proving in Higher Order Logics, Springer, 2001, pp. 136–153.

[Day77] B.J. Day, *Note on compact closed categories*, J. Austral. Math. Soc. Ser. A **24** (1977), no. 3, 309–311.

[Day11] ———, *Monoidal functor categories and graphic Fourier transforms*, Theory Appl. Categ. **25** (2011), No. 5, 118–141.

[DLL19] I. Di Liberti and F. Loregian, *On the unicity of formal category theories*, preprint arXiv:1901.01594 (2019).

[Dol58] A. Dold, *Homology of symmetric products and other functors of complexes*, Annals of Mathematics 68 (1958), 54–80.

[DS11] D. Dugger and D. I. Spivak, *Mapping spaces in quasi-categories*, Algebr. Geom. Topol. **11** (2011), no. 1, 263–325.

[Dub70] E.J. Dubuc, *Kan extensions in enriched category theory*, Lecture Notes in Mathematics, Vol. 145, Springer-Verlag, Berlin-New York, 1970. MR 0280560

[EK66a] S. Eilenberg and G.M. Kelly, *Closed categories*, Proc. Conf. Categorical Algebra (La Jolla, Calif., 1965), Springer, New York, 1966, pp. 421–562.

[EK66b] ———, *A generalization of the functorial calculus*, J. Algebra **3** (1966), 366–375.

[Elm83] A.D. Elmendorf, *Systems of fixed point sets*, Trans. Amer. Math. Soc. **277** (1983), no. 1, 275–284.

[EP08] P.J. Ehlers and T. Porter, *Ordinal subdivision and special pasting in quasicategories*, Advances in Mathematics **217** (2008), no. 2, 489–518.

[FGHW18] M. Fiore, N. Gambino, M. Hyland, and G. Winskel, *Relative pseudomonads, Kleisli bicategories, and substitution monoidal structures*, Selecta Mathematica 24 (2018), 2791–2830.

[Gam10] N. Gambino, *Weighted limits in simplicial homotopy theory*, J. Pure Appl. Algebra **214** (2010), no. 7, 1193–1199.

[GAV72] A. Grothendieck, M. Artin, and J.L. Verdier., *Théorie des topos et cohomologie étale des schémas. Tome 1: Théorie des topos*, Lecture Notes in Mathematics, Vol. 269, Springer-Verlag, Berlin-New York, 1972.

[Gen15] F. Genovese, *Quasi-functors as lifts of Fourier-Mukai functors: the uniqueness problem*, Ph.D. thesis, Università degli studi di Pavia, 2015.

[Get09] E. Getzler, *Operads revisited*, Algebra, Arithmetic, and Geometry, Progr. Math., vol. 269, Birkhäuser Boston, 2009, pp. 675–698.

[GF16] R. Garner and I.-L. Franco, *Commutativity*, Journal of Pure and Applied Algebra **220** (2016), no. 5, 1707–1751.

[GHN15] D. Gepner, R. Haugseng, and T. Nikolaus, *Lax colimits and free fibrations in ∞-categories*, preprint arXiv:1501.02161 (2015).

[GJ09] P.G. Goerss and J.F. Jardine, *Simplicial homotopy theory*, Modern Birkhäuser Classics, vol. 174, Birkhäuser Verlag, Basel, 2009, Reprint of the 1999 edition [MR1711612]. MR 2840650

[GJ17] N. Gambino and A. Joyal, *On operads, bimodules and analytic functors*, Mem. Amer. Math. Soc. **249** (2017), no. 1184, v + 110.

[GL12] R. Garner and S. Lack, *Lex colimits*, Journal of Pure and Applied Algebra **216** (2012), no. 6, 1372–1396.

[Gra80] J.W. Gray, *Closed categories, lax limits and homotopy limits*, Journal of Pure and Applied Algebra **19** (1980), 127–158.

[Gro13] M. Groth, *Derivators, pointed derivators and stable derivators*, Algebraic & Geometric Topology **13** (2013), no. 1, 313–374.

[Gui80] R. Guitart, *Relations et carrés exacts*, Ann. Sci. Math. Québec **4** (1980), no. 2, 103–125.

[Hir03] P.S. Hirschhorn, *Model categories and their localizations*, vol. 99, Mathematical Surveys and Monographs, no. 99, American Mathematical Society, Providence, RI, 2003.

[Hov99] M. Hovey, *Model categories*, Mathematical Surveys and Monographs, vol. 63, American Mathematical Society, Providence, RI, 1999.

[HP07] M. Hyland and J. Power, *The category theoretic understanding of universal algebra: Lawvere theories and monads*, Electronic Notes in Theoretical Computer Science **172** (2007), 437–458.

[Ide89] M. Idel, *Language, Torah, and Hermeneutics in Abraham Abulafia*, SUNY Series in Judaica, Hermeneutics, Mysticism and Religion, State University of New York Press, 1989.

[IK86] G.B. Im and G.M. Kelly, *A universal property of the convolution monoidal structure*, Journal of Pure and Applied Algebra **43** (1986), no. 1, 75–88.

[Isa09] S.B. Isaacson, *A note on unenriched homotopy coends*, Online Preprint: `http://www-home.math.uwo.ca/~sisaacso/PDFs/coends.pdf` (2009).

[Joy] A. Joyal, *The theory of quasi-categories I,II, in preparation.*

[Joy86] ______, *Foncteurs analytiques et espèces de structures*, Combinatoire énumérative, Lecture Notes in Math., vol. 1234, Springer, Berlin, 1986, pp. 126–159.

[Joy08a] ______, *Notes on quasicategories*, `https://www.math.uchicago.edu/~may/IMA/Joyal.pdf`, 2008.

[Joy08b] ______, *The theory of quasi-categories and its applications*, Proceedings of the IMA Workshop "n-Categories: Foundations and Applications", Citeseer, 2008, Lectures at the CRM (Barcelona). `https://mat.uab.cat/~kock/crm/hocat/advanced-course/Quadern45-2.pdf`.

[JSV96] A. Joyal, R. Street, and D. Verity, *Traced monoidal categories*, Mathematical Proceedings of the Cambridge Philosophical Society **3** (1996), 447–468.

[JT07] A. Joyal and M. Tierney, *Quasi-categories vs Segal spaces*, Categories in algebra, geometry and mathematical physics, Contemp. Math., vol. 431, Amer. Math. Soc., Providence, RI, 2007, pp. 277–326.

[Kel74] G.M. Kelly, *Doctrinal adjunction*, Category Seminar (Proc. Sem., Sydney, 1972/1973), Lecture Notes in Math., Vol. 420, Springer, Berlin, 1974, pp. 257–280.

[Kel80] ______, *A unified treatment of transfinite constructions for free algebras, free monoids, colimits, associated sheaves, and so on*, Bulletin of the Australian Mathematical Society **22** (1980), no. 01, 1–83.

[Kel89] ______, *Elementary observations on 2-categorical limits*, Bulletin of the Australian Mathematical Society **39** (1989), 301–317.

[Kel05a] ———, *Basic concepts of enriched category theory*, Repr. Theory Appl. Categ. **64** (2005), no. 10, vi+137, Reprint of the 1982 original [Cambridge Univ. Press, Cambridge; MR0651714]. MR 2177301

[Kel05b] ———, *On the operads of J. P. May*, Repr. Theory Appl. Categ. **13** (2005), no. 13, 1–13.

[KL14] A. Kuznetsov and V.A. Lunts, *Categorical resolutions of irrational singularities*, International Mathematics Research Notices **2015** (2014), no. 13, 4536–4625.

[Kme18] E. Kmett, *lens library, version 4.16*, Hackage `https://hackage.haskell.org/package/lens-4.16`, 2012–2018.

[Krö07] R. Krömer, *Tool and object: a history and philosophy of category theory*, Science networks historical studies 32, Birkhäuser, 2007.

[KS74] G.M. Kelly and R. Street, *Proceedings Sydney Category Theory Seminar 1972/1973*, ch. Review of the elements of 2-categories, pp. 75–103, Springer Berlin Heidelberg, 1974.

[Law63] F.W. Lawvere, *Functorial semantics of algebraic theories*, Proceedings of the National Academy of Sciences of the United States of America **50** (1963), no. 5, 869–872.

[Law73] ———, *Metric spaces, generalised logic, and closed categories*, Rendiconti del Seminario Matematico e Fisico di Milano, vol. 43, Tipografia Fusi, Pavia, 1973.

[Law07] ———, *Axiomatic cohesion*, Theory and Applications of Categories **19** (2007), no. 3, 41–49.

[Leh14] M.C. Lehner, *Kan extensions as the most universal of the universal constructions*, Master's thesis, Harvard College, 2014.

[Lei] T. Leinster, *Coend computation*, MathOverflow, `https://mathoverflow.net/q/20451` (v2010-04-06).

[Lei99] ———, **fc**-*multicategories*, arXiv preprint math/9903004 (1999).

[Lei04] ———, *Higher operads, higher categories*, London Mathematical Society Lecture Note Series, vol. 298, Cambridge University Press, Cambridge, 2004.

[Lei06] ———, *Are operads algebraic theories?*, Bulletin of the London Mathematical Society **38** (2006), no. 2, 233–238.

[Lei14] ———, *Basic category theory*, vol. 143, Cambridge University Press, 2014.

[Lor18] F. Loregian, *A Fubini rule for ∞-coends*, Preprints of the MPIM (2018), no. 68.

[LR11] S. Lack and J. Rosický, *Notions of Lawvere theory*, Applied Categorical Structures **19** (2011), no. 1, 363–391.

[Lur09] J. Lurie, *Higher Topos Theory*, Annals of Mathematics Studies, vol. 170, Princeton University Press, Princeton, NJ, 2009.

[LV12] J.-L. Loday and B. Vallette, *Algebraic operads*, Grundlehren der Mathematischen Wissenschaften [Fundamental Principles of Mathematical Sciences], vol. 346, Springer, Heidelberg, 2012.

[May72] J.P. May, *The geometry of iterated loop spaces*, Springer-Verlag, Berlin-New York, 1972, Lectures Notes in Mathematics, Vol. 271.

[Mil] B. Milewski, *Profunctor optics: The categorical view,* https://bartoszmilewski.com/2017/07/07/profunctor-optics-the-categorical-view/ *blog post.*

[ML70] S. Mac Lane, *The Milgram bar construction as a tensor product of functors*, The Steenrod Algebra and its Applications (Proc. Conf. to Celebrate N. E. Steenrod's Sixtieth Birthday, Battelle Memorial Inst., Columbus, Ohio,1970), Lecture Notes in Mathematics, Vol. 168, Springer, Berlin, 1970, pp. 135–152.

[ML98] ______, *Categories for the working mathematician*, second ed., Graduate Texts in Mathematics, vol. 5, Springer-Verlag, New York, 1998. MR 1712872

[MLM92] S. Mac Lane and I. Moerdijk, *Sheaves in geometry and logic: A first introduction to topos theory*, Universitext, vol. 13, Springer, 1992.

[Moe95] I. Moerdijk, *Classifying spaces and classifying topoi*, Lecture Notes in Mathematics, vol. 1616, Springer-Verlag, Berlin, 1995.

[Moo09] D. Moody, *The "physics" of notations: toward a scientific basis for constructing visual notations in software engineering*, IEEE Transactions on software engineering **35** (2009), no. 6, 756–779.

[MSS02] M. Markl, S. Shnider, and J. Stasheff, *Operads in algebra, topology and physics*, Mathematical Surveys and Monographs, vol. 96, American Mathematical Society, Providence, RI, 2002.

[MZ10] J.-P. Marquis and E.N. Zalta, *What is category theory*, What is category theory, 2010, pp. 221–255.

[nLa21] nLab authors, *Profunctor*, http://ncatlab.org/nlab/show/profunctor, March 2021.

[PGW17] M. Pickering, J. Gibbons, and N. Wu, *Profunctor optics: Modular data accessors. the art*, Science, and Engineering of Programming **1** (2017), no. 2.

[Pie82] R.S. Pierce, *Associative algebras*, Graduate texts in mathematics, Springer, Berlin, 1982.

[PS08] C. Pastro and R. Street, *Doubles for monoidal categories*, Theory and applications of categories **21** (2008), no. 4, 61–75.

[Rez17] C. Rezk, *Stuff about quasicategories*, Unpublished notes, http://www.math.illinois.edu/rezk/595-fal16/quasicats.pdf (2017).

[Rie] E. Riehl, *Understanding the homotopy coherent nerve,* https://golem.ph.utexas.edu/category/2010/04/understanding_the_homotopy_coh.html.

[Rie14] ______, *Categorical homotopy theory*, New Mathematical Monographs, vol. 24, Cambridge University Press, Cambridge, 2014.

[Rie17] ______, *Category theory in context*, Courier Dover Publications, 2017.

[RV14] E. Riehl and D. Verity, *The theory and practice of Reedy categories*, Theory Appl. Categ. **29** (2014), 256–301.

[RV15] ______, *The 2-category theory of quasi-categories*, Advances in Mathematics **280** (2015), 549–642.

[RV17a] ______, *Fibrations and Yoneda lemma in an ∞-cosmos*, Journal of Pure and Applied Algebra **221** (2017), no. 3, 499–564.

[RV17b] ———, *Kan extensions and the calculus of modules for ∞-categories*, Algebraic & Geometric Topology **17** (2017), no. 1, 189–271.

[Sch13] D. Schäppi, *The formal theory of Tannaka duality*, Astérisque (2013), no. 357.

[Seg74] G. Segal, *Categories and cohomology theories*, Topology **13** (1974), 293–312.

[Sel10] P. Selinger, *A survey of graphical languages for monoidal categories*, New structures for physics (Bob Coecke, ed.), vol. 813, Springer, 2010, pp. 289–355.

[Shu06] M. Shulman, *Homotopy limits and colimits and enriched homotopy theory*, preprint arXiv:math/0610194 (2006), Preprint.

[Shu16] ———, *Contravariance through enrichment*, Theory Appl. Categ. (2016).

[Sim11] H. Simmons, *An introduction to category theory*, Cambridge University Press, 2011.

[Str72] R. Street, *The formal theory of monads*, J. Pure Appl. Algebra **2** (1972), no. 2, 149–168.

[Str74] ———, *Fibrations and Yoneda lemma in a 2-category*, Proceedings Sydney Category Theory Seminar 1972/1973 (G.M. Kelly, ed.), Lecture Notes in Mathematics, vol. 420, Springer, 1974, pp. 104–133.

[Str76] ———, *Limits indexed by category-valued 2-functors*, Journal of Pure and Applied Algebra **8** (1976), no. 2, 149–181.

[Str80] ———, *Fibrations in bicategories*, Cahiers de topologie et géométrie différentielle catégoriques **21** (1980), no. 2, 111–160.

[Str81] ———, *Conspectus of variable categories*, Journal of Pure and Applied Algebra **21** (1981), no. 3, 307–338.

[Str11] J. Strom, *Modern classical homotopy theory*, vol. 127, American Mathematical Society Providence, RI, USA, 2011.

[SW78] R. Street and R. Walters, *Yoneda structures on 2-categories*, J. Algebra **50** (1978), no. 2, 350–379.

[Szl12] K. Szlachányi, *Skew-monoidal categories and bialgebroids*, Advances in Mathematics **231** (2012), no. 3-4, 1694–1730.

[Tam06] D. Tambara, *Distributors on a tensor category*, Hokkaido mathematical journal **35** (2006), no. 2, 379–425.

[Toë05] B. Toën, *Vers une axiomatisation de la théorie des catégories supérieures*, *K*-Theory **34** (2005), no. 3, 233–263.

[Tri] T. Trimble, *Towards a doctrine of operads.* `http://ncatlab.org:8080/toddtrimble/published/Towards+a+doctrine+of+operads`.

[Ulb90] K.-H. Ulbrich, *On Hopf algebras and rigid monoidal categories*, Israel Journal of Mathematics **72** (1990), no. 1-2, 252–256.

[Wei94] C.A. Weibel, *An introduction to homological algebra*, Cambridge Studies in Advanced Mathematics, vol. 38, Cambridge University Press, Cambridge, 1994.

[Woo82] R.J. Wood, *Abstract proarrows I*, Cahiers de topologie et géometrie différentielle categoriques **23** (1982), no. 3, 279–290.

[Yon60] N. Yoneda, *On Ext and exact sequences*, J. Fac. Sci. Univ. Tokyo, Sect. I **8** (1960), 507–576 (1960).

Index

Printed in the United States
by Baker & Taylor Publisher Services

Printed in the United States
by Baker & Taylor Publisher Services